Jelena Krmpotić-Nemanić
Wolfgang Draf · Jan Helms

Surgical Anatomy of Head and Neck

With 233 Illustrations, Mostly in Color

Springer-Verlag
Berlin Heidelberg NewYork
London Paris Tokyo

Professor Dr. Jelena Krmpotić-Nemanić
Zavod za Anatomiju "Drago Perović", Šalata 11, Zagreb/Jugolslavia

Professor Dr. Wolfgang Draf
Städtische Kliniken, Klinik für HNO-Krankheiten und Plastische Gesichtschirurgie,
Pacelliallee 4, 6400 Fulda, Federal Republic of Germany

Professor Dr. Jan Helms
HNO-Klinik der Universität,
Josef-Schneider-Straße 11, 8700 Würzburg, Federal Republic of Germany

Title of the German Edition:
Chirurgische Anatomie des Kopf-Hals-Bereiches

ISBN-13:978-3-642-71814-4 e-ISBN-13:978-3-642-71812-0
DOI: 10.1007/978-3-642-71812-0

Softcover reprint of the hardcover 1st edition 1988

2122/3130-543210

Foreword to the German Edition

Surgeons and clinicians of the head and neck have long recognized the need for a properly developed anatomic text and illustrations for swift and accurate orientation. There are of course a number of excellent anatomical atlases and texts available, but rapid orientation – as for example in defining intraoperative structures – is often made difficult by the detailed minutia of the anatomical text. The result is that such texts often fail to meet clinical demands.
The principal author of this book is indeed fortunate to have had a clinical and surgical education at an outstanding ENT institution in Sercer, in addition to her anatomical training and teaching experience. This has enabled her to design the illustrations with the problems of the surgeon in mind, with better information on the special features of approaches to deeper, more vital structures. Her two coauthors are both well-known surgeons in their own rights, and have been able to offer additional assistance and advice. The text has been deliberately kept brief to render orientation swift and to ensure general clarity. Nevertheless, all clinically essential information on the relevant structures of the head and neck and their variations is given. Moreover, this material is so presented that it portrays the often difficult anatomy of the head and neck to both the clinican and the surgeon in a way that is graphic and readily understood.
I can wholeheartedly recommend this book not only to inexperienced trainees, but also to experienced physicians. I am sure that the text will achieve the wide distribution it so richly deserves, and will find great acceptances in the various specialties which can most benefit from it.

Heidelberg H.-J. DENECKE

Preface

The original idea for this book came from Professor H.-J. Denecke of Heidelberg, who saw the need for a contemporary text on the surgical anatomy of the head and neck to fill an existing gap in the available literature. Thus, our aim has been to show the anatomy of this region from the perspective of the surgeon. However, to better illustrate this anatomy, the skin and other structures overlying the operative field are often not shown. In other illustrations, dotted lines show the spatial relationships of structures lying deep to the operative field. A large number of the figures used were drawn specifically for this volume to correspond with various surgical approaches currently employed. In addition, some excellent informative illustrations were taken from works previously published by Denecke, Guleke, Lang, Kempe, and other authors.

We have tried to include all significant topographical information in our narrative text, although understandably this material cannot replace careful study of more detailed surgical and anatomical literature. Additionally, we have described what we consider to be the most surgically important anatomic structures present. In so doing, we have also included a selected bibliography that should prove useful for further review.

In order to meet the clinical requirements of the surgeon, we were unable to use all of the most current anatomical terms. Unfortunately, anatomical nomenclature is still not absolutely settled, and the incorporation of new descriptive terms into the everyday language of surgery remains a task for the future. Where appropriate, we have placed alternative terms for certain structures in parentheses.

Our new figures were drawn with great care and precision by the medical artist Edita Schubert, to whom we are very grateful. We should also stress that each illustration used was the subject of careful discussion between the authors and the artist. We trust that the reader shall find the fruits of our labors to be more than justified by our efforts.

We have developed our book for all clinicians engaged in otorhinolaryngology, maxillofacial surgery and neurosurgery. It is our hope that the text will simplify the learning and execution of surgical procedures in a region of the body presenting special topographical and anatomical difficulties.

We would like to thank the staff of Springer-Verlag in Heidelberg and Dr. Alan Kornblut in Washington for their understanding and assistance.

Zagreb
Fulda
Würzburg

J. Krmpotić-Nemanić
W. Draf
J. Helms

Contents

1 The Neck and Thorax

1.1 (Plates 1–3) Topography of the Upper Mediastinum

The use of horizontal sections to provide topographic orientation to the upper or superior mediastinium is important not only for computed tomographic (CT) examinations, but also for doing mediastinoscopy. Plate 1: Topographic orientation at the level of T_1–T_2, corresponding to the region of the jugular notch. Plate 2: Section at the level of T_3 about 5 cm caudal to the manubrium sterni. Plate 3: Topographic relations at the level of T_5 at the tracheal bifurcation.

Surgical Importance

Mediastinoscopy, surgical exploration of the upper mediastinum

Reference

1. Pernkopf E and Ferner H (1963)

Plates 1, 3 modified from Pernkopf [1]

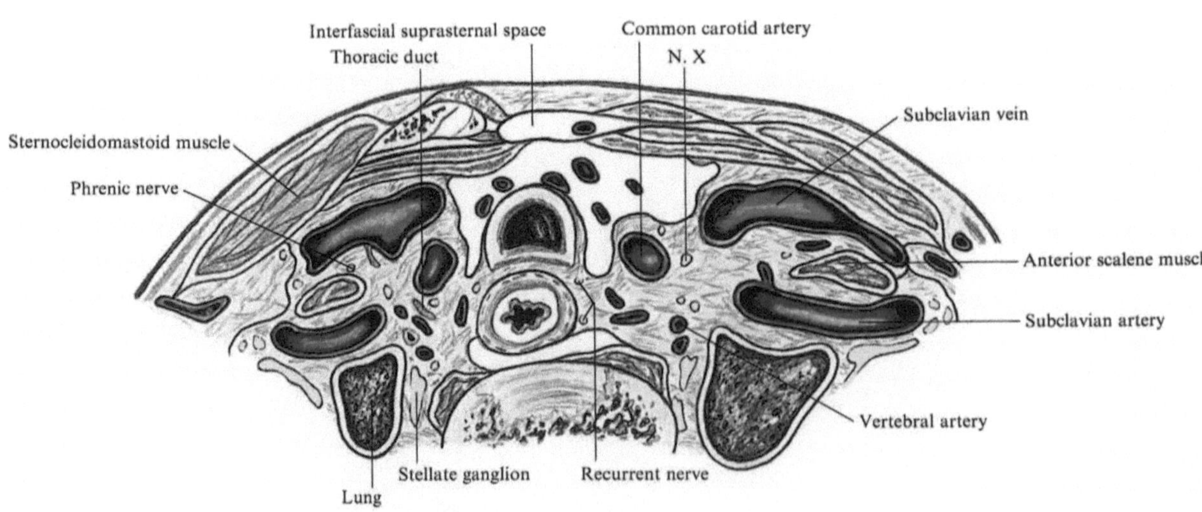

1

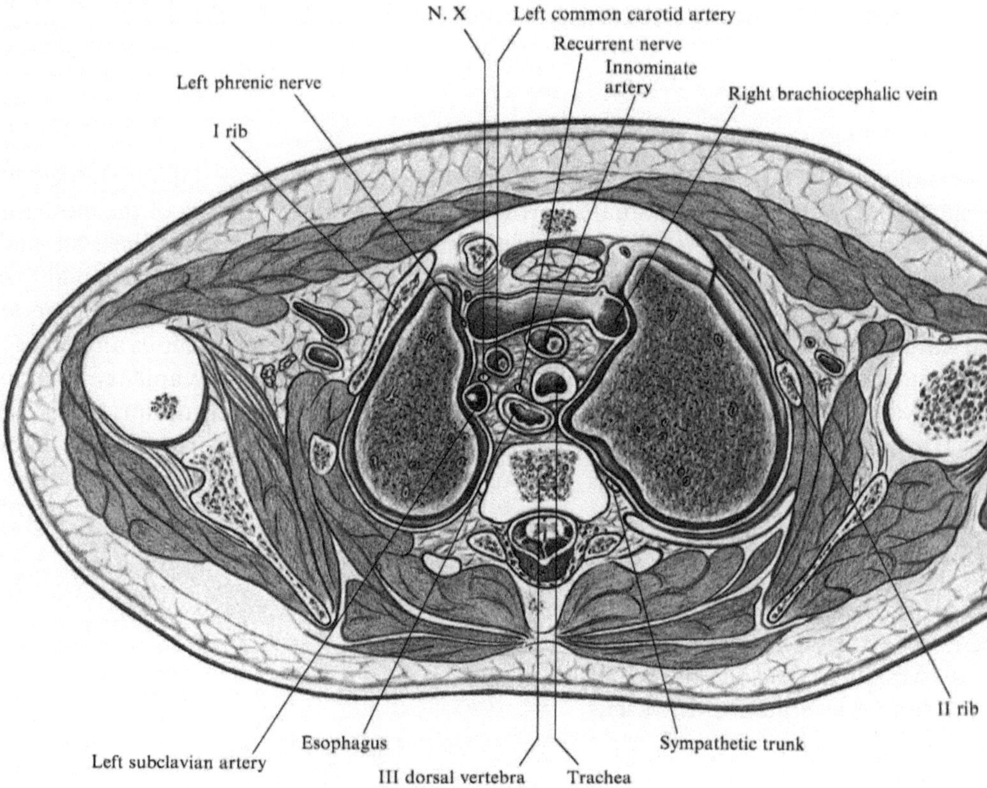
N. X
Left common carotid artery
Recurrent nerve
Innominate artery
Left phrenic nerve
Right brachiocephalic vein
I rib
II rib
Left subclavian artery
Esophagus
III dorsal vertebra
Trachea
Sympathetic trunk

2

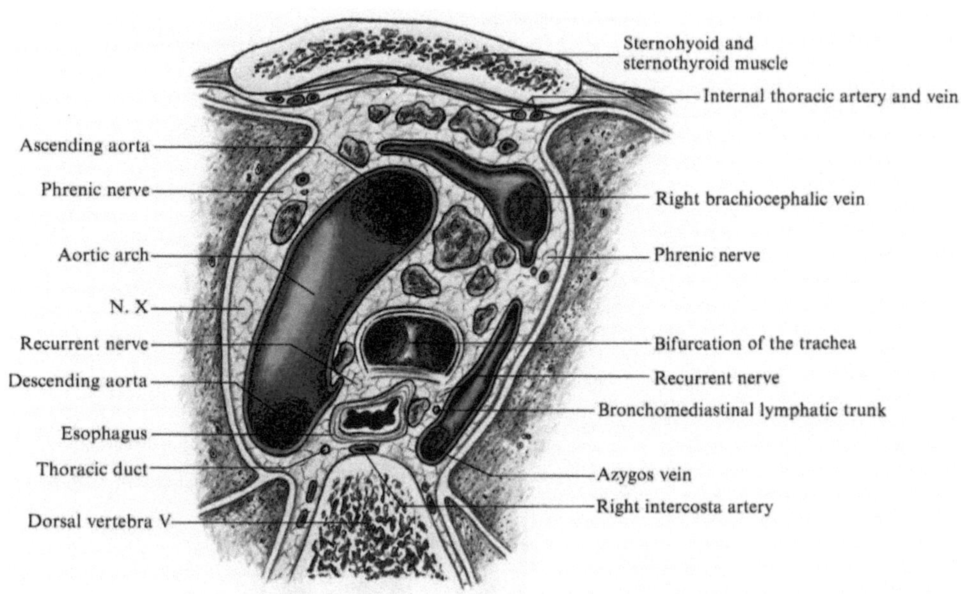
Sternohyoid and sternothyroid muscle
Internal thoracic artery and vein
Ascending aorta
Phrenic nerve
Right brachiocephalic vein
Aortic arch
Phrenic nerve
N. X
Recurrent nerve
Bifurcation of the trachea
Descending aorta
Recurrent nerve
Bronchomediastinal lymphatic trunk
Esophagus
Thoracic duct
Azygos vein
Dorsal vertebra V
Right intercosta artery

3

1.2 Normal Development of the Branches of the Aortic Arch

The aortic arch normally develops from the artery of the left IVth branchial arch, while the proximal part of the right subclavian artery evolves from that of the right IVth arch [1]. The external carotid arteries form from the so-called ventral aortae, while the internal carotid arteries arise from the dorsal aortae and the artery of the IIIrd branchial arch. The segments of the ventral aortae between the arteries of the IIIrd and IVth arches give rise to the common carotid arteries. The initial part of the VIth branchial arch artery forms the branches of the pulmonary artery. The ductus arteriosus of *Botalli* develops on the left side of the dorsal part of the VIth branchial arch artery and normally disappears after birth. If, however, certain segments of the vessels joining the branchial arch arteries do not disappear, a number of vascular anomalies can develop, causing such symptoms as dysphagia or dyspnea.

Selective and superselective angiographies have provided a better understanding of the mechanism giving rise to these anomalies [2]. These anomalies can be classified as active and inactive malformations. Examples of *active malformations* are arteriovenous fistulae and arteriovenous angiomas. *Inactive malformations* include capillary angiomas and cavernomas. This classification is useful when specific treatments are considered, such as embolization or vessel resection and reconstruction.

References

1. Krmpotić-Nemanić J (1978)
2. Merland JJ et al. (1980)

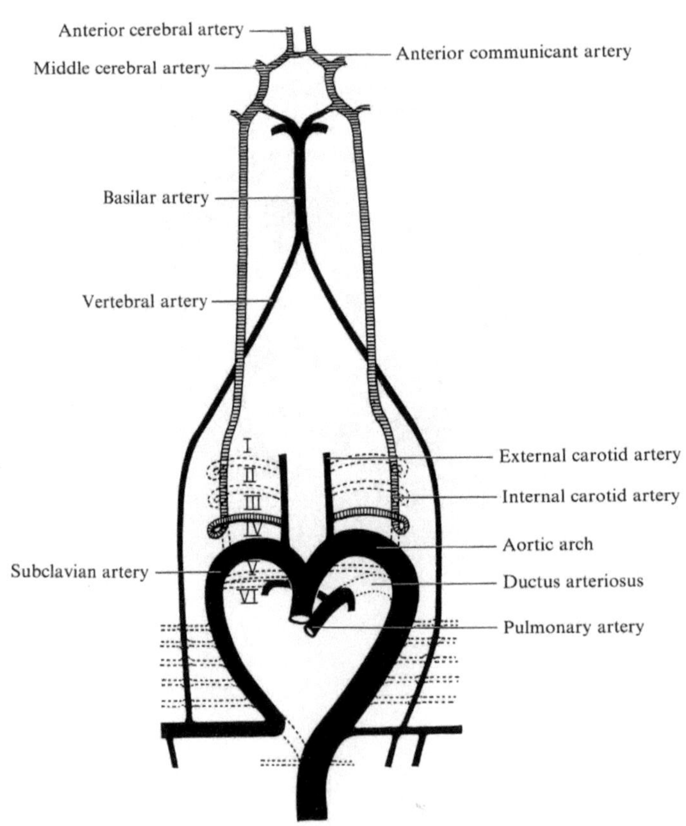

As shown in the first Plate, a double aortic arch can cause compression of the esophagus [2, 4]. A right-sided aortic arch represents another possible malformation, and can be associated with either a rare form of mirror-image branching or with an aberrant left subclavian artery that leaves the aortic arch close to the descending aorta (Plate 2). A further anomaly is a right aortic arch with a left retroesophageal innominate artery, or *truncus brachiocephalicus* [3], or a patent ductus arteriosus (Plate 4). In these deformities, the aortic arch can also have a coarctation either at its origin or in the region of the ductus Botalli.
Desnos et al. [1] have further described the various anomalies of the aortic arch. As a consequence of vascular compression against the developing lower respiratory tract, they represent an important cause of tracheobronchial stenosis with stridor in the neonate.

Surgical Importance

Bronchoscopy, esophagoscopy, mediastinoscopy, vascular causes of dysphagias.

References

1. Desnos J et al. (1981)
2. Leonardi HK et al. (1980)
3. Locatelli E et al. (1977)
4. Prot D et al. (1979)
5. Rowe LD et al. (1978)

Plates 1–4 adapted from Rowe et al. [5]

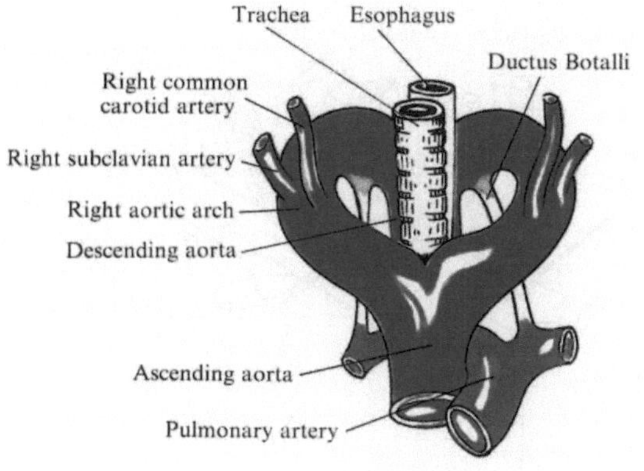

1

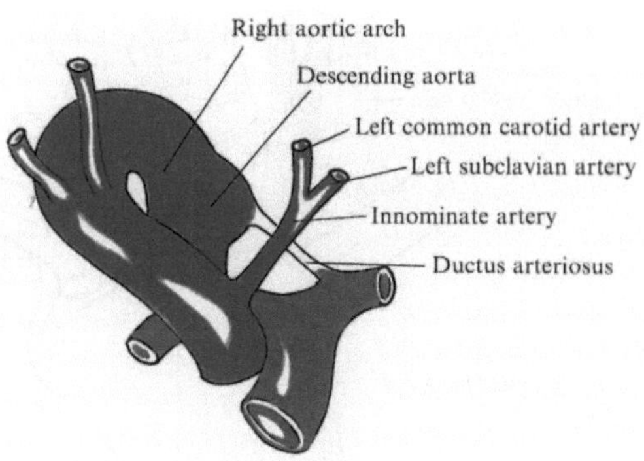

2

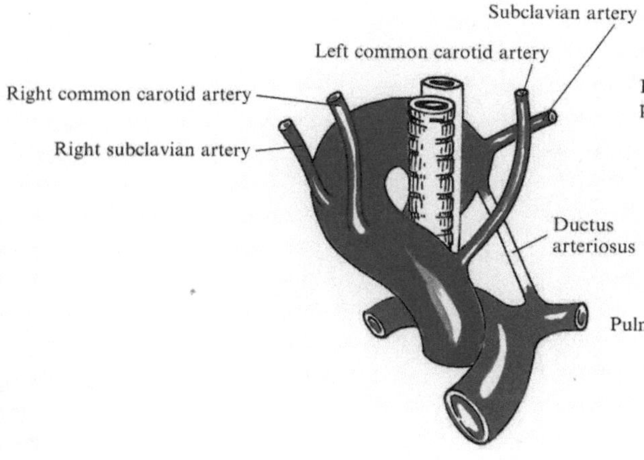

3

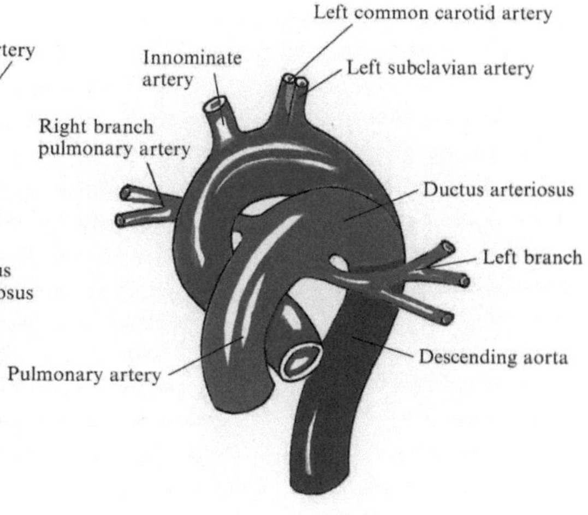

4

1.4 (Plates 1–2) Superior Displacement of the Aortic Arch

The aortic arch can be displaced in the region of the jugular notch and together with an enlarged innominate vessel runs either horizontally (Plate 1) or obliquely (Plate 2) across the trachea [1–4]. In either case, the relation of the trachea and the esophagus to the origin of the aortic branches can vary. Such variations should be kept in mind in tracheoscopy or mediastinoscopy, as well as in thyroid and parathyroid surgery. If a tracheotomy is indicated in these patients, Denecke [1] has recommended protecting the vessels with a pectoral flap at the time of surgery.

Surgical Importance

Surgical exploration of the neck, tracheotomy, thyroid and parathyroid surgery, mediastinoscopy

References

1. Denecke HJ (1980)
2. Krmpotić-Nemanić J (1978)
3. Nitzsche H (1980)
4. Tiraboschi R et al. (1980)

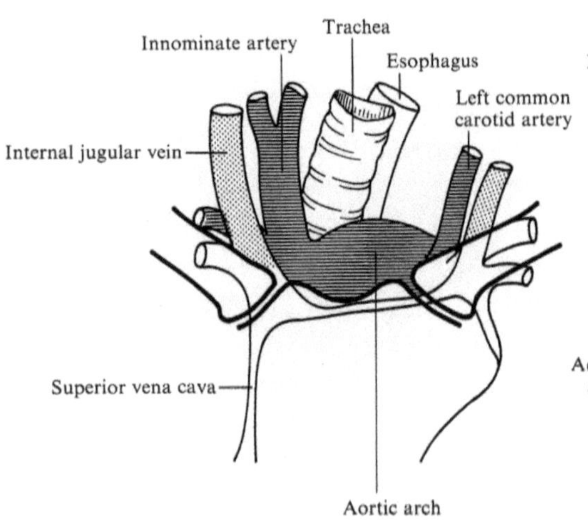

1

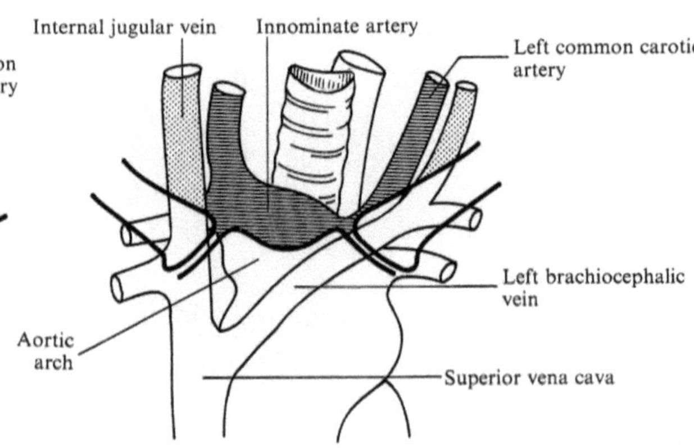

2

1.5 (Plates 1–2) The Cervical Aortic Arch

The aortic arch and its branches can be displaced upward and laterally so as to lie in the lower neck [1–3]. In one such pattern described by Tiraboschi et al. [3], the aortic arch is displaced toward the right side and is connected by a narrowed isthmus to the descending aorta (Plate 1). In another type of displacement, the arch is located on the left side and gives rise to an aberrant subclavian artery. In the case of a stenotic vessel or isthmus connecting the aortic arch to the descending aorta, the latter artery may undergo dilatation, as shown in Plate 2 [2, 3].

Surgical Importance

Exploration of the neck, thyroid and parathyroid surgery, tracheotomy, mediastinoscopy.

References

1. Krmpotić-Nemanić J (1978)
2. Nitzsche H (1980)
3. Tiraboschi R et al. (1980)

Plates 1–2 adapted from Tiraboschi et al. [3]

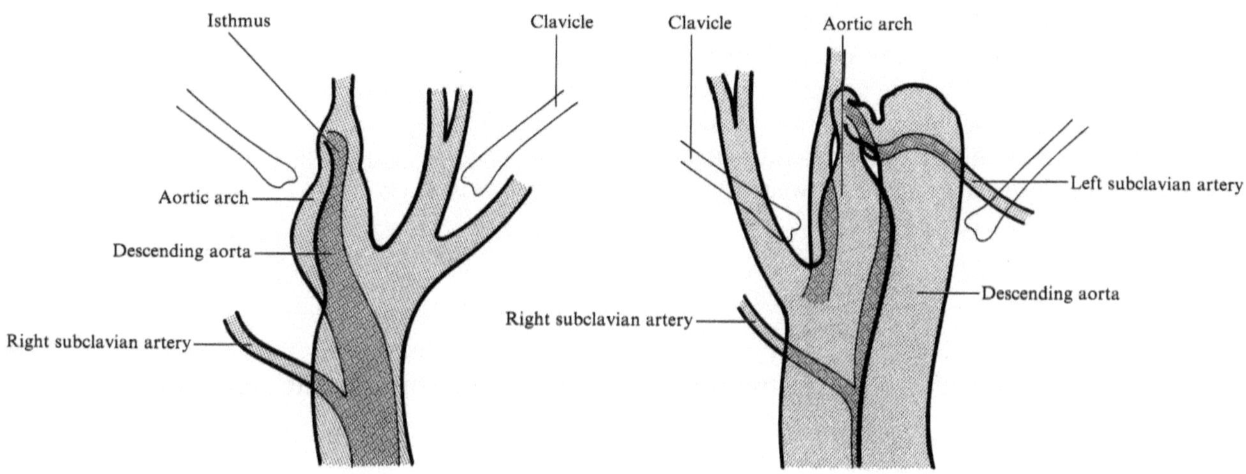

1 2

1.6 (Plates 1–10) Variations and Malformations of the Aortic Arch

The normal sequence from right to left of the branches of the aortic arch is shown in Plate 1 [4], i.e. the innominate artery (*truncus brachiocephalicus, arteria anonyma*), which branches off into the common carotid artery and the right subclavian artery, the left common carotid artery, and the left subclavian artery. The vertebral arteries originate from the subclavian arteries. Although there are many patterns of variation with respect to both the number and sequence of vessel origin, ten of these patterns are most common.

The innominate and the left common carotid arteries can have a common origin (Plate 2). The vertebral artery can originate as a separate vessel from the aortic arch medial to the left subclavian artery (Plate 3). The left vertebral artery and the left subclavian artery can share a common origin, with the remaining vessels also coming from a common source (Plate 4). The vertebral artery can ascend from the aortic arch lateral to the left subclavian artery (Plate 5). Both left carotid arteries can originate from the innominate artery (Plate 6). The right subclavian artery can be the lowermost branch of the aortic arch, originating at the commencement of the descending aorta and reaching the right superior mediastinum by passing behind the esophagus, by taking a path between the trachea and the esophagus or by passing in front of the trachea [5] (Plate 7). Kemmetmüller [2] has described the following variations, as depicted from right to left in Plate 8: innominate artery, thyroidea ima artery, left common carotid artery, medial vertebral artery, and lateral vertebral artery. The latter two vessels are situated on the left side. The lateral vertebral artery may give off the thyroidea ima artery (Plate 8), which can also originate between the innominate artery and the left common carotid artery (Plate 9). Finally, the right vertebral artery can originate as the lowermost branch of the aortic arch [1–3] (plane 10).

Surgical Importance

Exploration of the neck, thyroid and parathyroid surgery, tracheotomy, mediastinoscopy.

References

1. Holzapfel G (1899)
2. Kemmetmüller H (1911)
3. Krause W (1905)
4. Krmpotić-Nemanić J (1978)
5. Locatelli E et al. (1977)

1.7 (Plates 1–2) Anomalous Locations of the Innominate Artery

The innominate artery is situated in the midline in 52% of cases, to the left of the trachea in 43%, and to the right of the trachea in 5% [2]. Even in its normal position in front of the trachea, surgical injury to the artery can cause severe bleeding, especially during a tracheotomy [8]. As previously noted in Sect. 1.4, this artery should be protected by either a tracheal flap [1] or a pectoral flap [3] when a tracheotomy is performed.

In certain cases, the innominate artery can produce tracheal compression with stridor [6, 7]. The bifurcation of the artery is generally situated above the jugular notch. The artery can undergo an upward displacement in the neck, passing over the trachea obliquely upwards from left to right and covering the trachea either partially (Plate 1) or completely (Plate 2). In such instances, the bifurcation is situated close to the thyroid gland. These variations can be the cause of serious complications in neck injuries or during tracheotomy, mediastinoscopy, as well as partial or total laryngectomies.

Surgical Importance

Explorations of the mid-neck, tracheotomy, mediastinoscopy, thyroid and parathyroid surgery.

References

1. Björk V (1969)
2. Conso JF et al. (1974)
3. Denecke HJ (1980)
4. Draf W (1978)
5. Krmpotić-Nemanić J (1978)
6. Narcy P et al. (1979)
7. Schlemmer M (1979)
8. Zanolli L and Viada J (1979)

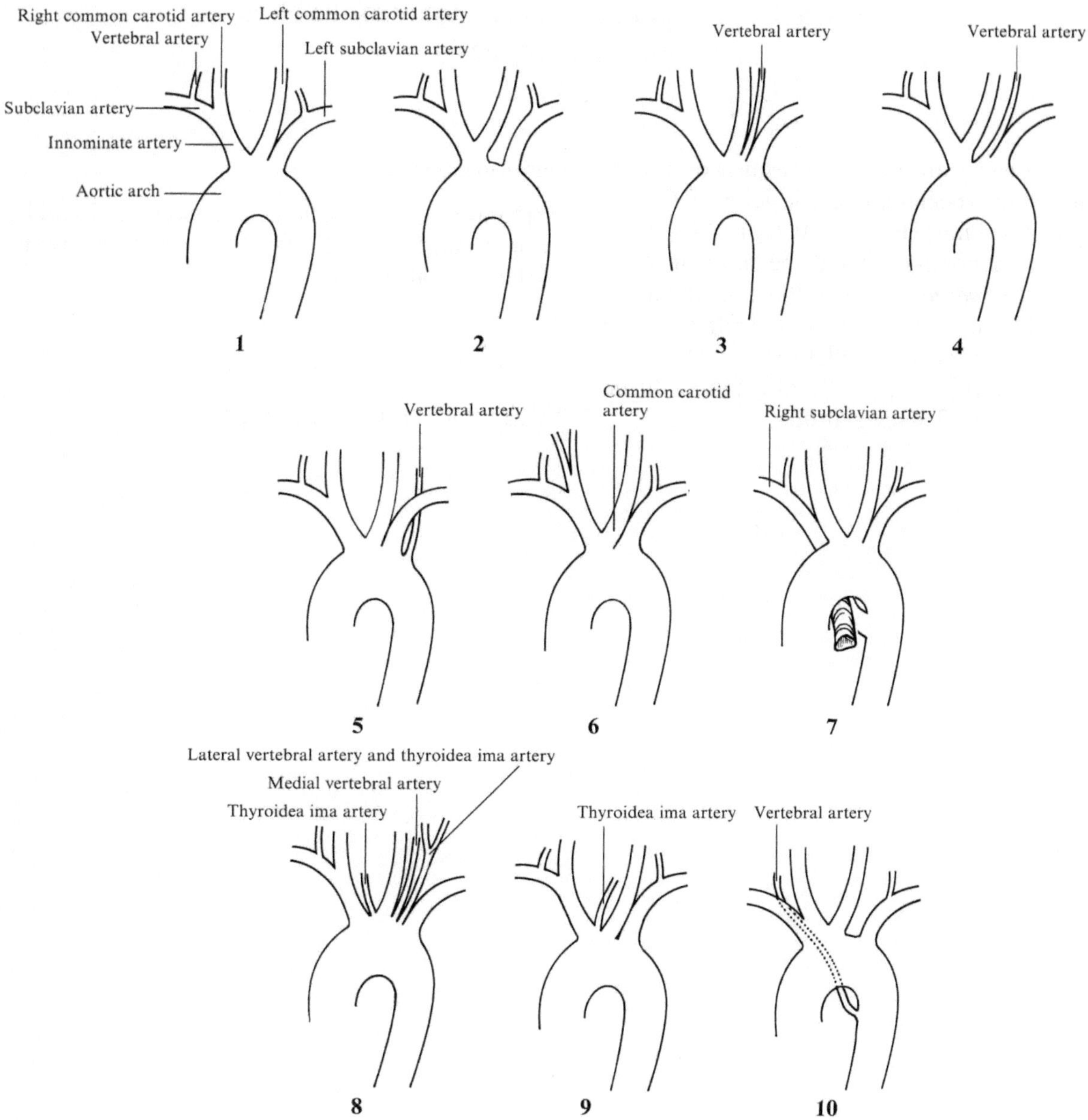
Right common carotid artery
Vertebral artery
Left common carotid artery
Left subclavian artery
Subclavian artery
Innominate artery
Aortic arch
1
2
Vertebral artery
3
Vertebral artery
4
Vertebral artery
5
Common carotid artery
6
Right subclavian artery
7
Lateral vertebral artery and thyroidea ima artery
Medial vertebral artery
Thyroidea ima artery
8
Thyroidea ima artery
9
Vertebral artery
10

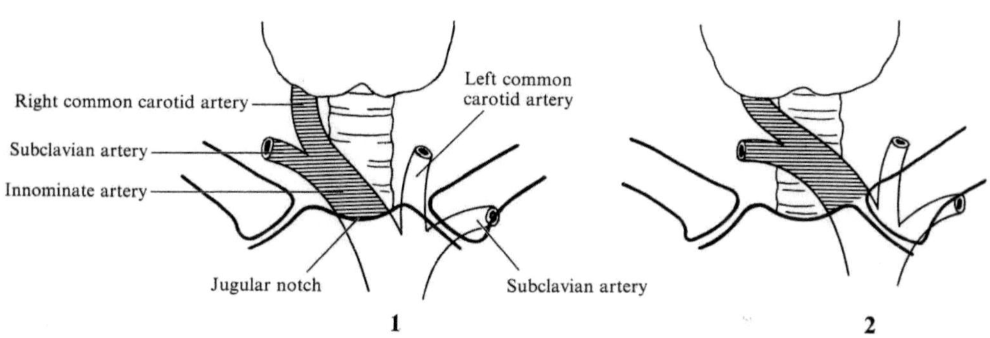
Right common carotid artery
Subclavian artery
Innominate artery
Left common carotid artery
Jugular notch
Subclavian artery
1
2

1.8 (Plates 1–3) Topographical Anatomy of the Neck (Horizontal Sections)

The Plates show an ascending sequence of horizontal sections beginning at the level of C_7 and the thyroid isthmus (Plate 1), at C_2 and C_3 at the level of the hypopharynx and the superior laryngeal aperture (*aditus laryngis*) (Plate 2), and at the level of the dens axis and atlas, with a section through the oropharynx (Plate 3) [1]. These sections are useful for anatomic orientation during ultrasound and CT examinations.

Surgical Importance

Explorations of the neck, neck dissection, thyroidectomy, surgery of the larynx, tracheotomy, fusion of the cervical spine

Reference

1. Pernkopf E and Ferner H (1963)

Plates 1–3 adapted from Pernkopf [1]

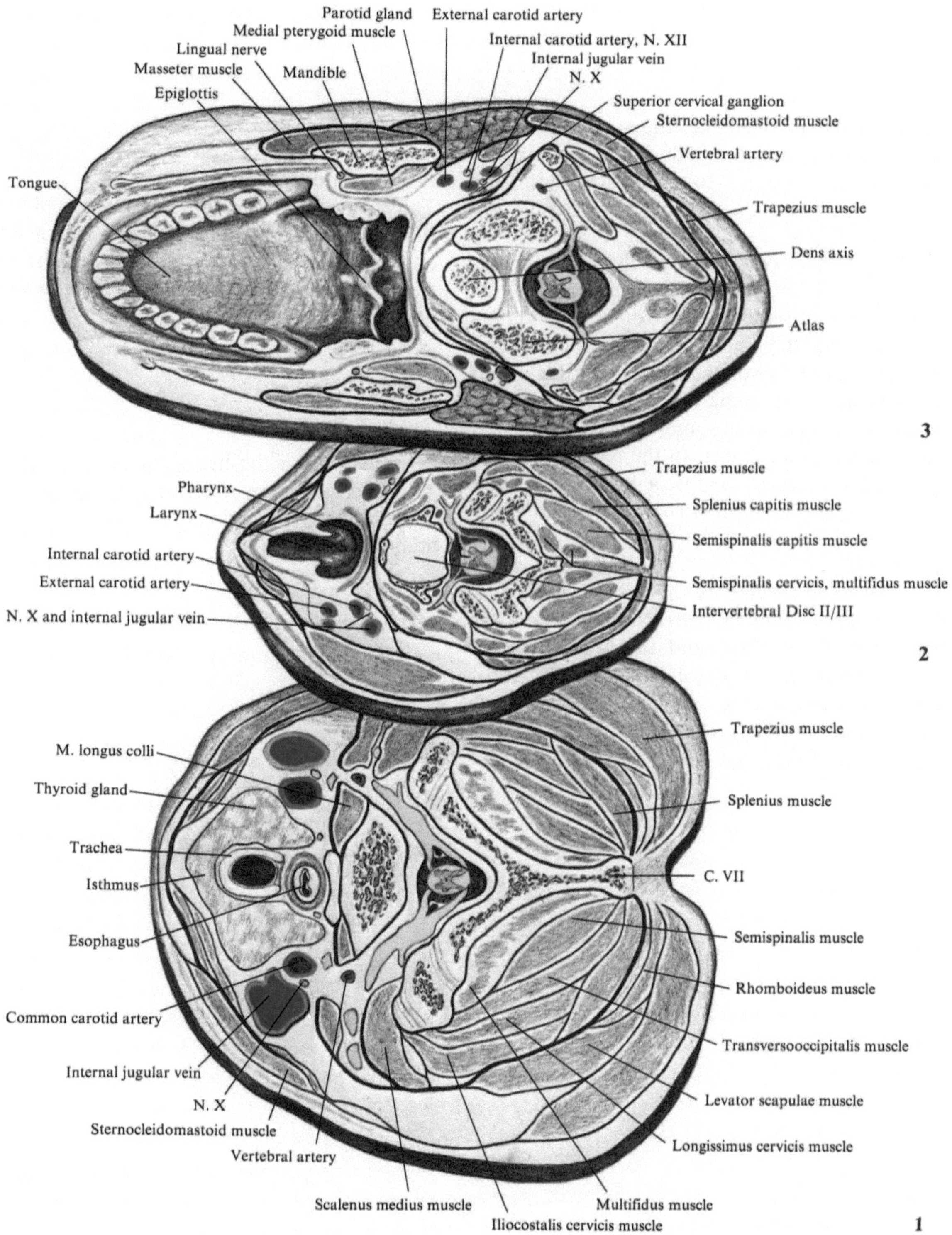
Parotid gland
External carotid artery
Medial pterygoid muscle
Internal carotid artery, N. XII
Lingual nerve
Internal jugular vein
Masseter muscle
Mandible
N. X
Epiglottis
Superior cervical ganglion
Sternocleidomastoid muscle
Vertebral artery
Tongue
Trapezius muscle
Dens axis
Atlas
3
Trapezius muscle
Pharynx
Splenius capitis muscle
Larynx
Semispinalis capitis muscle
Internal carotid artery
Semispinalis cervicis, multifidus muscle
External carotid artery
N. X and internal jugular vein
Intervertebral Disc II/III
2
Trapezius muscle
M. longus colli
Thyroid gland
Splenius muscle
Trachea
C. VII
Isthmus
Semispinalis muscle
Esophagus
Rhomboideus muscle
Common carotid artery
Transversooccipitalis muscle
Internal jugular vein
Levator scapulae muscle
N. X
Sternocleidomastoid muscle
Longissimus cervicis muscle
Vertebral artery
Scalenus medius muscle
Multifidus muscle
Iliocostalis cervicis muscle
1

1.9 The Cervical Fascia

As illustrated in the Plate, the superficial fascia (*lamina superficialis*) extends from the anterior border of the lower jaw to the clavicle and manubrium sterni. In its course, it ensheaths the sternocleidomastoid and trapezius muscles.

The middle layer of the cervical fascia (superficial layer of the *lamina pretrachealis*) ensheaths the suprahyoid muscles and extends from the hyoid bone to the superior thoracic aperture, enclosing the infrahyoid muscles over the larynx, thyroid gland, and trachea. This fascia is fixed caudally to the posterior sternal surface and extends laterally to each clavicle and scapula. In the region of the laryngeal prominence, the fascia fuses with the superficial layer. The suprasternal cavity (*spatium suprasternale*) lies between the split layers of the superficial fascia.

The trachea is covered by the middle layer of the cervical fascia (lamina pretrachealis). The space between the superficial and the middle layer is called the pretracheal space. The thyroid gland is ensheathed by its own connective tissue capsule (see Sect. 2.11). The pretracheal layer is situated at the level of the large vessels anterior to the larynx and extends from the hyoid bone over the laryngeal cartilage to the posterior lamella of the fibrous pericardium.

The posterior prevertebral layer is located between the spine and the pharyngeal constrictor muscles and esophagus [2]. This fascia covers the scalene muscles and also ensheaths the sympathetic trunk and the phrenic nerve, as later discussed.

Surgical Importance

Explorations of the neck, partial laryngeal resections, laryngectomy, tracheotomy, coniotomy, mediastinoscopy, excisions of the floor of the mouth.

References

1. Abrahams P and Webb P (1978)
2. Feneis H (1982)

Plate modified from Abrahams and Webb [1]

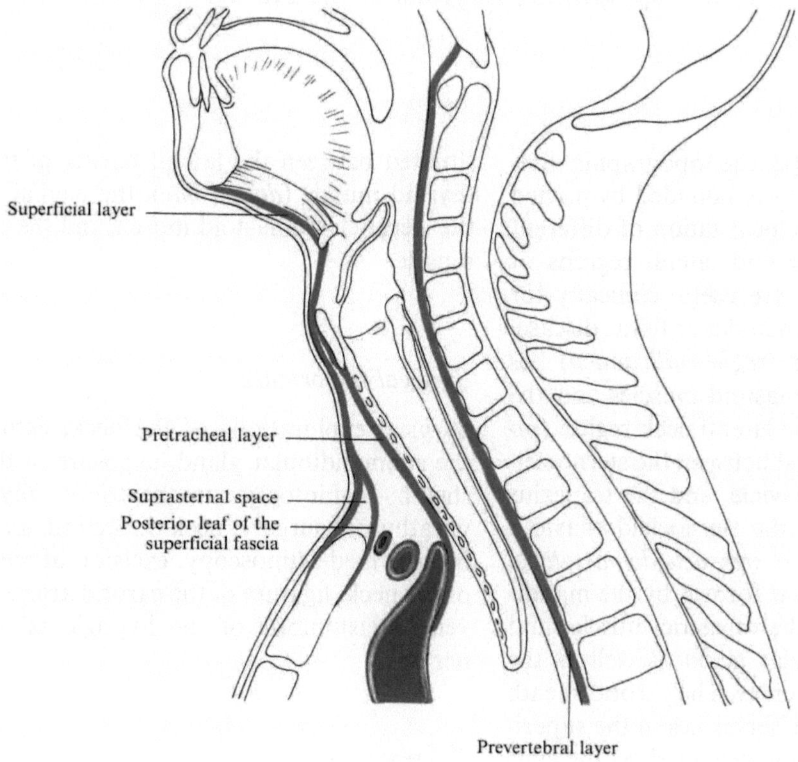
Superficial layer
Pretracheal layer
Suprasternal space
Posterior leaf of the superficial fascia
Prevertebral layer

1.10 Topographic Regions of the Neck

As illustrated by Guleke [1], the topographic division of the neck into triangles bounded by named muscles greatly facilitates localization of different structures in the anterior and lateral regions of the neck. These triangles are useful clinically for defining locations of lymph nodes or tissue disease. The anterior neck region (*regio colli antica*) lies between the sternocleidomastoid muscles and the body of the mandible. The lateral neck region (*regio colli lateralis*) is situated between the sternocleidomastoid muscle, the clavicle, and the trapezius muscle. The region over the sternocleidomastoid muscle is called the *regio sternocleidomastoidea.* The submandibular fossa is formed by the mandible and both bellies of the digastric muscle, and contains the submandibular gland as well as the lingual and hypoglossal nerves. The carotid sheath with its major vessels and nerves lies in the superior carotid triangle, which is delimited by the sternocleidomastoid muscle, the posterior belly of the digastric muscle, and the superior belly of the omohyoid muscle. The inferior carotid triangle is situated between the lateral border of the sternothyroid muscle (*dotted line*), the medial border of the sternocleidomastoid muscle and the omohyoid muscle.

Surgical Importance

Surgical explorations of the neck, extirpation of the submandibular gland, exposure of the carotid sheath, coniotomy, tracheotomy, thyroid and parathyroid surgery, neck dissection, scalene node biopsy, mediastinoscopy, excision of median cysts of the neck, ligature of the carotid artery or jugular vein, anastomosis of the hypoglossal and facial nerves

Reference

1. Guleke N (1953)

Plate from Guleke [1]

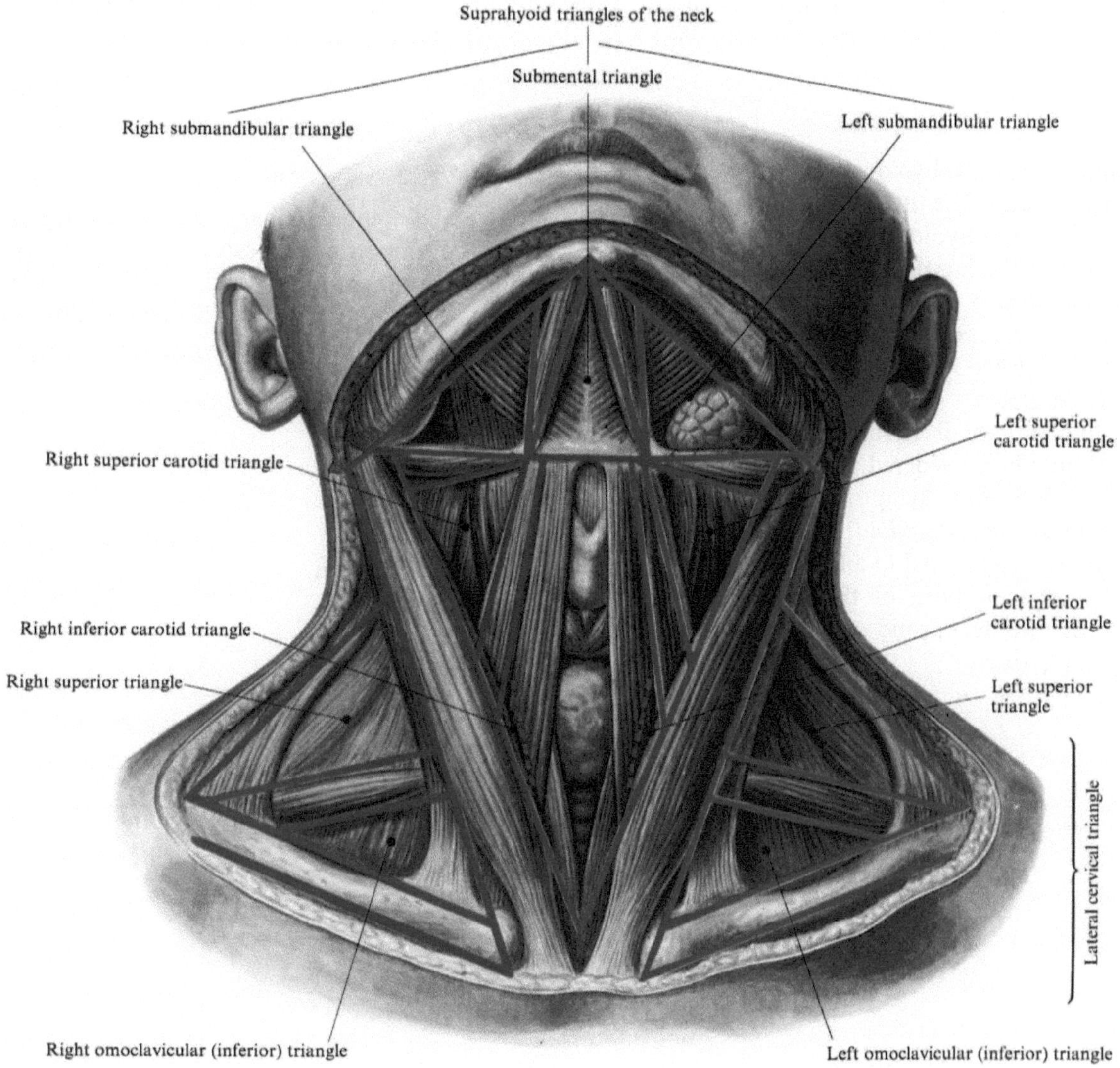
Suprahyoid triangles of the neck
Submental triangle
Right submandibular triangle
Left submandibular triangle
Right superior carotid triangle
Left superior carotid triangle
Right inferior carotid triangle
Left inferior carotid triangle
Right superior triangle
Left superior triangle
Lateral cervical triangle
Right omoclavicular (inferior) triangle
Left omoclavicular (inferior) triangle

1.11 Topographic Relationships of the Pharyngeal Muscles to the Facial Skeleton, the Larynx and the Vertebral Column

The supra- and infrahyoid muscles are attached to the hyoid bone. The thyrohyoid muscles together with the thyrohyoid membrane connect the hyoid bone to the larynx. Surgical division of the suprahyoid muscle makes possible a caudal displacement of the larynx when clinically indicated (as following sleeve resections of subglottic or tracheal stenoses).

The styloid process is normally 2.5 cm in length [1] and should be considered pathological if longer than 3 cm. However, symptoms due to nerve or vessel compression ("*styloid process syndrome*") are not necessarily linked to a long styloid process, since the superior horn of the thyroid cartilage can also irritate the carotid artery to produce cervical pain [2, 4].

Surgical Importance

Resection of the lateral pharyngeal wall, surgical correction of pharyngeal paralysis with dysphagia [3], lateral approach to the medial skull base, surgery of the floor of the mouth, styloid process syndrome, esophagoscopy, laryngotracheoscopy, surgery of median neck cysts.

References

1. Corell RW et al. (1979)
2. Counter RT (1980)
3. Denecke HJ (1980)
4. Moffat DB (1962)

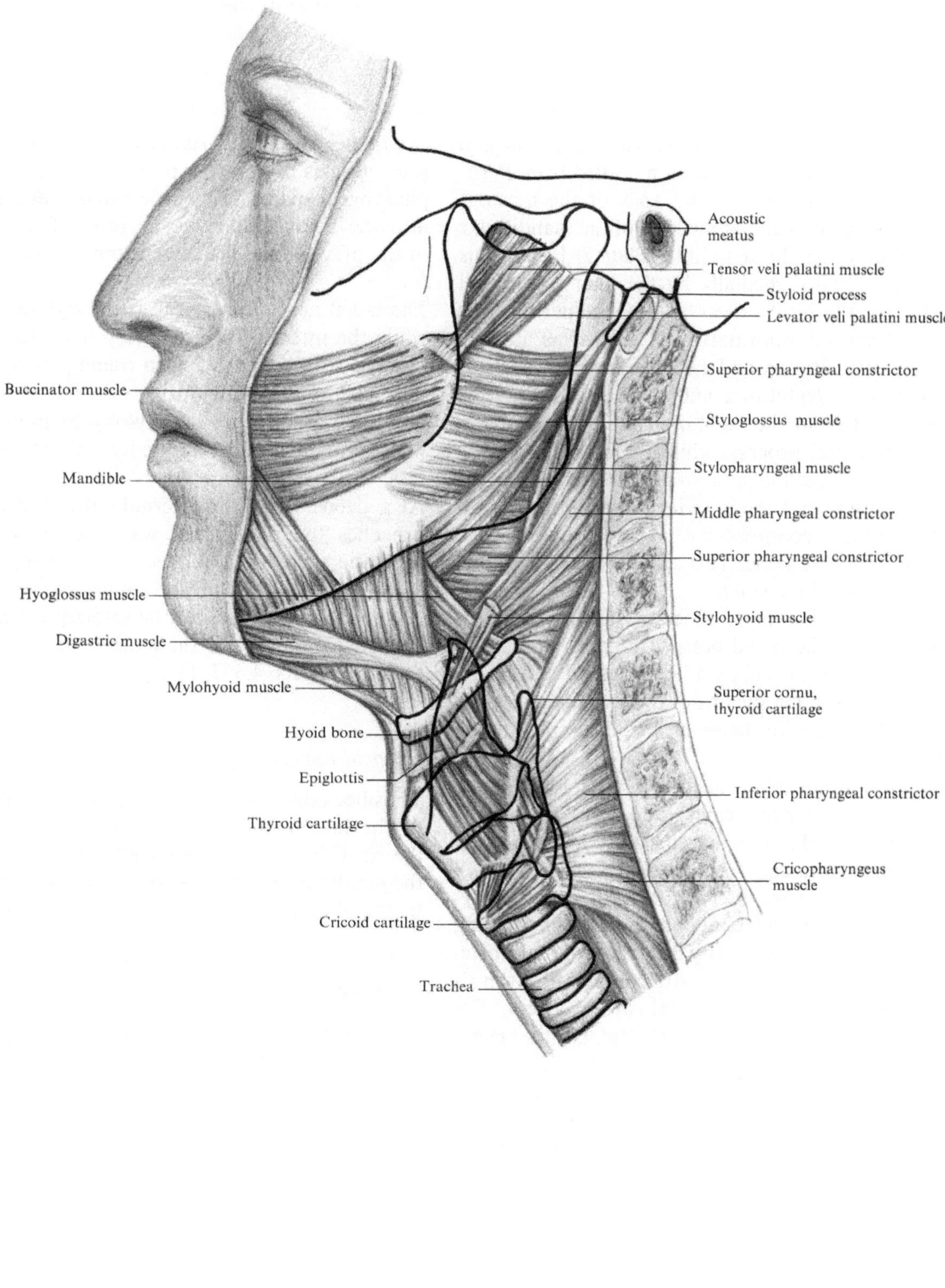
Acoustic meatus
Tensor veli palatini muscle
Styloid process
Levator veli palatini muscle
Superior pharyngeal constrictor
Buccinator muscle
Styloglossus muscle
Stylopharyngeal muscle
Mandible
Middle pharyngeal constrictor
Superior pharyngeal constrictor
Hyoglossus muscle
Stylohyoid muscle
Digastric muscle
Mylohyoid muscle
Superior cornu, thyroid cartilage
Hyoid bone
Epiglottis
Inferior pharyngeal constrictor
Thyroid cartilage
Cricopharyngeus muscle
Cricoid cartilage
Trachea

1.12 (Plates 1–3) The Parapharyngeal Space

The *parapharyngeal* or *lateral pharyngeal space* is shown in a horizontal section modified from Pernkopf [3] at the level of the apex of the mastoid process (Plate 1) and at the level of the mandibular nerve entering the mandible (Plate 2) [1–4]. This space is limited medially by the pharyngeal wall and laterally by the deep layer of the parotid fascia. It can be more narrowly defined as a space filled with fibrous and fatty tissues between the muscles of deglutition and mastication [4]. It is separated from the *carotid space* by the muscles of the styloid process, while the carotid space is separated from the *paraspinal space* by the longus colli muscle and the anterior scalene muscle. All these spaces comprise the parapharyngeal space in its broadest sense. This space extends upwards to the base of the skull.

The lateral pharyngeal space extends inferiorly to the level of the hyoid bone where it is limited by the sheath of the submandibular gland and its attachements to the sheaths of the stylohyoid muscle and the posterior belly of the digastric muscle [2].

The styloid process is located in the middle of the superior pharyngeal cavity and gives rise to the insertions of the stylopharyngeal, styloglossal, and stylohyoid muscles (*the muscle group of Riolan*). The stylopharyngeal septum extends from the styloid process to the lateral pharyngeal wall, dividing the parapharyngeal cavity into anterior and posterior parts. The anterior part contains the tonsils and branches of the external carotid artery. The posterior part includes the internal carotid artery, peripheral trunks of cranial nerves IX–XI, the sympathetic trunk, and the internal jugular vein. The parapharyngeal space is separated from the *retropharyngeal space* by a septum of thin connective tissue. The *retrovisceral space*, or *retro-esophageal space*, is in its upper part known as the retropharyngeal space. This space extends downwards into the mediastinum to the level of the fusion of the prevertebral fascia with the esophageal fascia.

The facial nerve lies lateral to the styloid process while the internal cartoid artery is medial to the process. Consequently, when tracing the artery to the base of the skull during surgery, it is often necessary to displace the stylohyoid muscle forwards and downwards and remove the styloid process.

At a deeper level the internal carotid artery approaches the pharyngeal wall and is separated from the tonsil by a connective tissue septum. In cases where the internal carotid artery forms a vascular loop, the vessel can lie directly beneath the pharyngeal mucosa or the muscles of the tonsillar fossa (see Sects. 4.5–7, 11.2).

Surgical Importance

Tonsillectomy; parotidectomy; lateral approach to the lateral, middle, and posterior sections of the skull base; mandibular resection; surgery of the jugular foramen; styloid process syndrome; ultrasound and CT examinations of the neck.

References

1. Denker A and Kahler O (1925)
2. Hollinshead WH (1961)
3. Pernkopf E (1963)
4. Silver AJ et al. (1984)

Plate modified from Pernkopf [3]

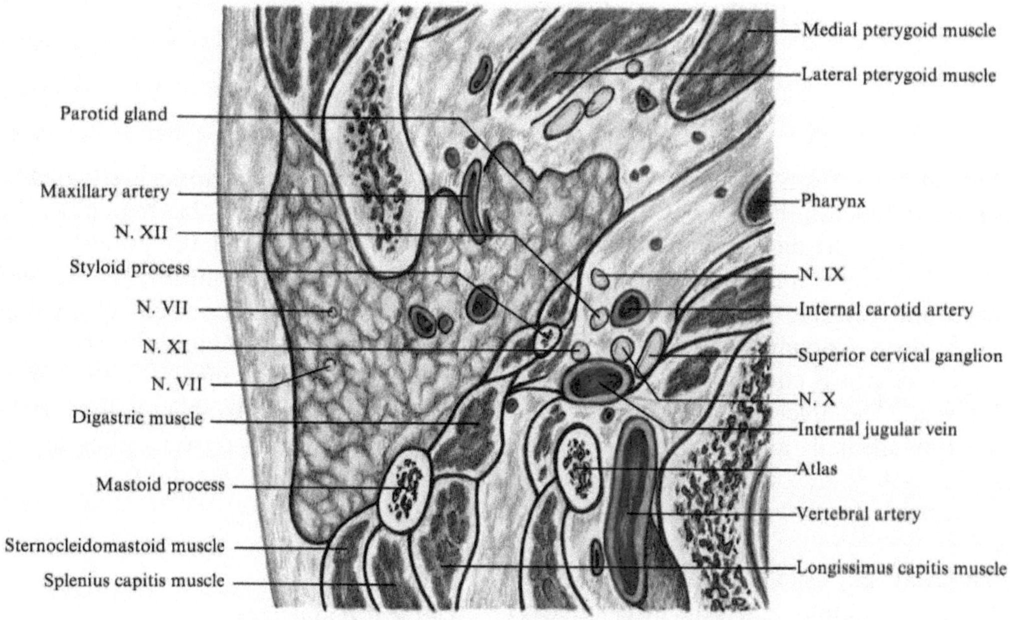
Medial pterygoid muscle
Lateral pterygoid muscle
Parotid gland
Maxillary artery
Pharynx
N. XII
Styloid process
N. IX
N. VII
Internal carotid artery
N. XI
Superior cervical ganglion
N. VII
N. X
Digastric muscle
Internal jugular vein
Atlas
Mastoid process
Vertebral artery
Sternocleidomastoid muscle
Longissimus capitis muscle
Splenius capitis muscle

1

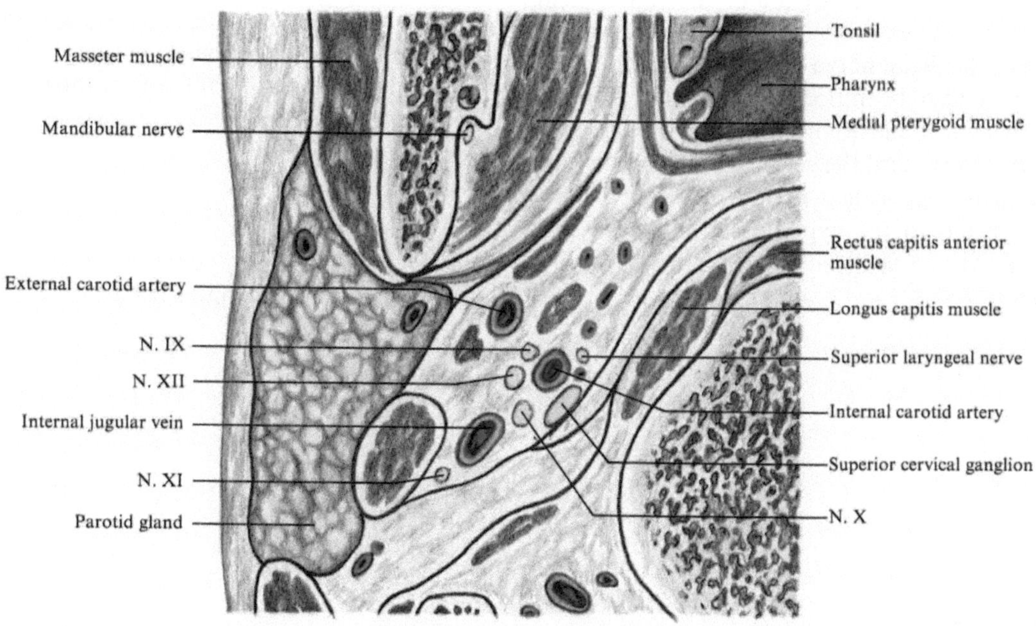
Tonsil
Masseter muscle
Pharynx
Mandibular nerve
Medial pterygoid muscle
Rectus capitis anterior muscle
External carotid artery
Longus capitis muscle
N. IX
Superior laryngeal nerve
N. XII
Internal carotid artery
Internal jugular vein
Superior cervical ganglion
N. XI
Parotid gland
N. X

2

1.13 Topography of the Submandibular and the Retromandibular Fossae

The submandibular fossa (*trigonum submandibulare*) is situated between the body of the mandible and both bellies of the digastric muscle. The medial wall of the fossa is formed by the mylohyoid, hyoglossus, and styloglossus muscles. The submandibular fossa contains the submandibular gland, which in turn is covered by the superficial cervical fascia. The duct of this gland (*ductus Warthoni*) runs medially and cranially toward the mylohyoid muscle to the orifice of the submandibular duct (*caruncula sublingualis, papilla salivaria sublingualis*).

While the facial vein usually crosses the posterior part of the gland laterally, the facial artery generally lies medially to it. When the submandibular gland is surgically extirpated, both vessels are commonly first exposed close to the point where the artery emerges above the gland. By ligating and displacing these vessels superiorly, the mandibular branch of the facial nerve (*ramus marginalis mandibulae*) will also be reflected superiorly from the gland. A neural anastomosis frequently passes between this nerve and the cervical plexus. The parotid gland is represented in the plate by a *dotted line*. Adjacent to the hypoglossal nerve are veins which are prone to bleed when the nerve is surgically exposed.

The retromandibular fossa is situated between the ascending ramus of the mandible, the external acoustic meatus, and the anterior border of the sternocleidomastoid muscle [1]. Its medial wall is formed by the posterior belly of the digastric muscle and the stylohyoid muscle. The external carotid artery normally lies medial to these muscles, while lateral and cranial to them is the facial nerve. The fossa contains the retromandibular vein and the parotid gland (see Sect. 1.19).

Surgical Importance

Excision of the submandibular gland; parotidectomy; lateral approach to the infratemporal and pterygoid fossae and pharyngeal wall; dorsocaudal approach to the lateral, posterior and medial sections of the skull base; surgery of the facial nerve; hypoglossal-facial anastomosis; exposure of the carotid sheath.

Reference

1. Lanz T v and Wachsmuth W (1955)

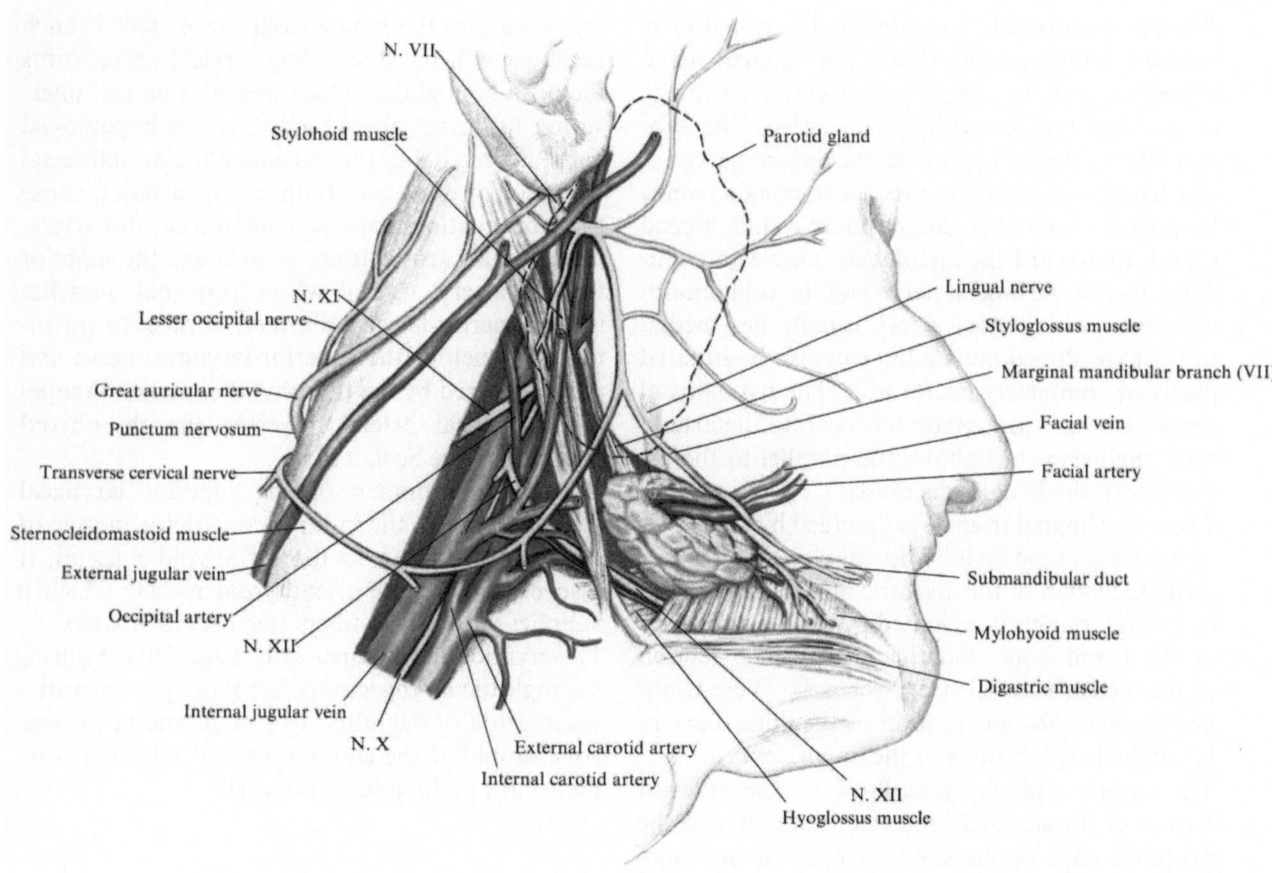
N. VII
Stylohoid muscle
Parotid gland
Lingual nerve
Styloglossus muscle
N. XI
Lesser occipital nerve
Marginal mandibular branch (VII)
Great auricular nerve
Facial vein
Punctum nervosum
Transverse cervical nerve
Facial artery
Sternocleidomastoid muscle
External jugular vein
Submandibular duct
Occipital artery
Mylohyoid muscle
N. XII
Digastric muscle
Internal jugular vein
N. X
External carotid artery
Internal carotid artery
N. XII
Hyoglossus muscle

1.14 Topography of the Carotid Triangle and the Submandibular Fossa After Removal of the Submandibular Gland

The parasympathetic submandibular ganglion is situated anatomically where the lingual nerve curves toward the tongue and receives secretory fibers from the chorda tympani nerve. The submandibular duct crosses over the lingual nerve.

The hyoglossus nerve and its accompanying veins lie lateral to the hyoglossus muscle and ascend superiorly toward the mylohyoid muscle and caudally toward the lingual nerve and the submandibular duct. The lingual artery usually lies medial to the hypoglossal muscle but can also be situated partly or completely lateral to it. The hypoglossal nerve can serve as a guide for locating the artery, since the nerve runs above and parallel to the artery above the hyoglossus muscle.

The upper lingual triangle is bounded by the posterior margin of the mylohyoid muscle and the intermediate tendon of the digastric muscle. The inferior lingual triangle is delimited by the greater cornu of the hyoid bone and the intermediate tendon of the digastric muscle (see Sect. 4.3). These triangles facilitate the localization of the lingual artery in surgical explorations of the upper neck.

The carotid triangle is outlined by the anterior border of the sternocleidomastoid muscle and the posterior edge of the superior belly of the omohyoid muscle (see Sect. 1.10). The contents of the triangle include (medially) the common carotid artery with its bifurcation into the external and internal carotid arteries and (laterally) the internal jugular vein. Between the two vessels in the carotid sheath and situated somewhat more dorsally are the vagus nerve and (ventrolaterally) the descending branch of the hypoglossal nerve. This branch together with the descending cervical nerve forms the ansa hypoglossi, which provides motor innervation to the infrahyoid muscles. The hypoglossal nerve lies medial to the retromandibular and facial veins, and crosses over both carotid arteries. Close to its bifurcation from the common carotid artery, the external carotid artery gives rise to the superior thyroid artery, one of whose principal branches is the superior laryngeal artery. Parallel to the internal branch of the superior laryngeal nerve and partly covered by the thyrohyoid muscle, the superior laryngeal artery traverses the thyrohyoid membrane (see Sect. 2.6).

The external branch of the superior laryngeal nerve innervates the inferior constrictor muscle of the pharynx. Deep to the infrahyoid muscles, it also extends to the cricothyroid muscle which it supplies with motor fibers (see Sect. 2.9, 2.16).

Preservation of the superior laryngeal nerve during supraglottic laryngectomy facilitates postoperative resumption of deglutition by maintaining the sensory supply of the endolarynx and allowing complete cord adduction to occur [1].

Surgical Importance

Extirpation of the submandibular gland, exploration of the carotid sheath, functional neck dissection, hypoglossal-facial nerve anastomoses, supraglottic laryngectomy, parotidectomy, lateral pharyngectomy, cricopharyngeal myotomy.

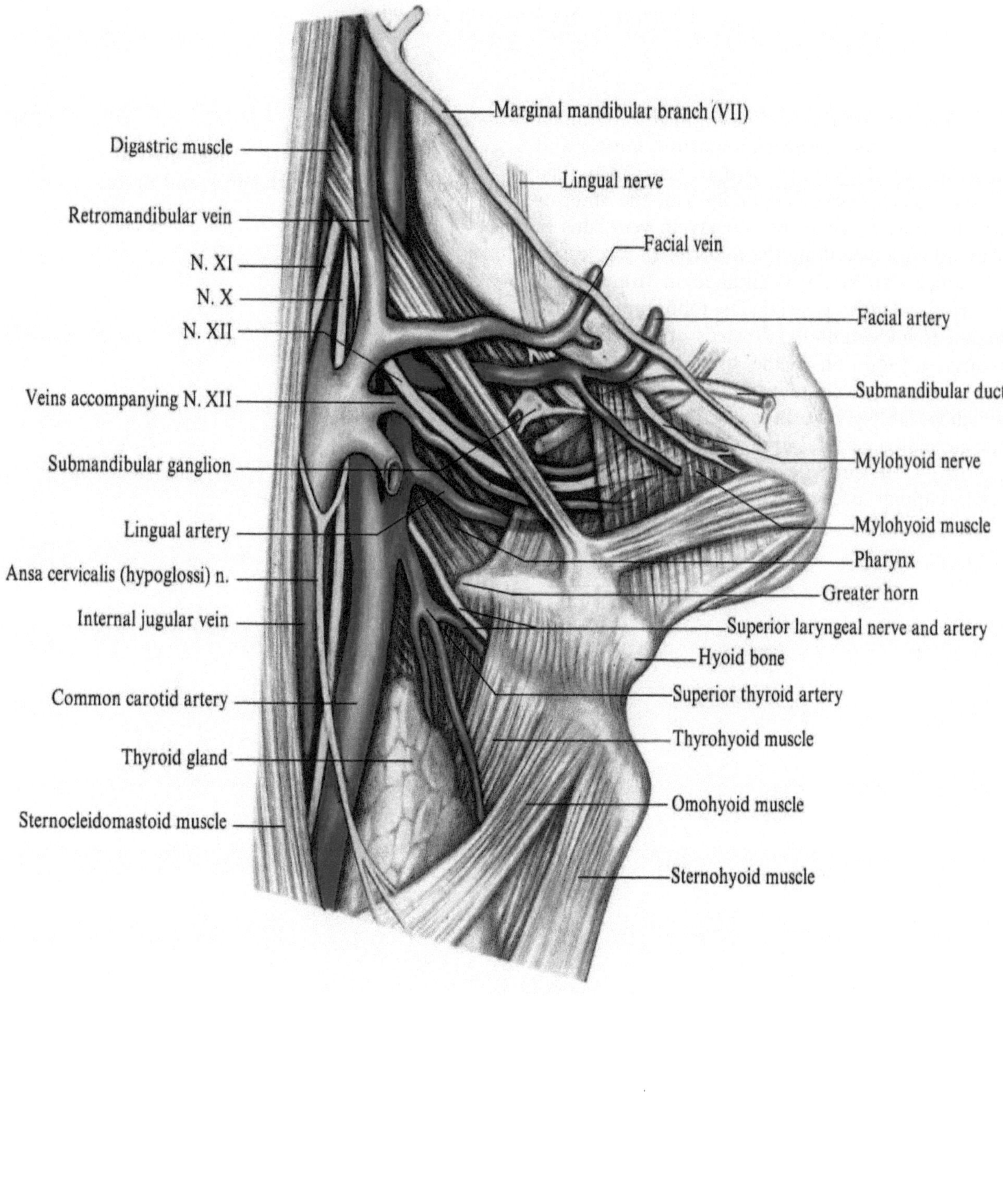
Marginal mandibular branch (VII)
Digastric muscle
Lingual nerve
Retromandibular vein
Facial vein
N. XI
N. X
Facial artery
N. XII
Veins accompanying N. XII
Submandibular duct
Submandibular ganglion
Mylohyoid nerve
Mylohyoid muscle
Lingual artery
Pharynx
Ansa cervicalis (hypoglossi) n.
Greater horn
Internal jugular vein
Superior laryngeal nerve and artery
Hyoid bone
Common carotid artery
Superior thyroid artery
Thyrohyoid muscle
Thyroid gland
Omohyoid muscle
Sternocleidomastoid muscle
Sternohyoid muscle

1.15 The Anterior Scalene Foramen (*Foramen Scalenorum Anterius*)

The so-called anterior scalene foramen is bounded anteriorly by the sternocleidomastoid muscle and inferiorly by the first rib [2] (see Sect. 2.10). The phrenic nerve passes diagonally over the anterior scalene muscle, while the subclavian vein runs in a shallow groove along the first rib. In rare cases, the subclavian artery is situated in front of the anterior scalene muscle [1]. The topography of this region is important for *Daniel*'s or scalene node biopsy. In addition to the structures mentioned, the foramen also contains the thyrocervical trunk, which variably gives off the inferior thyroid artery, the ascending cervical artery, the transverse cervical artery, the superficial cervical artery, and the suprascapular artery.

Surgical Importance

Scalene biopsy, neck dissection, thyroid surgery, myocutaneous flaps in the neck and thorax.

References

1. Adachi B (1928)
2. Guleke N (1953)

Plate from Guleke [2]

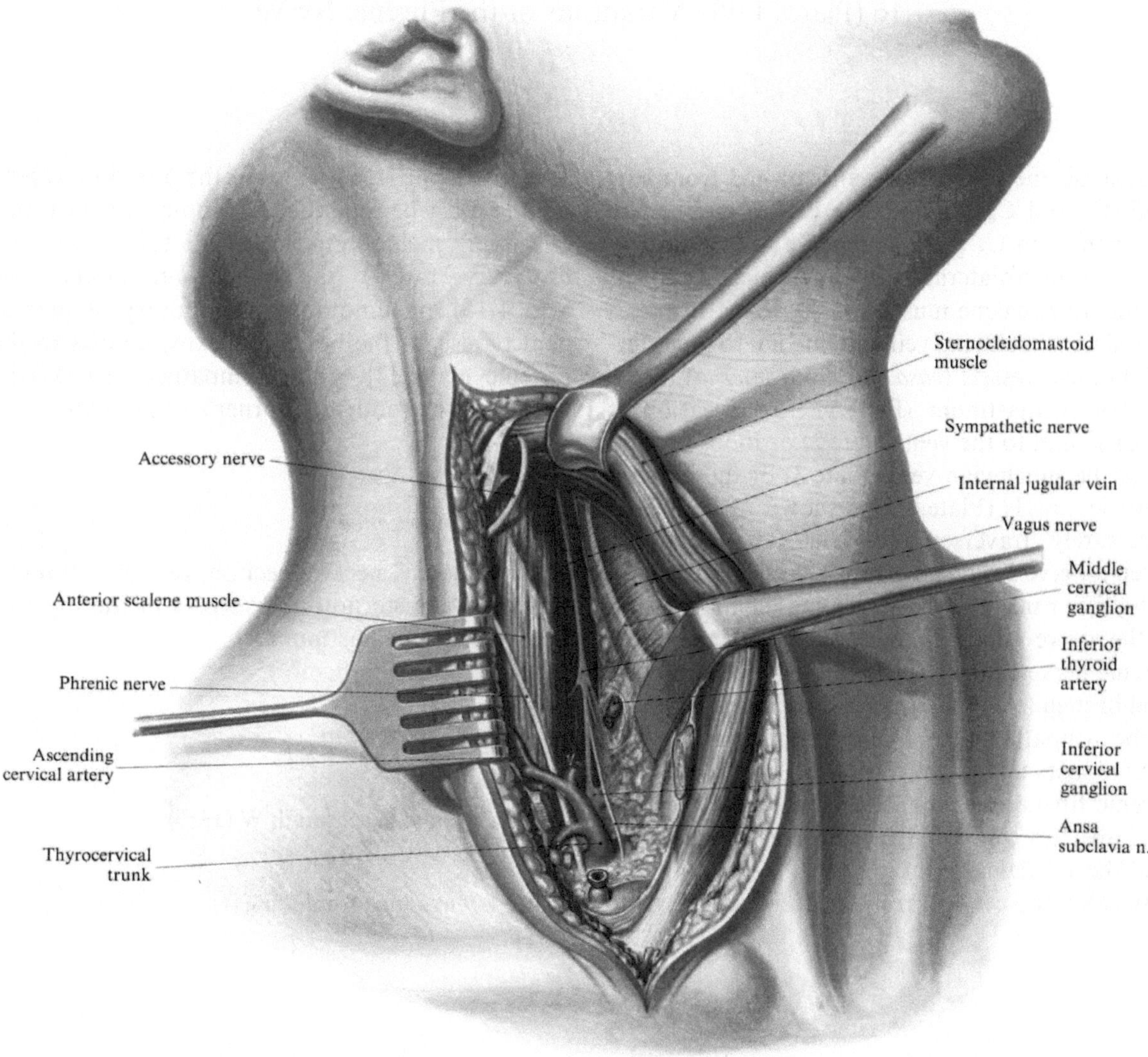
Sternocleidomastoid muscle
Sympathetic nerve
Internal jugular vein
Vagus nerve
Middle cervical ganglion
Inferior thyroid artery
Inferior cervical ganglion
Ansa subclavia n.
Accessory nerve
Anterior scalene muscle
Phrenic nerve
Ascending cervical artery
Thyrocervical trunk

1.16 (Plates 1–9) Variations of the Phrenic Nerve

In general, the phrenic nerve is formed from two rami (C_3 and C_4), although it can rarely include a branch from C_5 [1,2]. The nerve runs obliquely from the superolateral to the inferomedial part of the anterior scalene muscle. In so doing, it passes behind the subclavian vein and medial to the internal thoracic vessels (*vasa thoracica interna*) in the anterior mediastinum (Plate 1). On occasion it passes lateral to the vein (Plate 2) or loops around either the subclavian vein (Plate 3) or the internal thoracic vessels (Plate 4). As shown in Plate 5, it also rarely traverses the vein. The accessory phrenic nerve originates from C_8 or T_1 (Plate 6), and either runs separately or joins the main trunk of the nerve (Plate 7). Part of the fibers of the phrenic nerve can run through the descending cervical branch (*ramus cervicalis descendens*) (Plate 8) or the sympathetic trunk and the superior cervical cardiac nerve (Plate 9). In the latter case, the phrenic fibers leave the cardiac nerve at the thoracic aperture and join the main phrenic nerve behind the subclavian vein.

Surgically, the cervical ribs are best exposed from a ventral approach. As such, the anterior scalene muscle must be separated. Possible complications of surgery include injuries to the subclavian vein or (very rarely) the subclavian artery (see Sect. 1.15), pneumothorax, temporary or permanent paresis of the brachial plexus, injuries to the phrenic nerve [3] or the sympathetic trunk (the latter injury producing Horner's syndrome).

Surgical Importance

Scalene biopsy, neck dissection, cervical rib resections, median sternotomy, surgery of the superior and anterior mediastinum.

References

1. Lang J (1979)
2. Lanz T v and Wachsmuth W (1955)
3. Malin JP (1979)

Plates 1–9 modified from Lang [1]

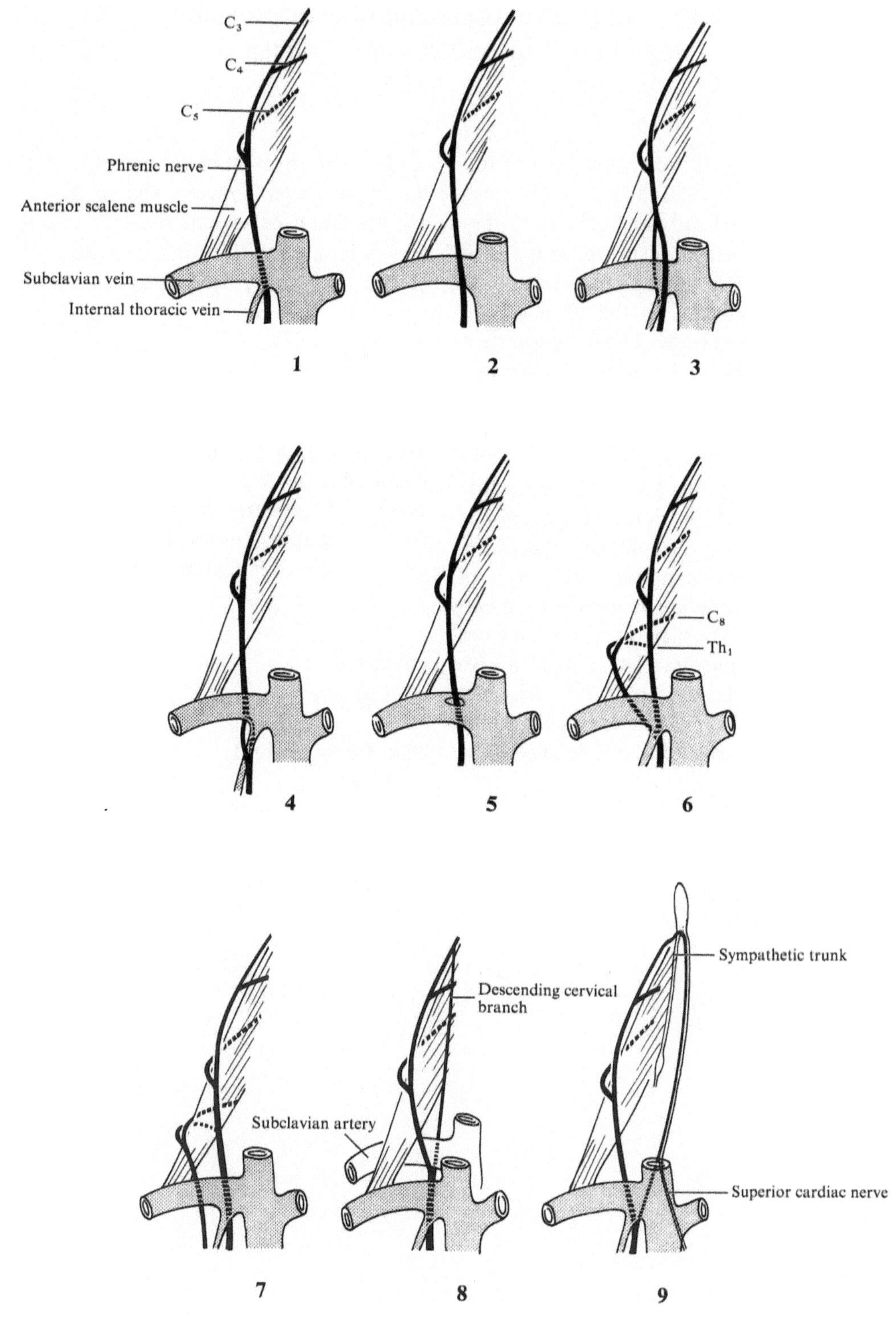

C_3
C_4
C_5
Phrenic nerve
Anterior scalene muscle
Subclavian vein
Internal thoracic vein
1
2
3
C_8
Th_1
4
5
6
Sympathetic trunk
Descending cervical branch
Subclavian artery
Superior cardiac nerve
7
8
9

1.17 The Stellate (Cervicothoracic) Ganglion and the Scalene-Vertebral Triangle

As described by Guleke [1], the sympathetic trunk is covered by the deep cervical fascia, and follows the longus muscles of the head and neck. The superior cervical ganglion has a fusiform shape and is approximately 2.5 cm long, begining at the base of the skull and descending to the level of the greater cornu of the hyoid bone. The ganglion then sends terminal branches to the internal, external, and common carotid plexuses, the pharyngeal and cardiac plexus, and the superior cervical cardiac nerve.

The middle cervical ganglion gives off the middle cervical cardiac nerve at the level of C_6, and is either ventral or dorsal to the inferior thyroid artery. The inferior cervical ganglion forms part of the stellate ganglion (*ganglion cervicothoracicum*) by fusing with the first thoracic ganglion and lies in front of the vertebral artery. This ganglion also lies dorsal to the subclavian artery on the head of the first rib. The sympathetic trunk loops around the subclavian artery (*ansa subclavia*). The scalene-vertebral triangle is medial to the anterior scalene muscle, between the longus colli muscle and the anterior scalene muscle. The vertebral artery is located within this triangle, which also includes the sympathetic plexus and the vertebral vein.

Surgical Importance

Stellate ganglion blockade, surgery of the cervical esophagus, resection of the sympathetic trunk, exposure of the vertebral artery, surgery of cervical chemodectomas (glomus tumors), tumors of the sympathetic trunk and brachial plexus.

Reference

1. Guleke N (1953)

Plate from Guleke [1]

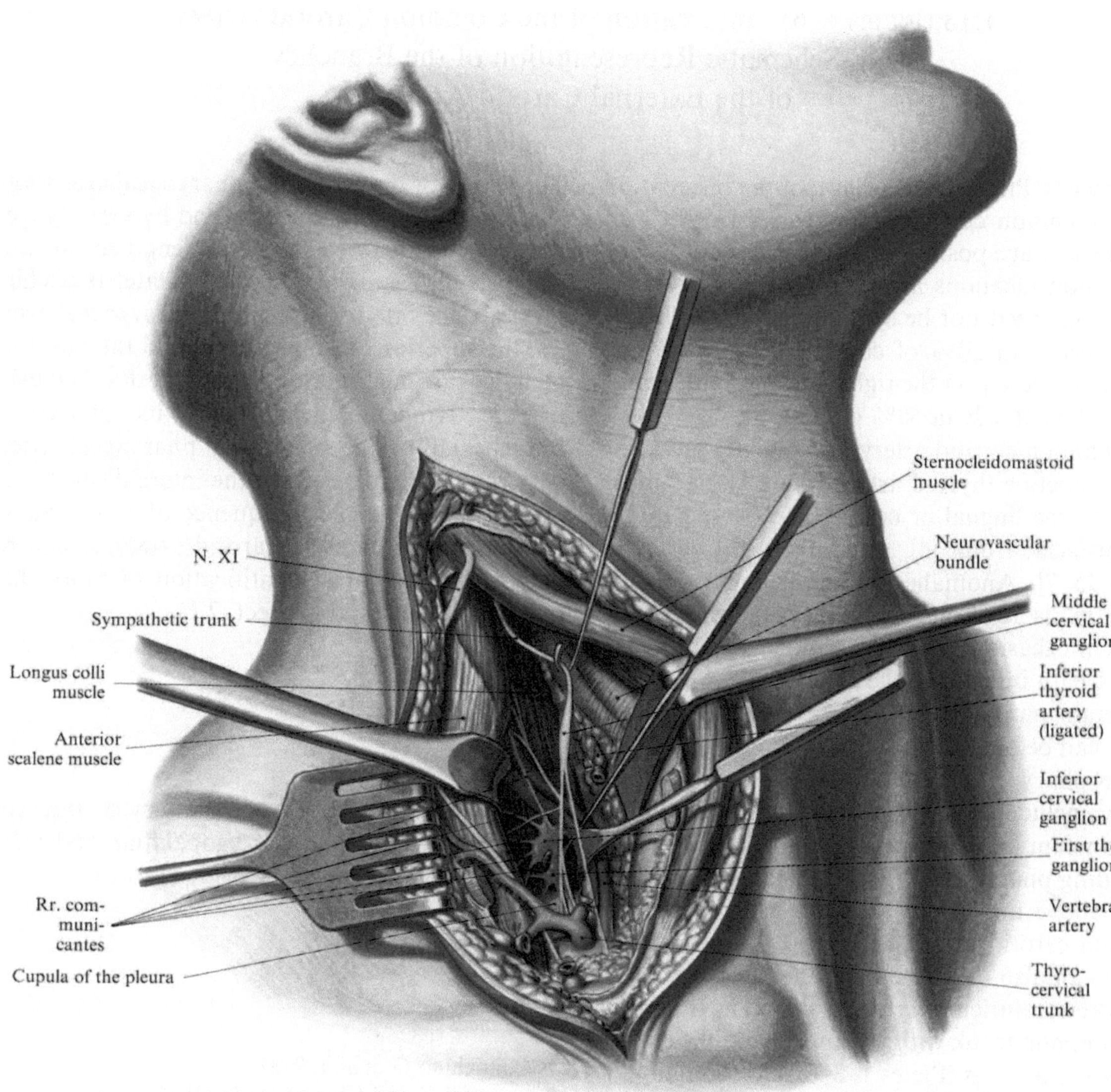

Sternocleidomastoid muscle
Neurovascular bundle
N. XI
Middle cervical ganglion
Sympathetic trunk
Longus colli muscle
Inferior thyroid artery (ligated)
Anterior scalene muscle
Inferior cervical ganglion
First thoracic ganglion
Rr. com- muni- cantes
Vertebral artery
Cupula of the pleura
Thyro- cervical trunk

1.18 (Plates 1–6) Bifurcation of the Common Carotid Artery and Schematic Representation of the Branches of the External Carotid Artery

As shown in Plates 1–3, the level of the bifurcation of the common carotid artery varies from C_6 to C_2. The average position is at the level of C_4 while uncommon locations are situated at the levels of T_1–T_3 and need not be symmetrical on each side of the neck. In 20% of cases, the bifurcation is located above C_4 on the right side and in the same site on the left side in 50% of cases [6].

The common carotid artery infrequently gives rise to the superior thyroid artery (Plate 4), and more rarely to the lingual or occipital arteries, a thyro-lingual-facial trunk [4], or to the right vertebral artery [5, 7]. Anomalies of the common carotid artery in its relationship to the vagus nerve are rare. On occasion, the nerve can lie within the carotid sheath in front of the artery.

In certain pathologies, the ascending pharyngeal artery can be the principle artery supplying tumors of the pharynx or lateral skull base. Such neoplasms include juvenile fibromas as well as glomus tympanicum and jugulare chemodectomas. The ascending pharyngeal artery originates at the level of or above the lingual artery as a branch of the external carotid artery. However, the pharyngeal artery can also originate from the facial, occipital or posterior auricular arteries, as well as from part of a common trunk with the lingual and the superior thyroid arteries (Plates 1, 3–6). Occasionally the ascending pharyngeal artery originates from the posterior side of the bifurcation (Plate 2) so that a conventional ligature of the external carotid artery to control bleeding will *not* include this vessel.

The length of the ascending pharyngeal artery varies. When it supplies the oro- and hypopharynges, the vessel tends to be short, but lengthens in cases where it also supplies the soft palate. Branching off it are the inferior and middle pharyngeal arteries, the superior tonsillar artery, and the posterior mengingeal and inferior tubal arteries (inclusive of the tympanic artery and its pharyngeal branches) [2]. The ascending pharyngeal artery very rarely originates from the internal carotid artery (Plate 1). As a consequence of this, planned ligation of the external carotid artery to control hemorrhage requires identification of more than one arterial branch (see Sect. 2.1).

Surgical Importance

Neck dissection; surgery of the larynx, pharynx, upper esophagus, and pterygopalatine and infratemporal fossae (see Sects. 1.12, 13, 6.23)

References

1. Chidichino G et al. (1978)
2. Guerrier Y and Charissoux G (1981)
3. Krmpotić-Nemanić J (1978)
4. Nikolić V (1961)
5. Palmer FJ (1977)
6. Smith D and Larsen JL (1979)
7. Tan WS and Spigos DG (1979)

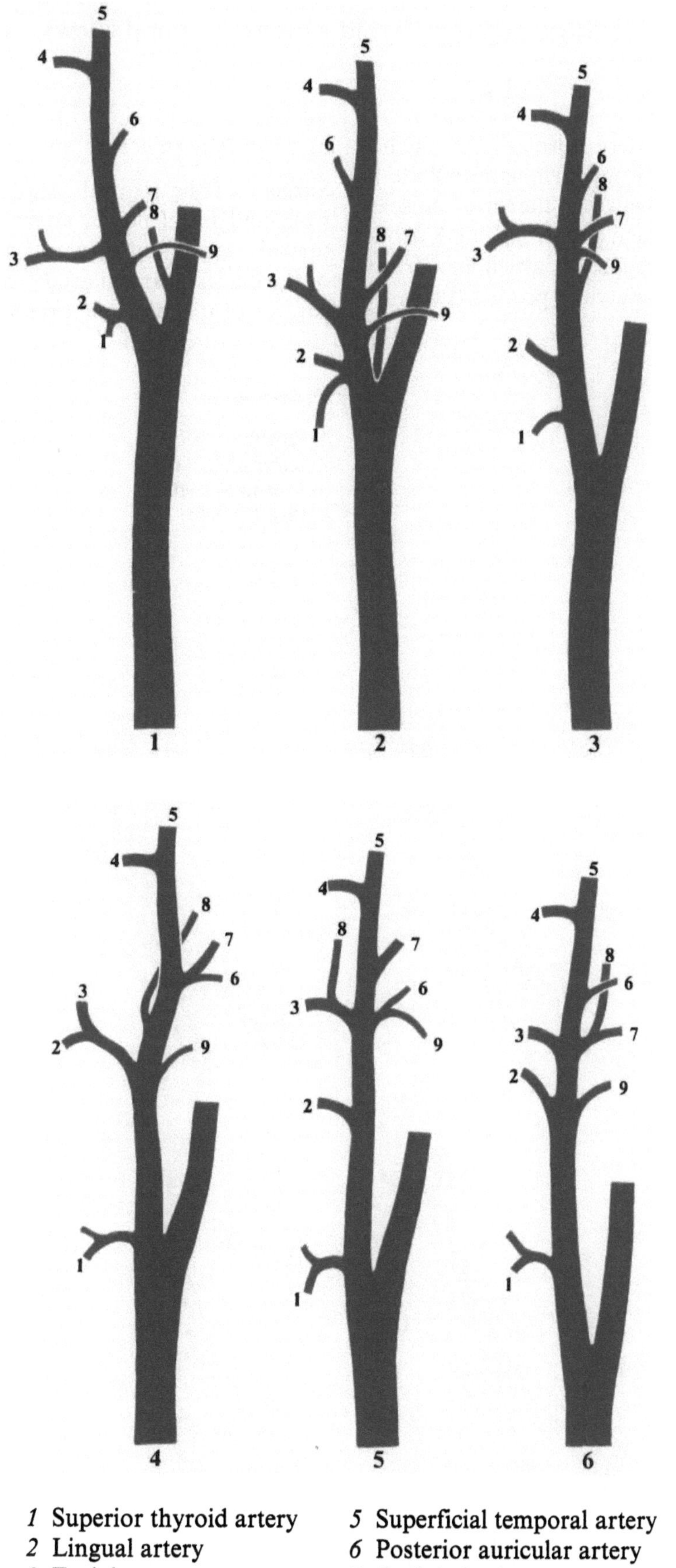

1 Superior thyroid artery
2 Lingual artery
3 Facial artery
4 Maxillary artery
5 Superficial temporal artery
6 Posterior auricular artery
7 Occipital artery
8 Ascending pharyngeal artery
9 Sternocleidomastoid artery

1.19 (Plates 1–4) Relationship of the External Carotid Artery to the Digastric and Stylohyoid Muscles (Lateral View)

The external carotid artery normally lies medial to the digastric and stylohyoid muscles (Plate 1) and less frequently between the two muscles (Plate 2) or lateral to them (Plate 3) [2–4]. Koebke [1] also describes dissections in which this artery forms a loop around the styloid process (Plate 4).

Surgical Importance

Surgery of the parotid gland, lateral pharyngeal wall, and facial nerve; lateral approaches to the medial, lateral and posterior skull base; exposure of the internal carotid artery and jugular foramen; surgery of the infratemporal fossa.

References

1. Koebke J (1976)
2. Krmpotić J (1946)
3. Morita S (1902)
4. Pisk E (1914)

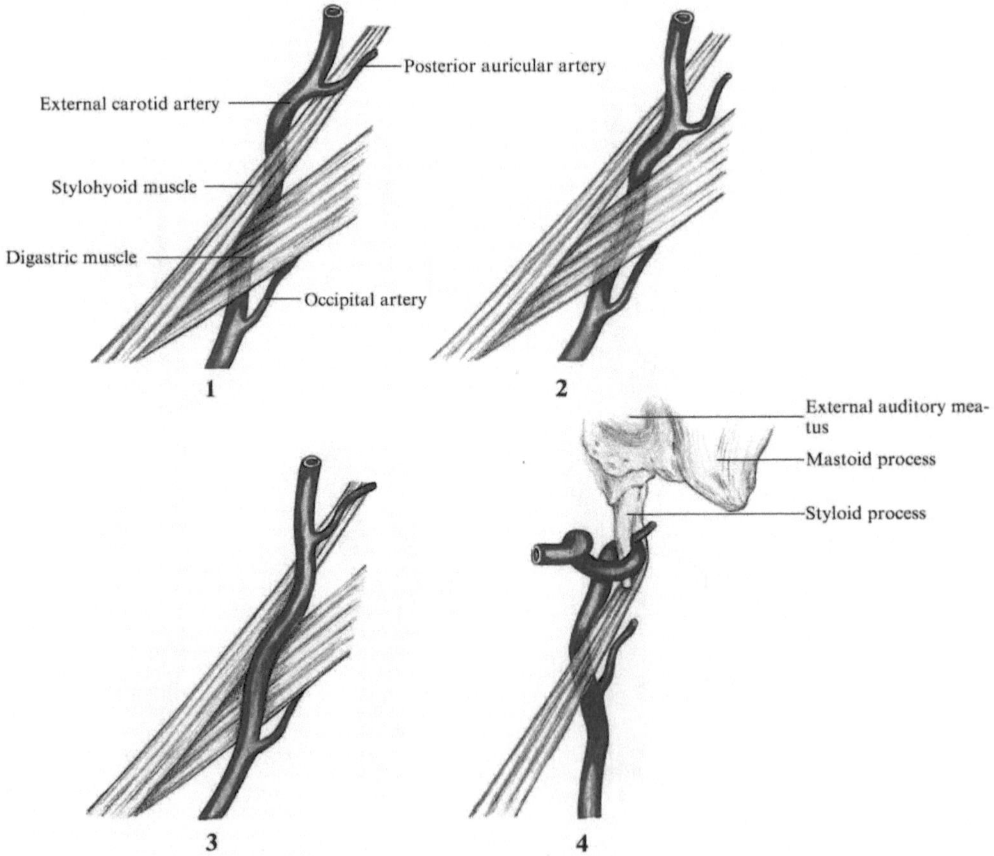

Plate 1 shows the classic pattern for branching of the external carotid artery and occurs in 32.5% of cases. However, many variations exist, as depicted in Plates 2–8 [1–3].

In clinical practice involving bleeding tumors of the neck and skull base (e.g. chemodectomas), it is important to perform angiography, if possible with subtraction techniques, *before* planning any surgery. The use of selective and superselective embolizations to occlude specific vessels and decrease tumor vascularity depends on the vessel anastomoses present.

In certain instances, the ascending pharyngeal artery is an important feeding vessel to skull base lesions. Clinically significant variations have been discussed in Sect. 1.8.

Surgical Importance

Surgery of lesions of the neck and skull base, inclusive of chemodectomas and nasopharyngeal fibromas (see Sects. 1.18, 19).

References

1. Aaron C and Chawaff AR (1967)
2. Krmpotić-Nemanić J (1978)
3. Lasjaunias P (1981, 1983)

Plates 1–8 adapted from Aaron and Chawaff [1]

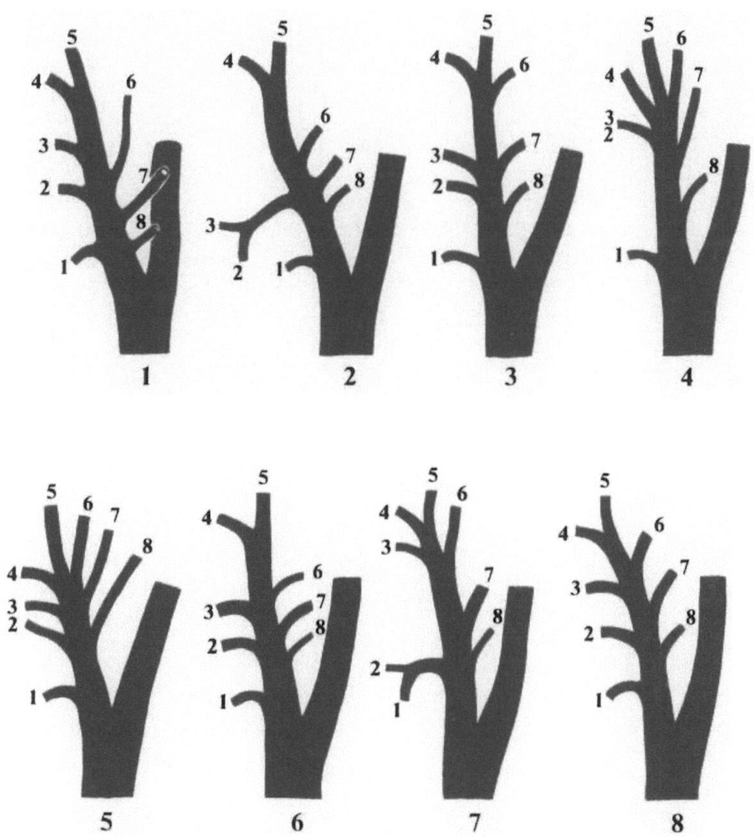

1 Superior thyroid artery
2 Lingual artery
3 Facial artery
4 Maxillary artery
5 Superficial temporal artery
6 Ascending pharyngeal artery
7 Posterior auricular artery
8 Occipital artery

1.21 (Plates 1–5) Variations of the Internal Jugular Vein

The tributaries of the internal jugular vein display many variations, as shown in Plates 1–5. The facial, lingual, or superior thyroid veins can join the jugular vein at separate locations (Plate 1) or form common trunks, including a thyrolingual-facial trunk (Plate 2), a thyrolingual trunk with a separate opening for the facial vein (Plate 3), a lingual-facial trunk with a separate communication for the superior thyroid vein (Plate 4), or a double thyrolingual-facial trunk (Plate 5).

In selected cases, venipuncture of the internal jugular vein in the newborn can be performed in *Sedillot's triangle*, which is located between the sternal and clavicular heads of the sternocleidomastoid muscle and the clavicle [2].

The relationship of the internal jugular vein or its bulb to the cranial nerves within the jugular foramen varies considerably, with multiple anastomoses possible (see Sect. 4.5).

Surgical Importance

Insertion of central venous catheters (subclavian or internal jugular vein), cervical node biopsies, neck dissection, dissection of the floor of the mouth.

References

1. Lang J (1979, 1983)
2. Seriat-Gautier B and Collet S (1981)

Plates 1–5 modified from Lang [1]

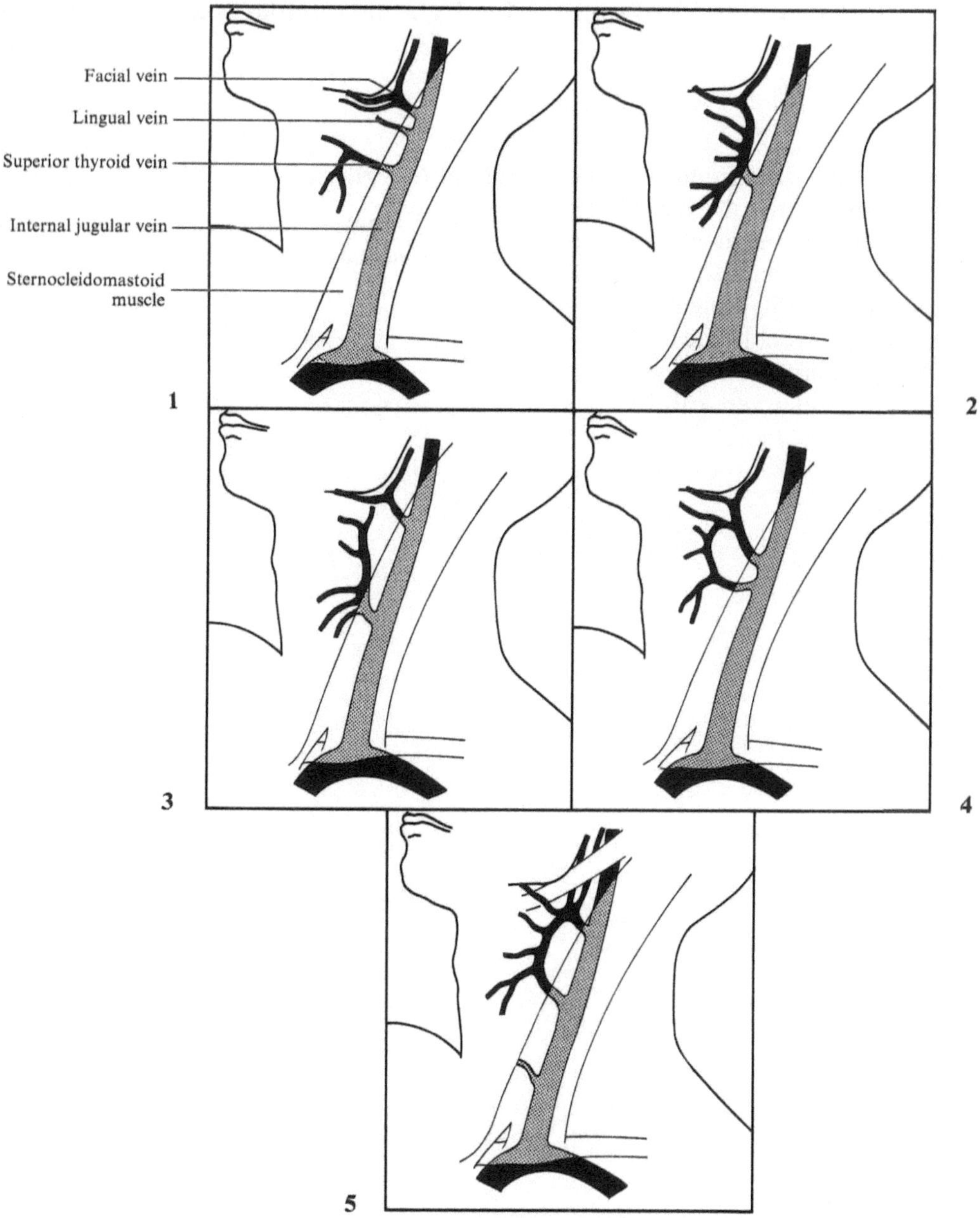

Facial vein
Lingual vein
Superior thyroid vein
Internal jugular vein
Sternocleidomastoid muscle
1
2
3
4
5

1.22 (Plates 1–8) Variations of the External Jugular Vein

The branching patterns and the opening of the extenal jugular vein are frequently variable [1]. The vein most commonly flows into the subclavian vein (Plates 1–5, 7, 8). However, it can also enter both the subclavian and the internal jugular veins (Plate 3) or only the latter vessel (Plate 6). In the presence of a very large external jugular vein, the diameter of the internal jugular vein can be relatively small.

Surgical Importance

Insertion of central venous catheter, neck explorations or dissection.

Reference

1. Krmpotić-Nemanić J (1978)

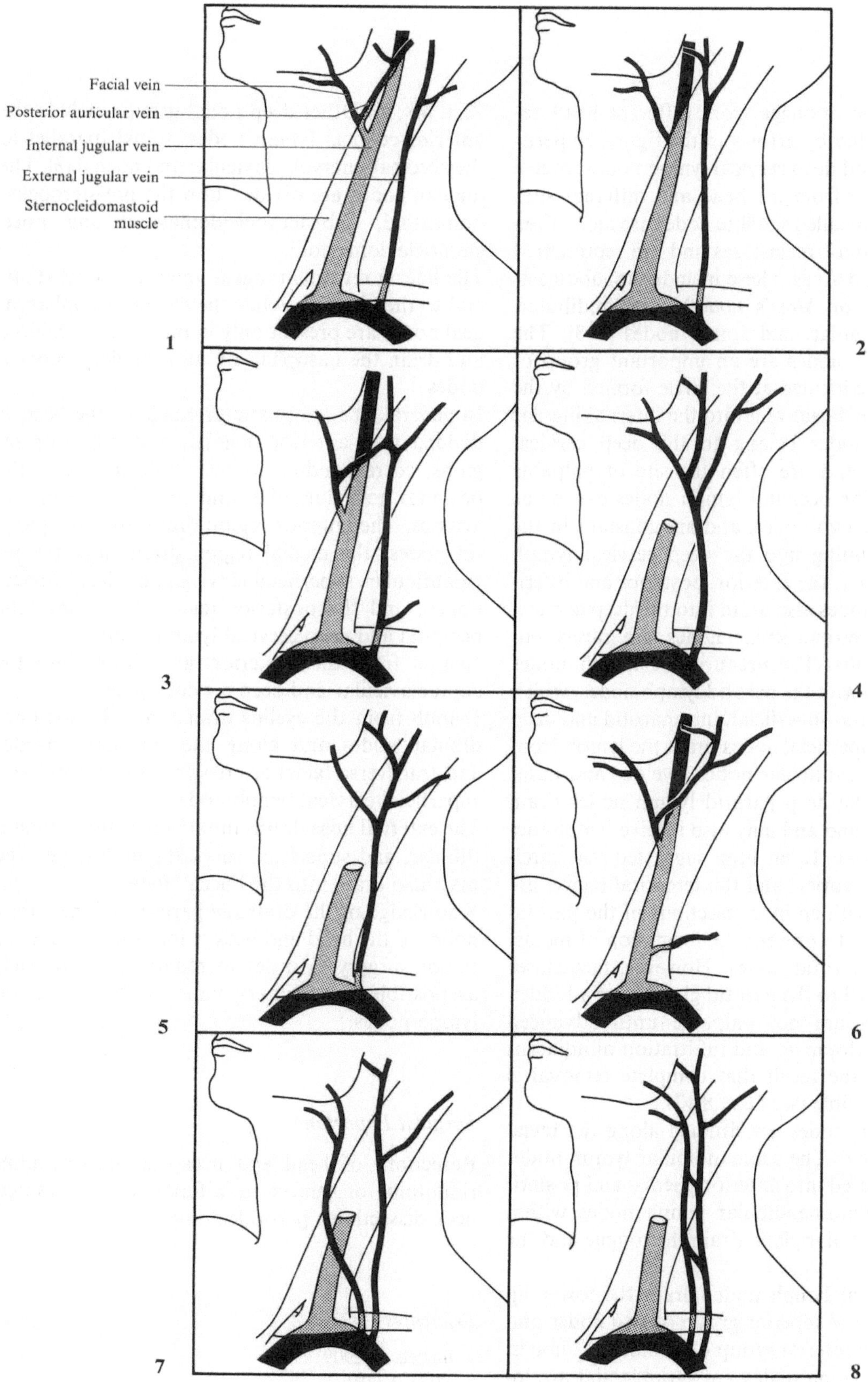

Facial vein
Posterior auricular vein
Internal jugular vein
External jugular vein
Sternocleidomastoid muscle
1
2
3
4
5
6
7
8

The lymphatic drainage from different head regions is depicted by arrows in the figure. Superficial cervical and deep cervical lymph nodes receive lymphatic flow from the head and different sites in the neck. So-called satellite nodes are most often the sites of tumor metastases and are represented as dark, shaded areas. These include the subdigastric (*Küttner's* or *Most's nodes*), submandibular, submental, jugular, and spinal nodes [1–3]. The jugulodigastric nodes are an important group of nodes that are located at the angle formed by the entrance of the facial vein into the internal jugular vein. These nodes belong to the deep cervical lymph nodes and are often the site of palpable metastases. The occipital lymph nodes can be as few as one or two nodes, and are constant in the newborn, draining into the deep cervical lymph nodes. Similarly, the anterior, posterior and inferior auricular nodes also drain into the deep cervical nodes. The retroauricular nodes are rarely enlarged in adults. The preauricular lymph nodes communicate with the parotid lymph nodes, which are grouped into superficial, intraparotid and deep nodes. The superficial nodes drain the lymph from the auricle, preauricular nodes, eyelids, and nasal soft tissue. The deep parotid lymph nodes drain the parotid gland and may also receive lymphatics from the tongue. It has been suggested that carcinomas of the auricle and the temporal region are best treated with en bloc resections of the parotid gland in order to prevent the formation of metastases in the parotid nodes. However, metastases situated medial to the parotid gland have a hidden location and are not palpable until advanced stages of development and infiltration of adjacent tissues, with the result that complete removal is no longer possible (see Sect. 8.17).

Buccal lymph nodes are situated along the facial artery and vein. The submandibular lymph nodes can be separated into anterior, medial and posterior groups. Paramandibular lymph nodes within the submandibular gland drain the tongue and the lips.

The submental lymph nodes drain the lower lip and constitute a superior group of two nodes and an inconstant inferior group of at most 1–2 nodes. The lingual lymph nodes are variable and are located along the lingual artery and vein (see Sect. 4.8, 9). Other deep nodal groups include the anterior cervical lymph nodes, spinal (parallel to the accessary nerve), clavicular and nuchal [3]. The jugular nodes are divided into the pre-sternocleidomastoid, sub-sternocleidomastoid and post-sternocleidomastoid.

The lateral retropharyngeal nodes are located lateral to the pharynx, while the medial retropharyngeal nodes are present only in infants and children and drain the nasopharynx into the deep cervical nodes.

In general, the lymphatic drainage of the head is divided into anterior, medial, and posterior regions, corresponding to the ramifications of the occipital, retroauricular, and superficial temporal arteries. The anterior region drains into the parotid nodes; the medial region drains into the retroauricular, superficial cervical, and deep cervical nodes; and the posterior region drains into the occipital and deep cervical lymph nodes.

Lymph from the posterior neck drains into the supraclavicular and deep cervical nodes.

Lymph from the eyelids drains into the submandibular nodes and along the zygomaticoorbital and transverse facial arteries into the parotid and superficial cervical lymph nodes.

The external nose drains into the parotid, submandibular, and superficial cervical lymph nodes and may also drain into the buccal nodes.

Knowledge of the drainage patterns of the lymph nodes of the head and neck is important in radical tumor surgery in order to remove as completely as possible any primary tumor with its relevant lymph nodes.

Surgical Importance

Resections of head and neck tumors (including rhinotomy or rhinectomy), functional and radical neck dissections, parotidectomy.

References

1. Bartels P (1909)
2. Most A (1917)
3. Taillens JP (1960)

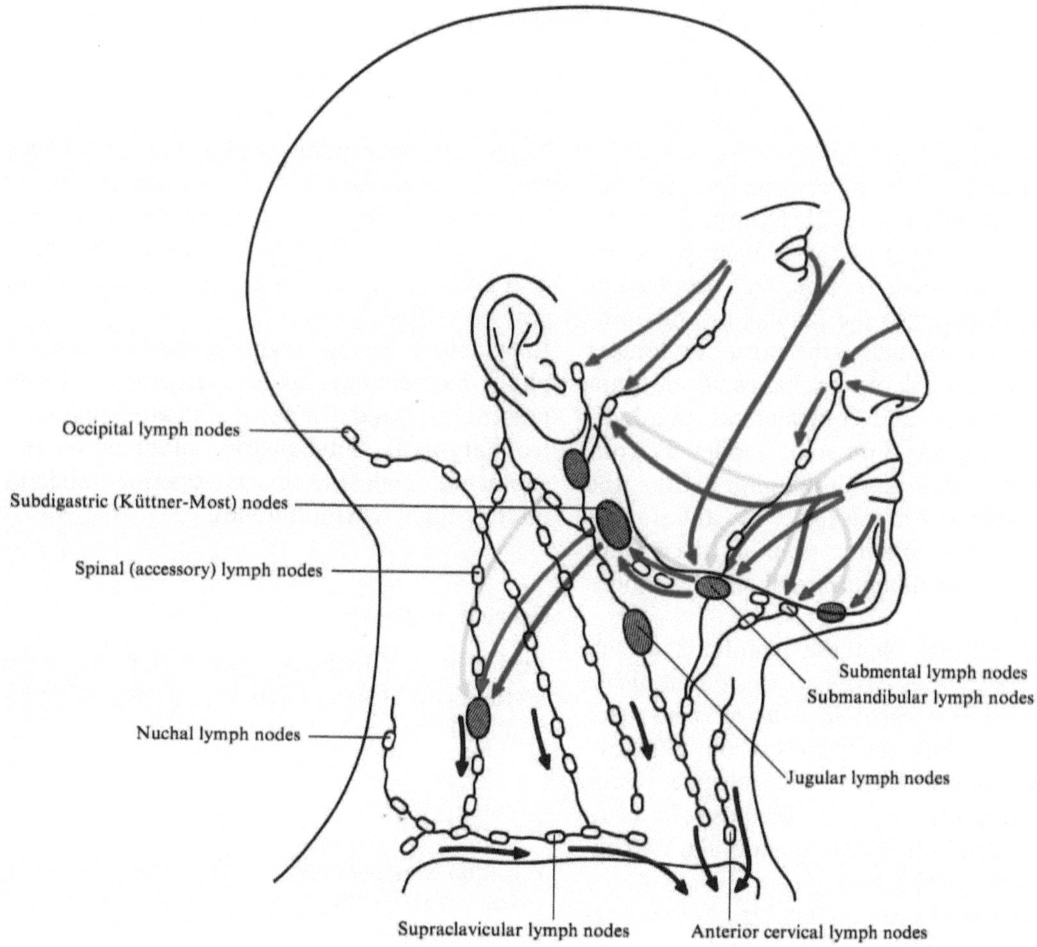

Occipital lymph nodes
Subdigastric (Küttner-Most) nodes
Spinal (accessory) lymph nodes
Nuchal lymph nodes
Submental lymph nodes
Submandibular lymph nodes
Jugular lymph nodes
Supraclavicular lymph nodes
Anterior cervical lymph nodes

1.24 Lymphatic Drainage of the Upper Respiratory Tract and the Esophagus

Lymphatic drainage from the nasal cavity and the nasopharynx drains into the retropharyngeal and upper jugular lymph nodes [2]. Lymph from the soft palate, tonsils, base of the tongue, posterior alveolar rim, retromolar trigone, and oropharyngeal walls all drains into the jugular or angulodigastric nodes. Additionally, the posterior tongue has a very rich network of lymphatics which drain into both ipsilateral and contralateral angulodigastric, medial jugular, and spinal nodes. In contrast, lymphatic drainage (and tumors) from the anterior two thirds of the tongue, the anterior alveolar rim and the anterior floor of the mouth drain to the submandibular and submental nodes (see Sects. 4.8, 9).

It should be stressed that the lymphatic vessels of the oral and nasal cavities can communicate through the incisive canal (*canalis incisivus*) (see Sect. 5.14). The lymphatic vessels of the olfactory region communicate with the epidural space, while those of the respiratory part of the nose run horizontally and drain into the submandibular and superficial cervical lymph nodes [1].

Tumors of the tonsils can metastasize to the parapharyngeal, the jugular, and the spinal accessory nodes. In tumors or inflammations of the retromolar triangle (*dentitio difficilis*), the subdigastric, jugular, and parapharyngeal lymph nodes become enlarged or are involved with metastatic disease.

The anterior laryngeal commissure and the subglottic space are drained into either the prelaryngeal nodes or the nodes along the recurrent laryngeal nerve (see Sect. 2.14).

In extensive nasopharyngeal disease, three lymphatic barriers have to be overcome by the lesion present [2]: Waldeyer's ring, the satellite nodes (retropharyngeal, subdigastric, submandibular and submental nodes), and the superior, medial and inferior jugular lymph nodes.

Surgical Importance

Localization of tumors and inflammatory diseases of the head and neck, tumor surgery of the head and neck

References

1. Cachin Y et al. (1969)
2. Taillens JP (1960)

Plate adapted from Cachin et al. [1]

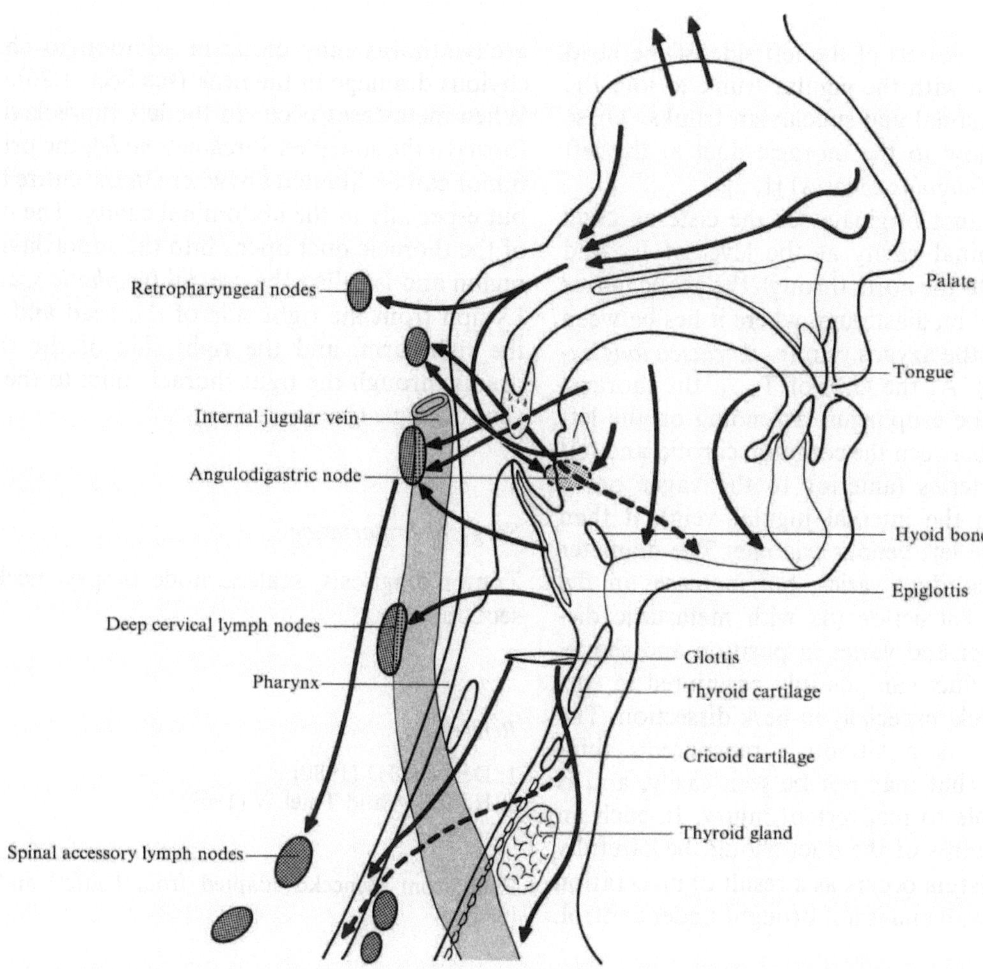
Retropharyngeal nodes
Palate
Tongue
Internal jugular vein
Angulodigastric node
Hyoid bone
Epiglottis
Deep cervical lymph nodes
Glottis
Pharynx
Thyroid cartilage
Cricoid cartilage
Thyroid gland
Spinal accessory lymph nodes

1.25 The Thoracic Duct and the Right Lymphatic Duct (*Truncus Lymphaceus Dexter*)

The lymphatic vessels of the left side of the head and neck unite with the jugular trunk to join the bronchomediastinal and subclavian trunks. These vessels then flow to the thoracic duct at the left venous angle (*angulus venosus*) [1, 2].

The thoracic duct originates as the cisterna chyli in the abdominal cavity at the level of L_2 and passes dorsal to the aorta through the aortic hiatus into the dorsal mediastinum, where it lies between the aorta and the azygos vein (*v. thoracica longitudinalis dextra*). At the level of T_{4-5}, the thoracic duct crosses the esophagus, ascending on the left side. Passing between the common carotid and left subclavian arteries (anterior to the vagus nerve and dorsal to the internal jugular vein), it then opens into the left venous triangle. The diameter of the thoracic duct varies, but increases in the presence of obstruction (as with metastatic disease). Its upper end varies in position and shape. The thoracic duct can possibly be injured in surgery of the neck, especially in neck dissection. The duct appears as a vitreous, transparent, thin-walled vessel that may not be seen easily, and is thus vulnerable to inadvertent injury. In such an event, the stumps of the duct should be carefully ligated. If a fistula occurs as a result of maceration of the duct's wall and is not brought under control, a chylothorax may ensue in addition to chronic chylous drainage in the neck (see Sect. 1.26).

When metastases occur in the left supraclavicular fossa (to the so-called *Virchow's node*), the primary tumor can be situated anywhere in the entire body, but especially in the abdominal cavity. The orifice of the thoracic duct opens into the supraclavicular region and is called the *central lymphatic space*.

Lymph from the right side of the head and neck, the right arm, and the right side of the thorax drains through the right thoracic duct to the right venous angle (see Sect. 2.10).

Surgical Importance

Tumor diagnosis, scalene node biopsy, neck dissection.

References

1. Denecke HJ (1980)
2. Hafferl A and Thiel W (1969)

Plate from Denecke adapted from Hafferl and Thiel [1, 2]

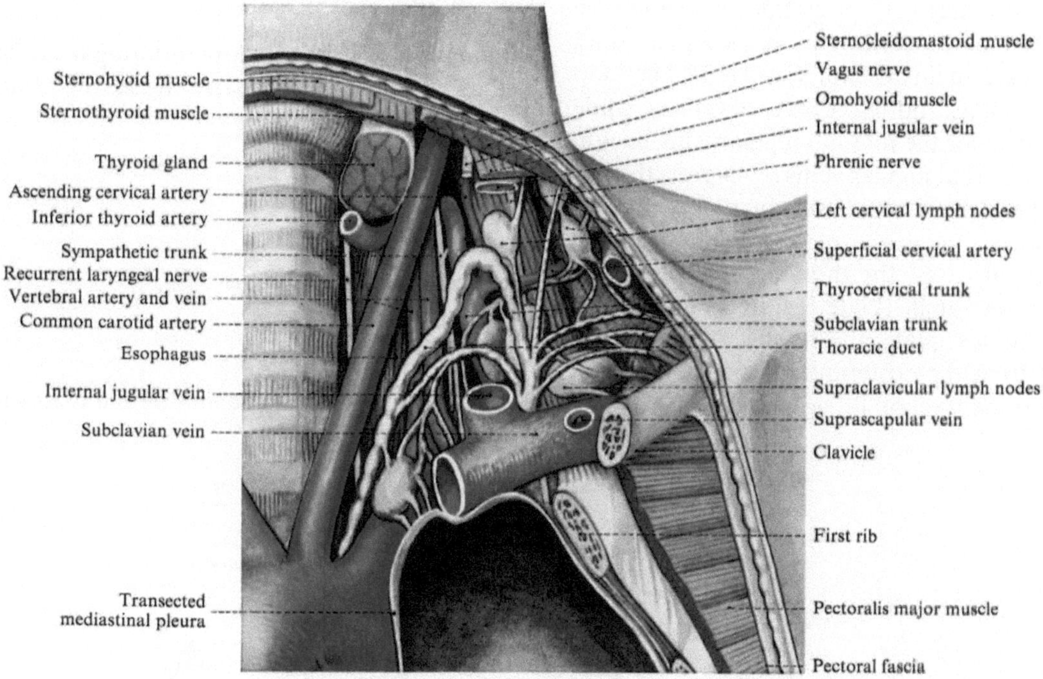

Sternohyoid muscle
Sternothyroid muscle
Thyroid gland
Ascending cervical artery
Inferior thyroid artery
Sympathetic trunk
Recurrent laryngeal nerve
Vertebral artery and vein
Common carotid artery
Esophagus
Internal jugular vein
Subclavian vein
Transected mediastinal pleura
Sternocleidomastoid muscle
Vagus nerve
Omohyoid muscle
Internal jugular vein
Phrenic nerve
Left cervical lymph nodes
Superficial cervical artery
Thyrocervical trunk
Subclavian trunk
Thoracic duct
Supraclavicular lymph nodes
Suprascapular vein
Clavicle
First rib
Pectoralis major muscle
Pectoral fascia

1.26 (Plates 1–12) Variations and Anomalies of the Orifice of the Thoracic Duct

As shown in Plates 1–11, variations in the course and location of the orifice of the thoracic duct are represented in an anterior view [1–3]. The loop of the duct can ascend to the level of C_7–C_6, while the duct can enter the internal jugular, subclavian, azygos, or external jugular veins. However, the latter two drainage patterns are not represented in these plates.

In exceptional cases, the thoracic duct can enter the right venous angle (Plate 12). In the case represented in the last plate, the right vertebral artery originates from the posterior wall of the descending aorta at the level of T_3.

Surgical Importance

Diagnosis of primary tumors resulting in supraclavicular metastases; scalene node or *Daniel's* biopsy, neck dissection.

References

1. Bartels P (1909)
2. Krmpotić-Nemanić J (1978)
3. Zdanov DA (1952)

Plates 1–12 adapted from Zdanov [3] and Bartels [1]

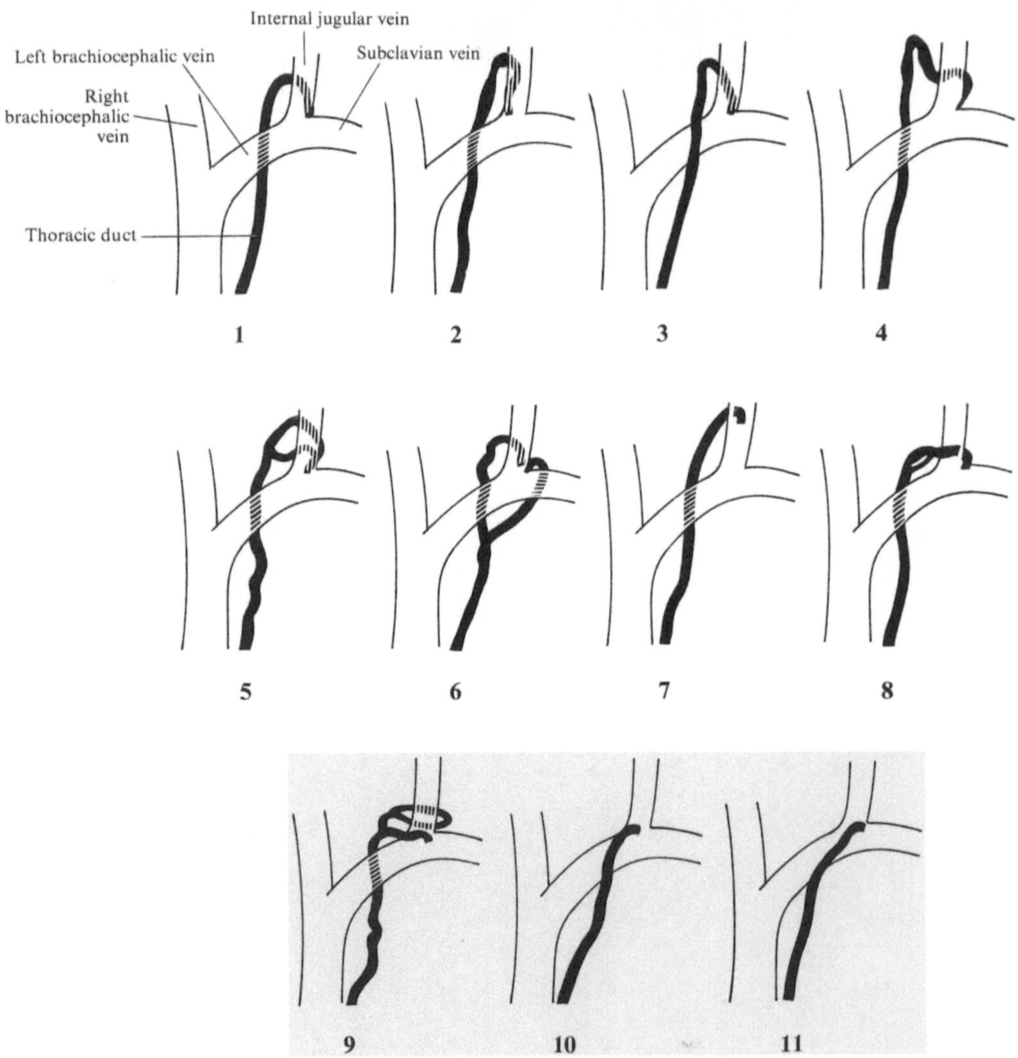

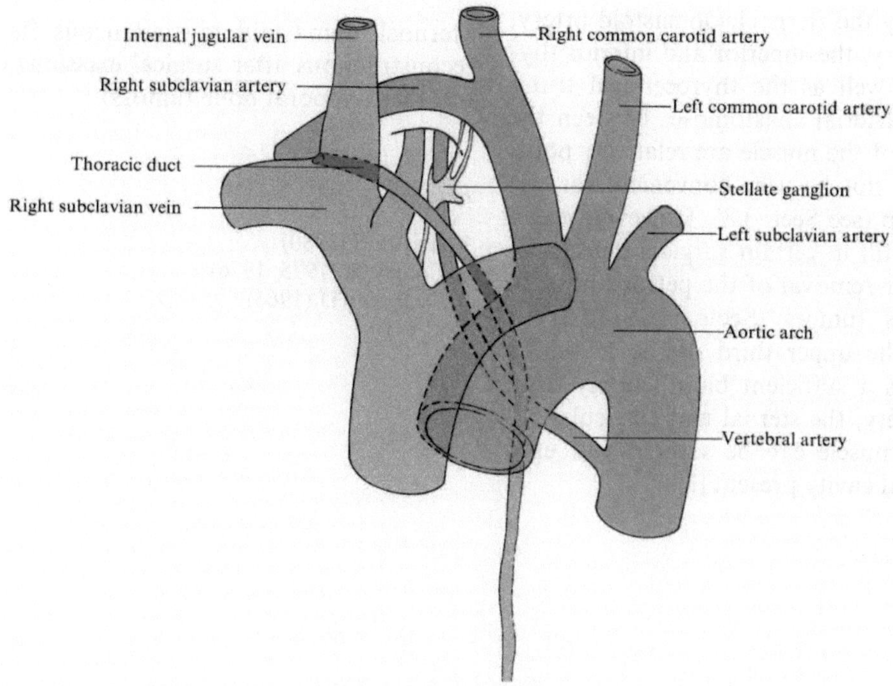
Internal jugular vein
Right common carotid artery
Right subclavian artery
Left common carotid artery
Thoracic duct
Stellate ganglion
Right subclavian vein
Left subclavian artery
Aortic arch
Vertebral artery

12

The blood supply of the sternocleidomastoid muscle is provided by the sternocleidomastoid artery, the occipital artery, the superior and inferior thyroid arteries, as well as the thyrocervical trunk [1–3]. Since the arterial anastomoses between the various sections of the muscle are relatively poor, this muscle may not be very convenient for use as a pedicled flap (see Sect. 3.8). However, it can provide bulk to fill in certain surgical defects, as might occur after removal of the petrous bone in resecting glomus tumors (Sects. 11.19, 12.2) or carcinomas. If the upper third of the muscle is pedicled and has a sufficient blood supply from the occipital artery, the sternal and clavicular insertions of the muscle can be severed and used to fill the surgical cavity present [3].

Surgical Importance

Sternocleidomastoid myocutaneous flaps (defect reconstructions after surgical excisions of glomus or other temporal bone tumors).

References

1. Ariyan S (1980)
2. Conley JJ (1975, 1976)
3. Denecke HJ (1966)

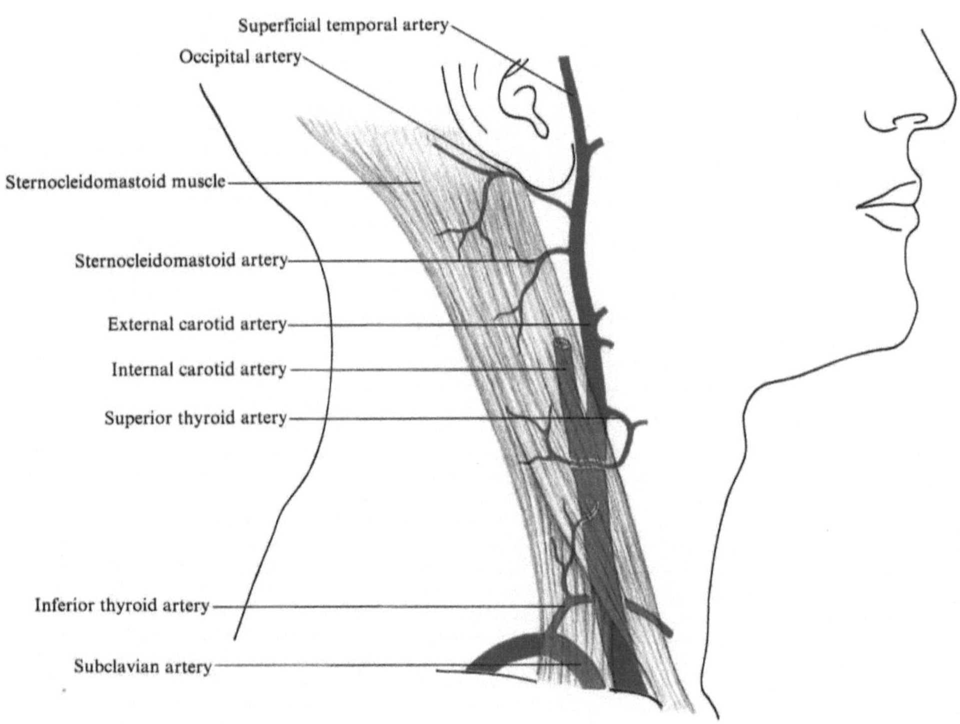

1.28 The Blood Supply of the Pectoralis Major Muscle (Anterior View)

The blood supply of the pectoralis major muscle is derived from the thoracoacromial and the lateral thoracic arteries, which are branches of the subclavian and the axillary arteries. The peripheral diameters of these arteries remain large so that the muscle is well vascularized and very suitable for use as a pedicled flap. Certain landmarks are of practical importance in locating the flap's major vessels. The thoracoacromial artery is located between the middle and lateral thirds of the clavicle at its inferior side. The sharp angle of *Mohrenheim's fossa* (Deltoideopectoral notch) under the clavicle serves as an orientation point. The lateral thoracic artery lies more laterally and caudally.

Surgical Importance

Myocutaneous pectoral flap [1], anterior pectoral flap [2].

References

1. Ariyan S (1979)
2. Conley JJ (1970, 1976)

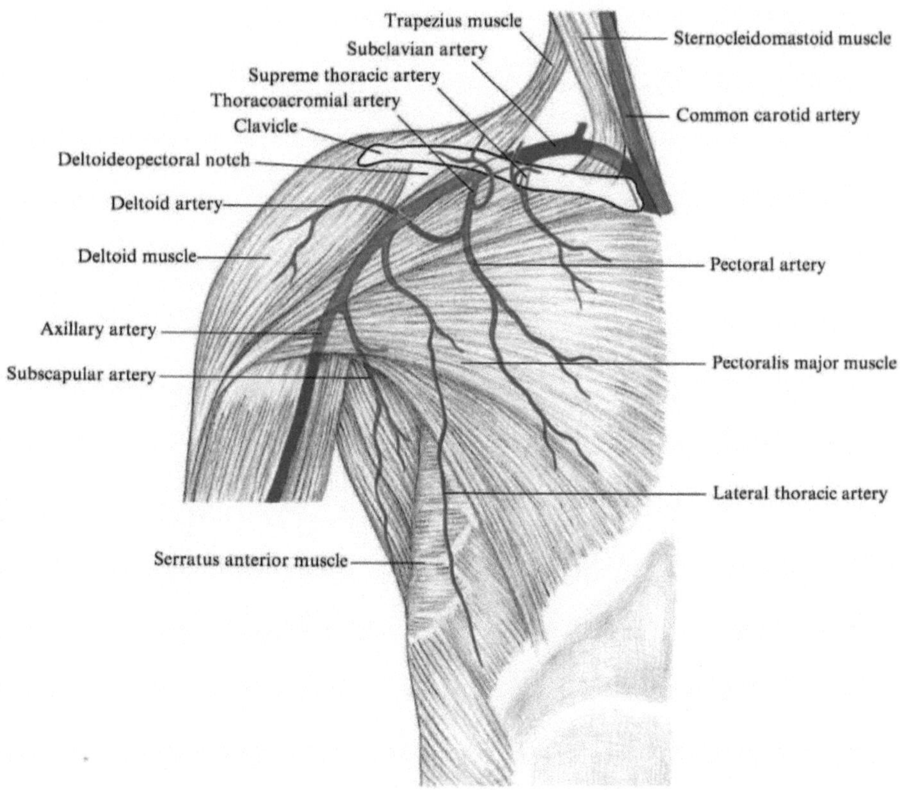

1.29 The Blood Supply of the Trapezius and Latissimus Dorsi Muscles (Posterior View)

The major arteries of the trapezius muscle are supplied by the occipital and transverse cervical arteries (*a. transversa colli*). If the transverse cervical artery is not cut during extirpation of adjacent lymph nodes, the artery – and especially its descending branch – can be followed peripherally for a considerable distance, making the trapezius muscle very suitable for use as a myocutaneous flap [1].

The nerves and vessels to the latissimus dorsi muscle bifurcate beneath the head of the humerus close to the rotation point of the corresponding myocutaneous flap. This allows the formation of two independent flaps, each with a complete neurovascular supply [2].

Surgical Importance

Trapezius and latissimus dorsi myocutaneous flaps, microvascular anastomoses.

References

1. Ariyan S (1980)
2. Tobin GR et al. (1980)

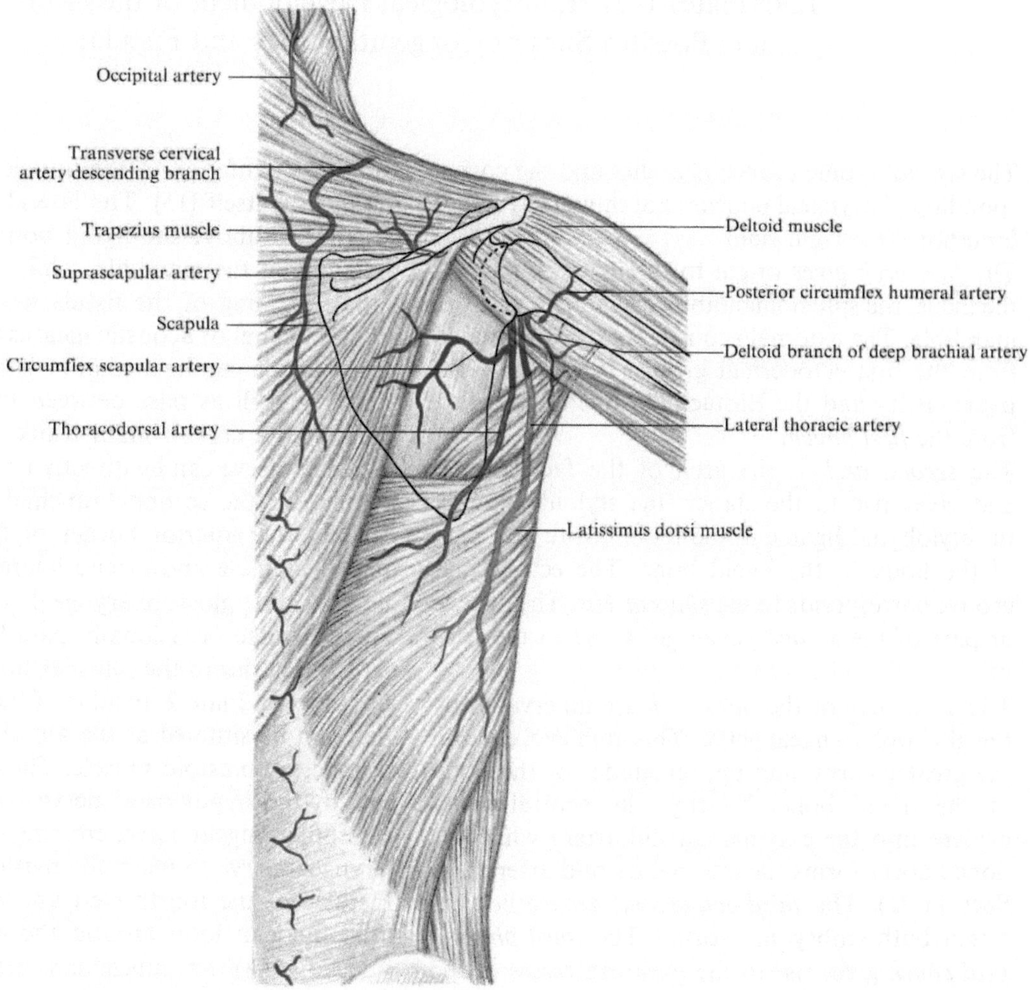
Occipital artery
Transverse cervical artery descending branch
Trapezius muscle
Suprascapular artery
Scapula
Circumflex scapular artery
Thoracodorsal artery
Deltoid muscle
Posterior circumflex humeral artery
Deltoid branch of deep brachial artery
Lateral thoracic artery
Latissimus dorsi muscle

1.30 (Plates 1–3) Embryological Development of the Neck and Possible Sites of Congenital Cysts and Fistulae

The six embryonic branchial arches and the corresponding pharyngeal pouches are shown in Plate 1 (anterior view, right side).

The *first arch* gives origin to the malleus, part of the incus, the sphenomandibular ligament, and the mandible. The external acoustic meatus originates from the first ectodermal groove, while the tympanic cavity and the Eustachian tube are derived from the *first pouch*.

The *second arch* is the arch of the facial nerve and gives rise to the stapes, the styloid process, the stylohyoid ligament, the lesser cornu and part of the body of the hyoid bone. The ectodermal groove corresponds to the *sinus of His*. The anterior part of the *second pouch* gives rise to the tonsil.

The structures of the *third arch* are innervated by the glossopharyngeal nerve. This arch evolves into the greater cornu and the remainder of the body of the hyoid bone. Nearby, the ventral aorta evolves into the external carotid artery while the dorsal aorta forms the internal carotid artery (see Sect. 11.21). The *third ectodermal groove* lies between both embryonal aortae. The *third pharyngeal pouch* gives rise to the pyriform recess of the larynx.

The *fourth arch* evolves into the thyroid cartilage while its nerve corresponds to the superior laryngeal nerve. The *fourth pouch* forms the upper part of the esophagus.

The *fifth arch* subsequently disappears, but the *fifth pouch* gives rise to the ultimobranchial bodies.

The *sixth arch* forms the cricoid cartilage; its nerve corresponds to the inferior laryngeal nerve.

In the 3–4 mm embryo, the thyroid gland forms as a ventral evagination from the pharynx in the region of the future foramen cecum of the tongue (*tuberculum impar His*). The gland grows caudally but remains fixed to the pharynx by the thyroglossal duct, the remnants of which can lead to the formation of median cervical cysts or fistulae (Plate 2) [6, 10]. When present, these lesions are usually situated caudal to the hyoid bone but can be found at any site in the upper mid neck. Thyroid tissue can also be found along the duct [14].

The formation of lateral cysts or fistulae can have different presentations, and may even be mistaken for parotid tumors [5, 7]. Fistulae of the first branchial cleft can show ectodermal doubling of the external acoustic meatus, with an orifice located anterior or inferior to the auricle or even within the meatus itself [15]. The lateral orifice can also be situated above the hyoid bone at the level of the angle of the mandible, while the superior (internal) opening of the fistula lies in the external auditory canal or acoustic meatus [3, 11]. The duct of the fistula can lie medial or lateral to the facial nerve, as well as pass between the primary divisions of the nerve's main trunk. In either event, the facial nerve can lie directly under the skin.

Fistulae of the second branchial cleft open caudally at the anterior border of the sternocleidomastoid muscle and extend laterally to the hypoglossal and glossopharyngeal nerves (Plate 2). These fistulae occasionally pass between the two carotid arteries to the tonsillar fossa [5].

As shown in Plate 2, fistulae of the third branchial cleft can be situated at the anterior border of the sternocleidomastoid muscle. They are usually lateral to the hypoglossal nerve but medial to the glossopharyngeal nerve, crossing over the superior laryngeal nerve to reach the pyriform recess.

Fistulae of the fourth cleft are relatively uncommon but can loop around the aortic arch (left-sided fistula) or subclavian artery (right-sided fistula) and open into the esophagus (Plate 3) [5, 8, 9, 12].

The congenital fistulae can be classified as *complete* when two openings are present and *incomplete* when there is only one internal or external opening [1]. Additionally, *nasopharyngeal fistulae* extend from the base of the skull to the tonsil and are of some surgical importance.

The congenital fistulae can also give rise to cysts having no recognizable internal or external openings.

In order to prevent recurrences, fistulae opening into the pharynx must be removed completely and include all orifices, which can be defined radiographically [2]. If possible, excision can be done by inverting the fistulous tract into the operative field and ligating the pharyngeal communications present.

In successfully eliminating median cervical cysts or fistulae, Sistrunk [13] has shown that removal of the body of the hyoid bone is necessary as well as a portion of the posterior tongue(at the site of the embryonic foramen cecum).

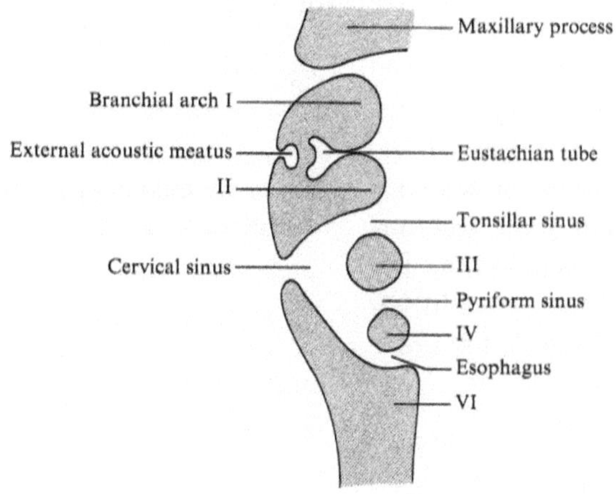

1

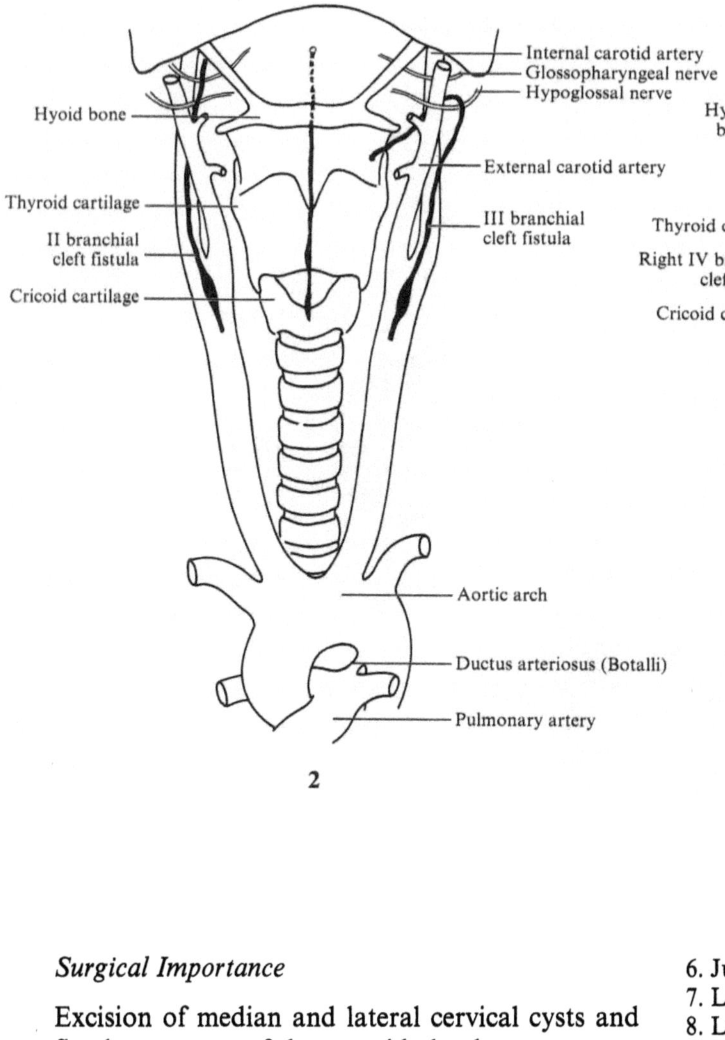

2

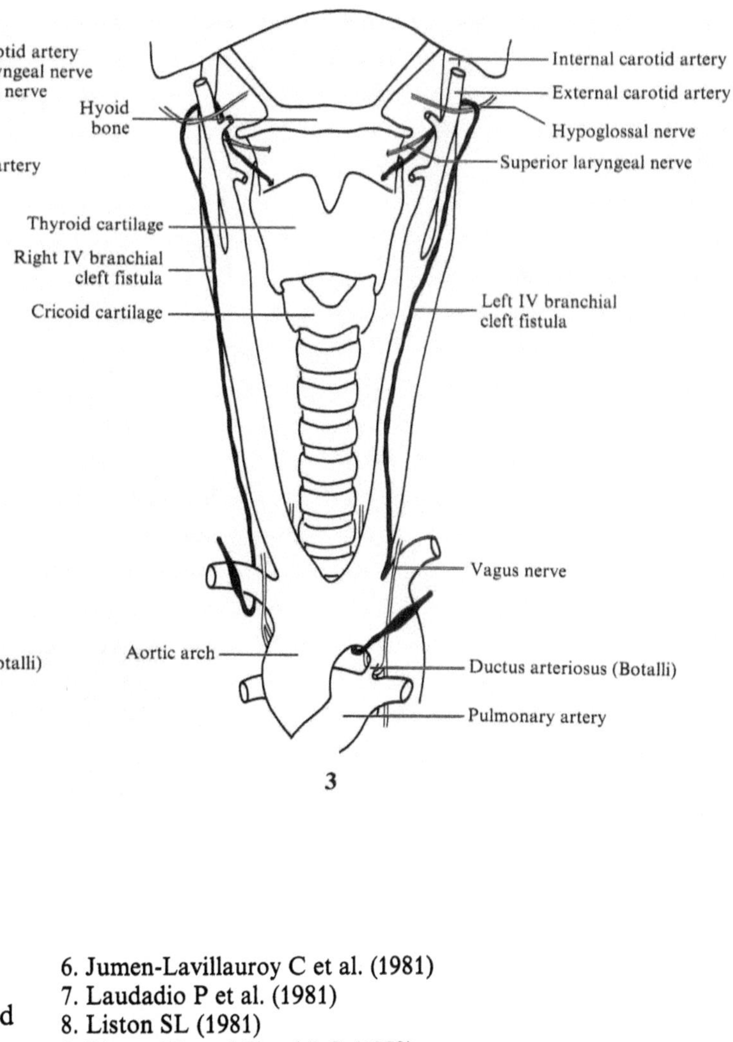

3

Surgical Importance

Excision of median and lateral cervical cysts and fistulae, surgery of the parotid gland.

References

1. Chandler JR and Mitchell B (1981)
2. Dupuch KM et al. (1981)
3. Gupta SK et al. (1979)
4. Haar JG et al. (1981)
5. Himalstein MR (1980)
6. Jumen-Lavillauroy C et al. (1981)
7. Laudadio P et al. (1981)
8. Liston SL (1981)
9. Liston SL and Siegel LG (1979)
10. Noyek A and Friedberg J (1981)
11. Olsen KD et al. (1980)
12. Shugar MA and Healy GB (1980)
13. Sistrunk WE (1920/1928)
14. Thomas JR (1980)
15. Tucker AG (1981)

Plates 1–3 adapted from Himalstein [5]

1.31 Laryngoceles

Laryngoceles represent malformations of *Morgagni's ventricle* and expand as a result of the accumulation of air and mucus within their lumina [1]. Internal laryngoceles can hinder respiration, especially when superinfection occurs. Laryngoceles can also develop externally (lateral to the thyrohyoid membrane) and expand by compression of the abdominal muscles. These lesions must also be differentiated from dilatation of the cervical veins or atypical cervical cysts.
In some cases microlaryngoscopy can be used to treat internal laryngoceles by their marsupialization. In contrast, external laryngoceles must be removed by means of an external approach in order to protect and retract vital structures of the neck.

Surgical Importance

Endolaryngeal microsurgery, thyrotomy, laryngocele extirpations.

Reference

1. Denecke HJ (1980)

Plate from Denecke [1]

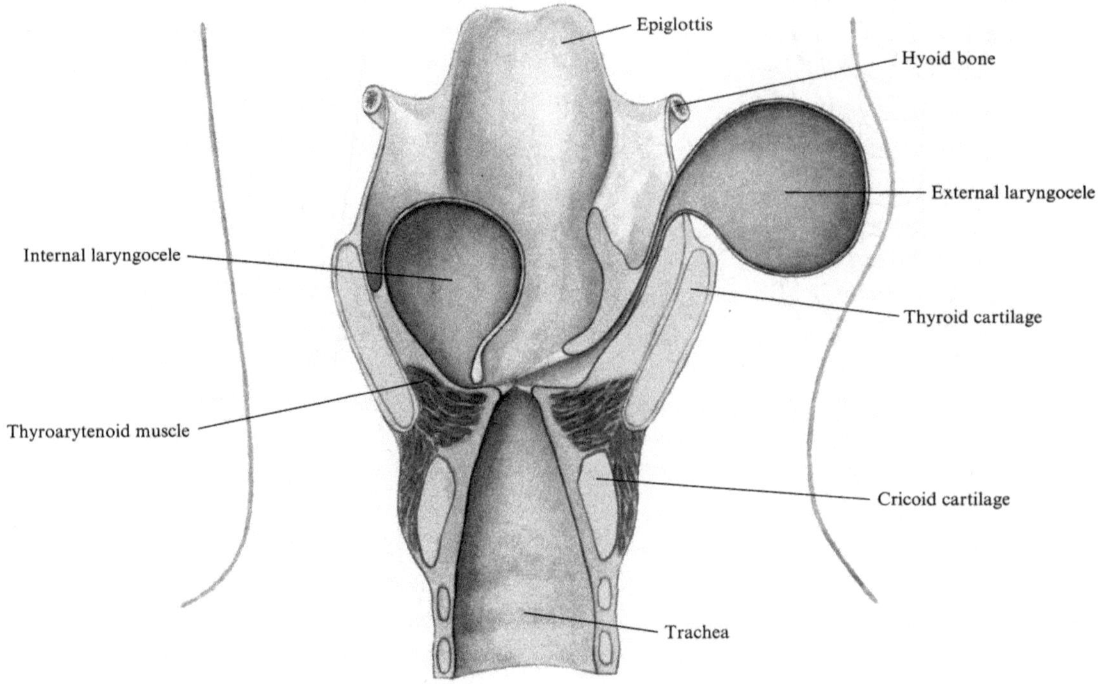

2 The Larynx and the Thyroid Gland

2.1 (Plates 1–2) Development of the Cartilaginous Laryngeal Skeleton

Plate 1 depicts the laryngeal skeleton in the newborn child, while the larynx in Plate 2 is that of a 10 year old adolescent.

The larynx of the newborn differs in shape, dimensions, position, and structure from the larynx of the adult. The size of the adult larynx is twice as large as that of the neonatal larynx. The female larynx is smaller than the male larynx, although the dimensions of the larynx remain nearly constant from the 3rd year of life to puberty.

The subglottic diameter of the newborn larynx is 6 mm and its connective tissue is much looser than in children or in adults. Its small dimensions and its more pronounced reaction to mechanical and inflammatory irritations (*laryngitis subglottica*) are the primary reasons why the pediatric glottis is reduced to a critical diameter of 50% of patency much sooner than is the adult larynx.

Reference

1. Peter K (1925)

Plates 1–2 adapted from Peter [1]

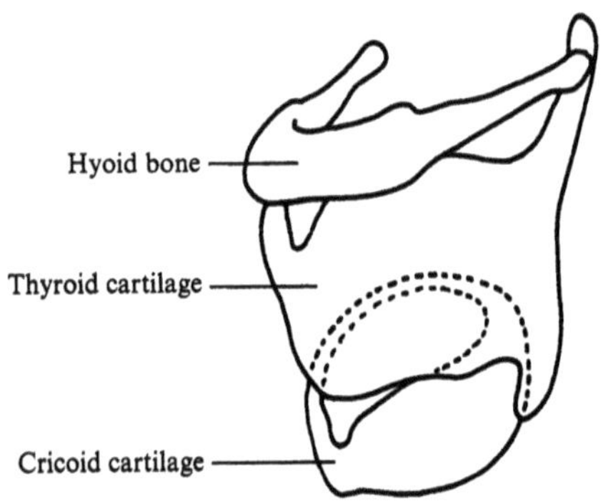

1

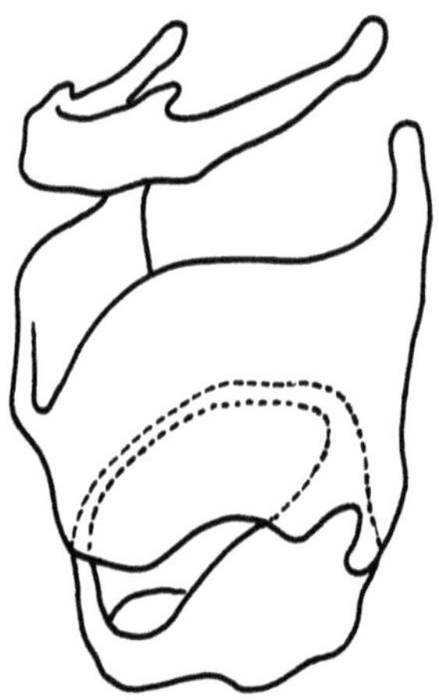

2

2.2 (Plates 1–3) The Position of the Larynx Relative to the Fifth Cervical Vertebra in the Neonate and the Adult

The position of the larynx is shown in relationship to the level of the fifth cervical vertebra in the neonate (Plate 1), the adult female (Plate 2), and the adult male (Plate 3) in the middle phase of expiration. The larynx of the newborn has a rather high position in the neck (see Sect. 2.1) and later descends as growth and development take place.

Surgical Importance

Intubation, laryngotracheoscopy, esophagoscopy.

Reference

1. Lanz T v and Wachsmuth W (1955)

Plates 1–3 adapted from Lanz-Wachsmuth [1]

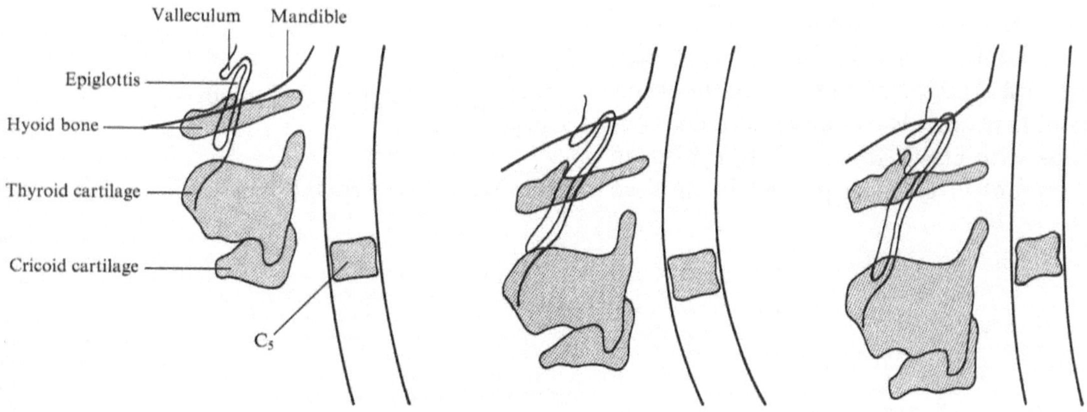

1 2 3

2.3 (Plates 1–2) The Aditus Laryngis in the Newborn and in the Adult

Due to the omega-shape of the epiglottis, the orifice of the larynx in the newborn has the shape of a vertical groove [2]. The epiglottis lies flat over the glottis and forms a sharp angle with the vocal cords. In contrast, the position of the epiglottis is more vertical in the adult than in the newborn and is unrolled, making the laryngeal orifice larger. The *plica accessoria* is a mucosal fold in the newborn that runs from the arytenoid to the thyroid cartilage below the vocal cords.
The neonatal larynx is covered by ciliated columnar epithelium, with the exceptions of the vocal cords and the lingual and (occasionally) the laryngeal surfaces of the epiglottis. Additionally, taste buds may be found in the lingual surfaces of the epiglottis. The stratified epithelium of the vocal cords in the neonate covers a narrower zone than occurs in the adult. There are 3–4 layers of cells in the vocal cords and six layers in the conus elasticus. The lamina propria is composed of loose connective tissue with elastic fibers. There is also an increased amount of collagen present in children to make this structure more resilient in protecting the larynx from traumatic injuries. Additionally, the laryngeal mucosa contains round and conical stretch receptors on the inferior side of the vocal cords and in the cricothyroid membrane [1].
The lymphatic system of the larynx develops after birth, but is at most sparse in the true vocal cords and more abundant in supra- and infraglottic tissues.

Surgical Importance

Intubation, laryngotracheobronchoscopy, esophagoscopy, endolaryngeal interventions.

References

1. Akazaku FK and Wyke B (1979)
2. Peter K (1925)

Plates 1–2 modified from Peter [2]

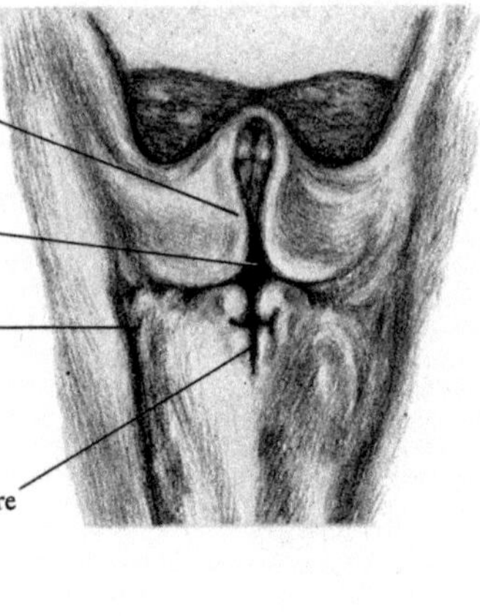

1

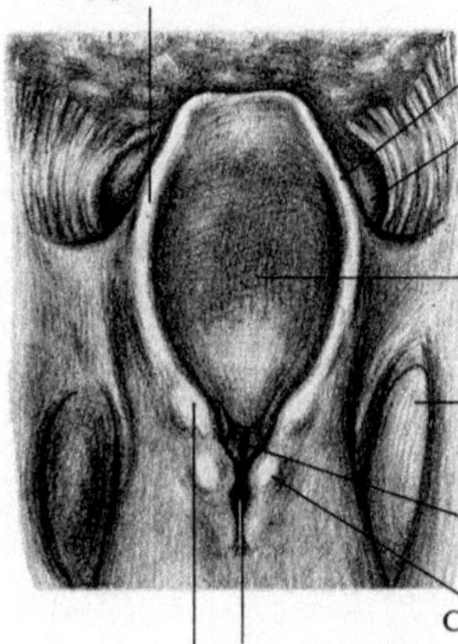

2

2.4 The External Laryngeal Muscles and the Muscles of the Tongue and Pharynx (Lateral View)

Although the supra- and infrahyoid muscles have their origins or insertions on the hyoid bone, partial or even total extirpation of the bone does not usually cause functional disturbances [2]. In the surgical treatment of hypopharyngeal diverticula (such as the *Zenker's diverticulum*), the fundiform portion of the cricopharyngeal muscle and the musculature of the upper part of the esophagus are divided. This procedure – or myotomy – is also useful in treating unilateral paralysis of the pharyngeal constrictor muscles, in addition to resection of a portion of the paralyzed pharyngeal wall with medialization of the affected vocal cord [1].

Surgical Importance

Interventions in the oral cavity, oro- and hypopharynx, cervical esophagus; medial and lateral pharyngotomy or pharyngectomy; removal of median cervical cysts and fistulae; coniotomy, tracheotomy; partial and total laryngectomies; laryngoplasty and tracheoplasty; resections of hypopharyngeal diverticula; partial or total thyroidectomy; surgical correction of pharyngeal paralysis.

References

1. Denecke HJ (1980)
2. Lesoine W (1966, 1970)

Plate adapted from Denecke [1]

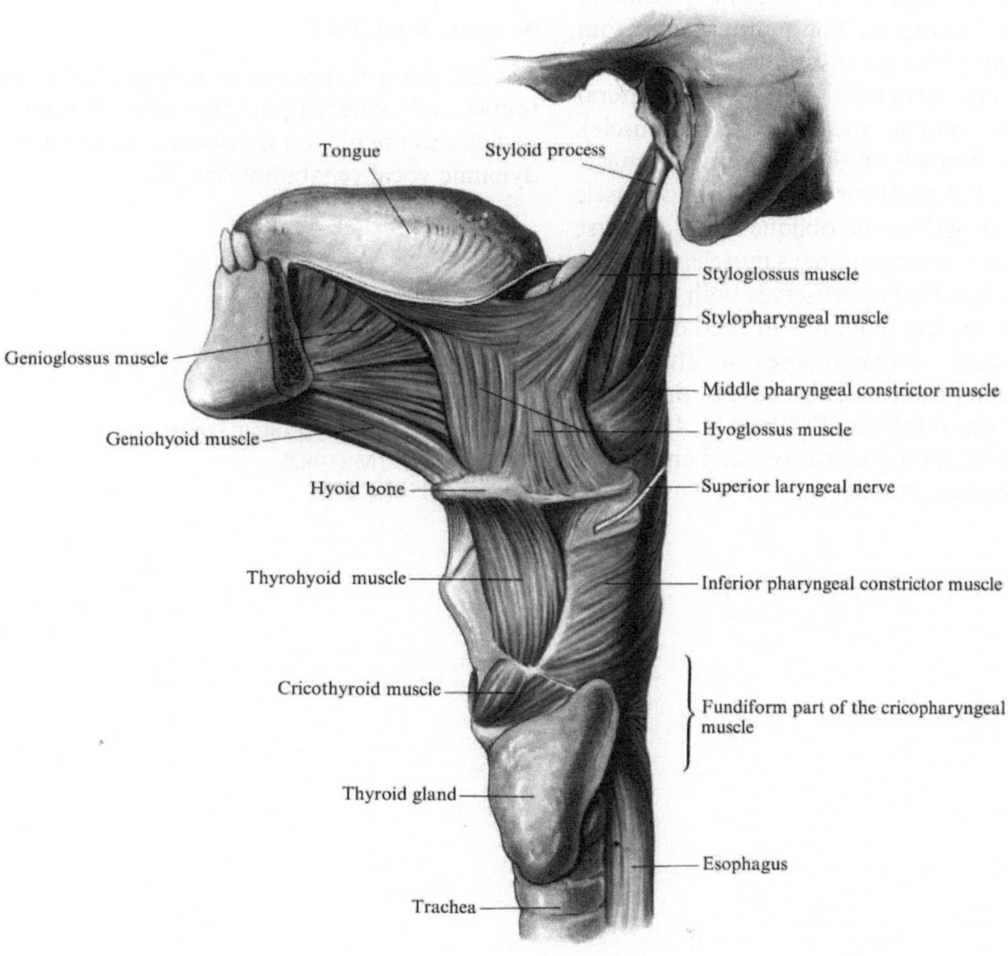

2.5 The Blood and Nerve Supply of the Intrinsic Laryngeal Muscles (Dorsal View)

The superior laryngeal artery passes through the thyrohyoid membrane together with the internal branch of the superior laryngeal nerve. The artery reaches the pyriform sinus and anastomoses beneath the mucosa of the sinus with the inferior laryngeal artery. This latter vessel first lies dorsal to the trachea and then pierces the inferior pharyngeal constrictor and the cricopharyngeal muscle to reach the pyriform sinus.

The internal ramus of the superior laryngeal nerve is sensory to the mucosa of the epiglottic vallecula and to the supraglottic larynx to the level of the vocal cords. This nerve forms the *ansa Galeni* with the inferior laryngeal nerve.

The inferior (recurrent) laryngeal nerve is normally divided into an anterior and a posterior branch, and can have a number of variations (see Sect. 2.19). The anterior branch of the nerve passes lateral to the cricothyroid joint. Because of its close relationship to this joint, the nerve can be involved in diseases affecting the joint's articulation or it can be injured as a result of trauma or surgery in this region. The motor fibers from the nerve supply the intrinsic laryngeal muscles (the lateral cricoarytenoid muscle, the lateral thyroarytenoid muscle and the vocalis muscle). The posterior branch of the recurrent laryngeal nerve supplies the posterior cricoarytenoid muscle (*m. posticus*), as well as the oblique and transverse portions of the interarytenoideus muscle. Since the recurrent laryngeal nerve subserves both abductor and adductor muscle functions in the larynx, the results of neural reanastomoses involving the nerve have been problematic because of the potential misdirection of the nerve's fibers [3, 4].

The cricothyroid muscle is the common antagonist of all the intrinsic muscles of the larynx and is supplied by the external branch of the superior laryngeal nerve.

Andrea [1] used angiography to study the microvasculature of the larynx and defined its blood supply through the superior, middle and inferior laryngeal arteries. The middle laryngeal artery is present in 97% of cases and usually has three branches piercing the cricothyroid ligament. An avascular zone exists in the anterior commissure, above which is a zone devoid of glands and vessels [1, 2]. This zone separates the anterior laryngeal commissure from the thyroepiglottic ligament, and is clinically important when doing horizontal laryngectomy.

The pharyngesophageal sphincter (or *pars cricopharyngea*) is innervated by an anastomosis of the external branch of the superior laryngeal nerve with the recurrent laryngeal nerve (*Ansa Dillworthi*) [5].

Surgical Importance

Thyroid and parathyroid surgery, partial laryngectomies, cricopharyngeal myotomy, nerve-muscle pedicles to the posterior cricoarytenoid muscle for dynamic vocal rehabilitations [6].

References

1. Andrea M (1981)
2. André M and Guerrier Y (1981)
3. Hiroto et al. (1968)
4. Malmgren L and Gacek T (1981)
5. Thomassin JM (1982)
6. Tucker HM (1978)

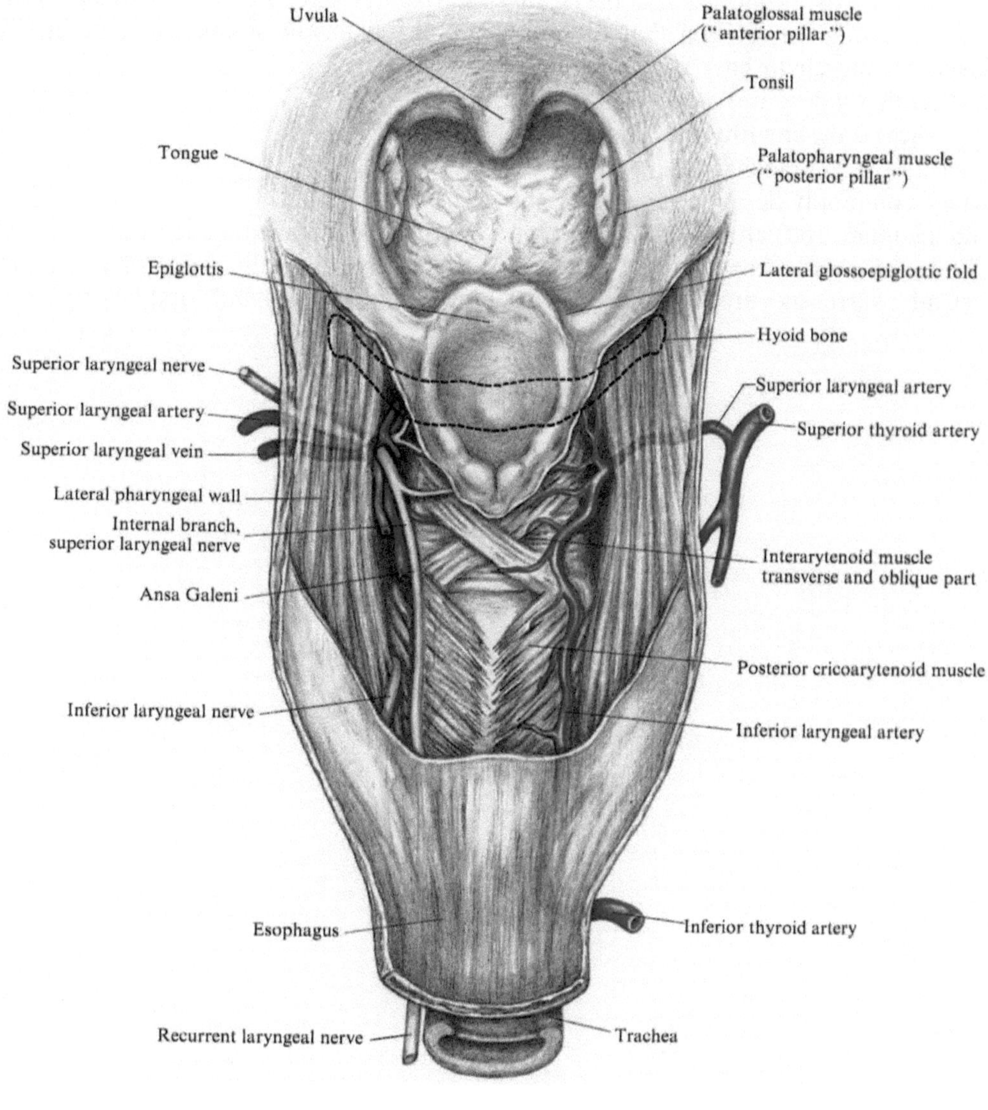

Uvula
Palatoglossal muscle
("anterior pillar")
Tonsil
Tongue
Palatopharyngeal muscle
("posterior pillar")
Epiglottis
Lateral glossoepiglottic fold
Hyoid bone
Superior laryngeal nerve
Superior laryngeal artery
Superior laryngeal vein
Superior laryngeal artery
Superior thyroid artery
Lateral pharyngeal wall
Internal branch,
superior laryngeal nerve
Interarytenoid muscle
transverse and oblique part
Ansa Galeni
Posterior cricoarytenoid muscle
Inferior laryngeal nerve
Inferior laryngeal artery
Esophagus
Inferior thyroid artery
Recurrent laryngeal nerve
Trachea

2.6 The External and Intrinsic Laryngeal Muscles With Their Arteries and Nerves (Anterior View)

The plate shows an artist's conception of the laryngeal muscles. It should be realized that the only external laryngeal muscle to have motor function to the vocal cords (and be important in the production of a voice) is the cricothyroid muscle (*anticus*) (see Sect. 2.5). The branches of the middle laryngeal artery commonly lie in the region of the cricothyroid ligament, and can be cut if coniotomy is done. Since the elastic fibers of the ligament have a vertical course and are transcended in a horizontal direction by the artery, any incisions on the ligament should be kept in a horizontal plane.

Surgical Importance

Extirpation of median cervical cysts and fistulae, thyroid and parathyroid surgery, coniotomy, tracheotomy, partial and total laryngectomy.

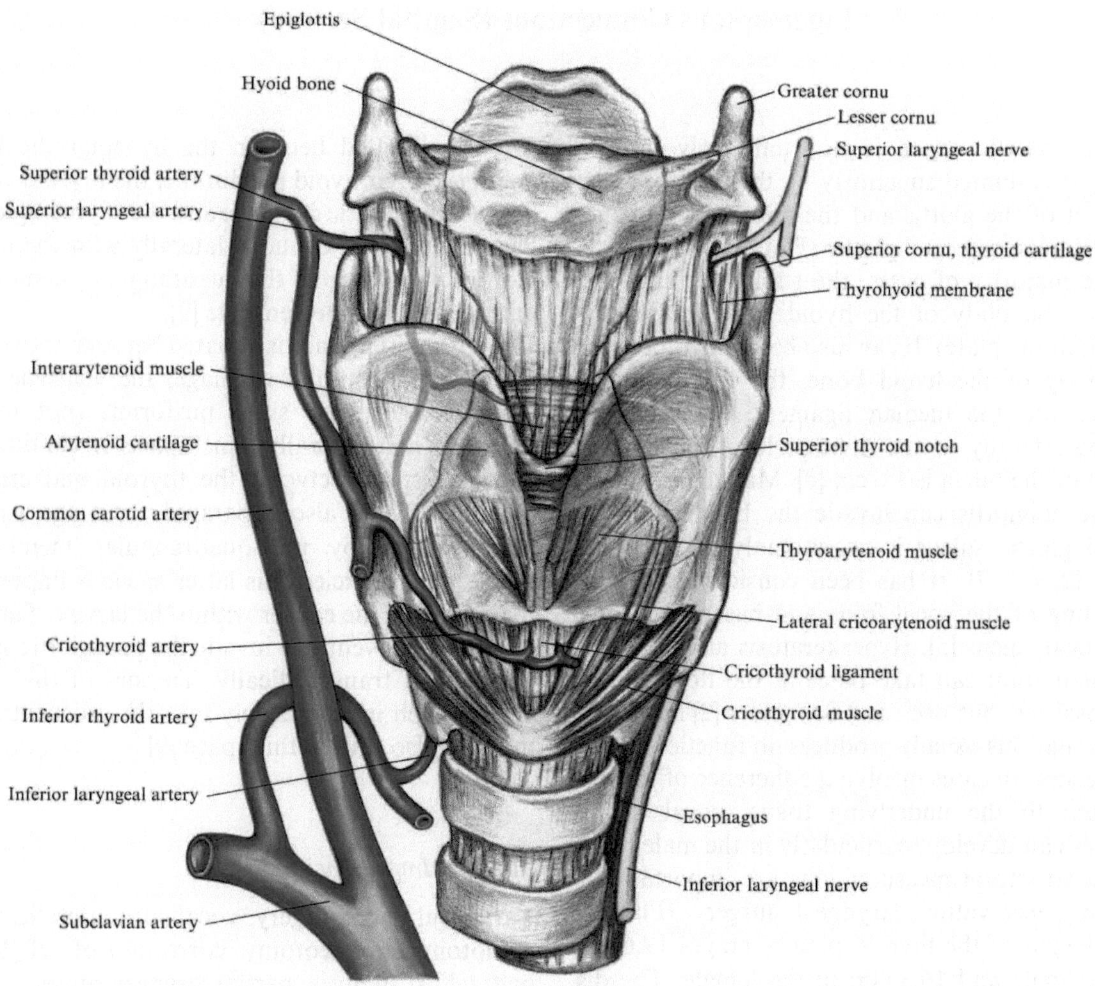
Epiglottis
Hyoid bone
Greater cornu
Lesser cornu
Superior laryngeal nerve
Superior thyroid artery
Superior laryngeal artery
Superior cornu, thyroid cartilage
Thyrohyoid membrane
Interarytenoid muscle
Arytenoid cartilage
Superior thyroid notch
Common carotid artery
Thyroarytenoid muscle
Lateral cricoarytenoid muscle
Cricothyroid artery
Cricothyroid ligament
Inferior thyroid artery
Cricothyroid muscle
Inferior laryngeal artery
Esophagus
Inferior laryngeal nerve
Subclavian artery

2.7 (Plates 1–2) The Laryngeal Cartilages and Their Ligamentous Connections (Sagittal Section)

The subglottic space is not uniformly defined [1, 3]. It is limited superiorly by the stratified epithelium of the glottis and the appearance of mucous glands in the subglottis (Plate 1).
In the majority of cases, the subhyoid bursa lies behind the body of the hyoid bone (not represented in the plate). It can also be situated between the body of the hyoid bone, the thyroid membrane, and the median ligament, or it can be shifted laterally to the right or left. The greatest width of the bursa is 1.6 cm [8]. Malignant tumors of the epiglottis can invade the bursa. The so-called glottic sulcus is an anomaly of the vocal folds [2, 4, 5, 7]. It has been considered to be a doubling of the vocal folds and has been named the vocal sulcus [5]. Hyperkeratosis and epithelial desquamation can take place in the floor of this mucosal sac, but Beck and Schneider [2] have indicated that this usually produces no functional consequences. In cases involving adherence of the epithelium to the underlying tissue, vocal disturbances can develop, particularly in the male [6].
Some structural measurements are important for doing conservation laryngeal surgery (Plate 2). The height of the thyroid plate averages 19.6 mm in the male and 16.3 mm in the female. The distance of the anterior commissure from the inferior margin of the thyroid cartilage is 9.6 mm in the male and 12.3 mm in the female. The distance of the commissure from the upper border of the cricoid is 18.9 mm in the male and 16.4 mm in the female. This distance from the lower margin of the cricoid cartilage is 26 mm in the male and 23.1 mm in the female.
The anterior ligamentous part of the glottis is 15.5 mm in the male and 11.5 mm in the female, while the posterior part is 7.5 mm in the male and 5.5 mm in the female. The pre-epiglottic space of *Bojer* is situated between the hyoepiglottic ligament, the thyrohyoid membrane, the thyroid cartilage, the epiglottic cartilage and the thyroepiglottic ligament. It is continuous laterally with the paraglottic space, deep to the quadrangular membrane and superior to the ventricle [9].
The paraglottic space is situated between the conus elasticus, the thyroid cartilage, the ventricle and the mucosa of the sinus piriformis (not represented). Inferolaterally, this space is continuous with the space between the thyroid and cricoid cartilages. It is also separated from the supraglottic space by the quadrangular membrane above the ventricle. This latter space is important for spread of the cancer within the larynx. Tumors involving the ventricle invade the paraglottic space and spread transglottically. Tumors of the vocal cords which invade deeply into the adjacent cord muscles also invade this space [9].

Surgical Importance

Laryngeal microsurgery, vocal cord lateralization, coniotomy, tracheotomy, correction of subglottic or tracheal stenosis, partial laryngectomies.

References

1. André M and Guerrier Y (1981)
2. Beck K and Schneider P (1926)
3. Carter DR and Meyers AD (1979)
4. Citelli (1906)
5. Lautenschläger A (1912)
6. Roch JB et al. (1981)
7. Salvi G (1901)
8. Sellars J (1982)
9. Silver CE (1981)

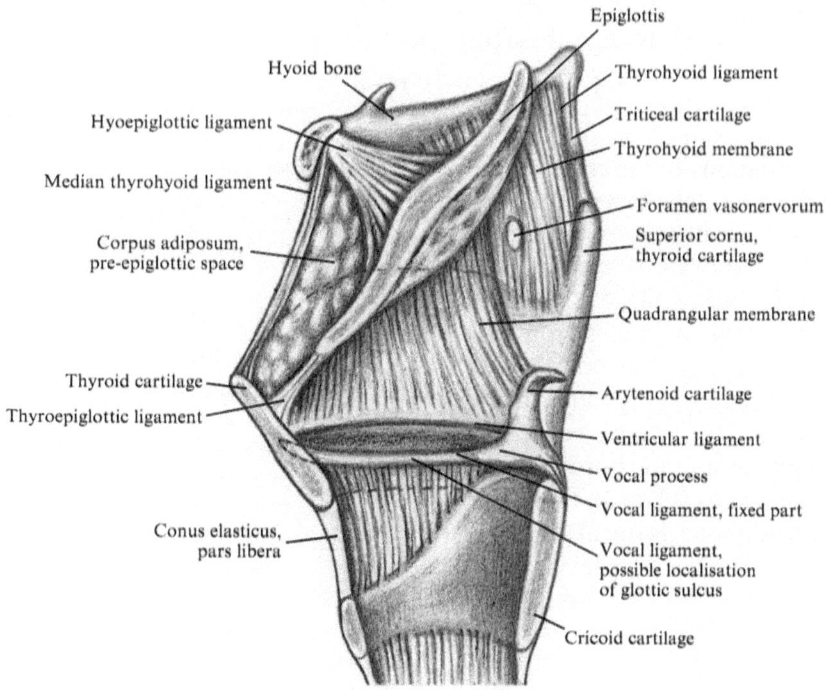

Epiglottis
Hyoid bone
Thyrohyoid ligament
Triticeal cartilage
Hyoepiglottic ligament
Thyrohyoid membrane
Median thyrohyoid ligament
Foramen vasonervorum
Corpus adiposum, pre-epiglottic space
Superior cornu, thyroid cartilage
Quadrangular membrane
Thyroid cartilage
Arytenoid cartilage
Thyroepiglottic ligament
Ventricular ligament
Vocal process
Vocal ligament, fixed part
Conus elasticus, pars libera
Vocal ligament, possible localisation of glottic sulcus
Cricoid cartilage

1

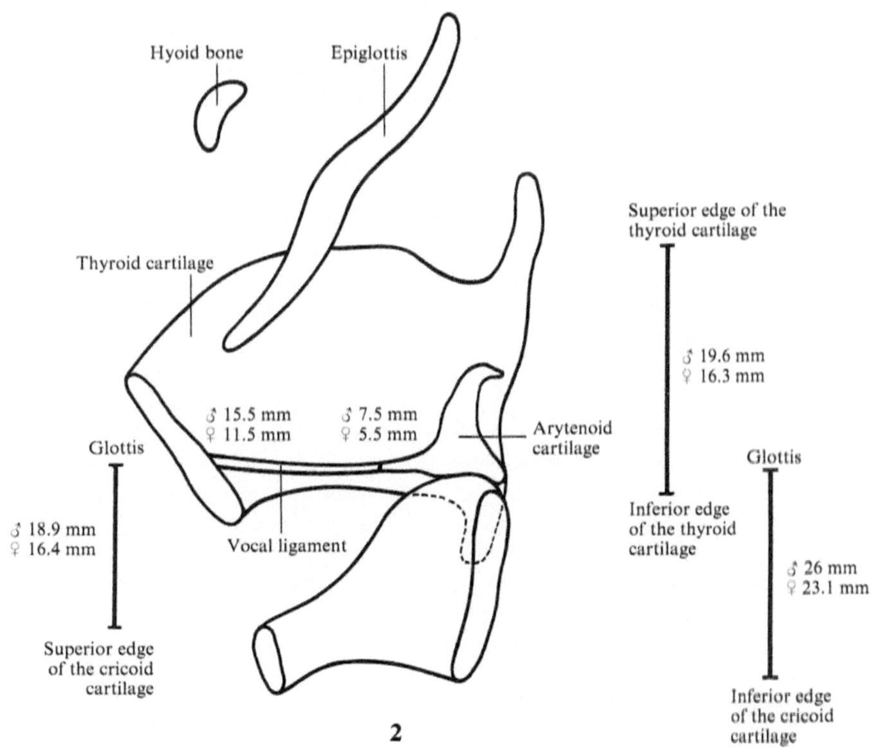

Hyoid bone
Epiglottis
Superior edge of the thyroid cartilage
Thyroid cartilage
♂ 19.6 mm
♀ 16.3 mm
♂ 15.5 mm
♀ 11.5 mm
♂ 7.5 mm
♀ 5.5 mm
Arytenoid cartilage
Glottis
Glottis
♂ 18.9 mm
♀ 16.4 mm
Vocal ligament
Inferior edge of the thyroid cartilage
♂ 26 mm
♀ 23.1 mm
Superior edge of the cricoid cartilage
Inferior edge of the cricoid cartilage

2

2.8 (Plates 1–3) Variations in the Entrance of the Superior Laryngeal Artery Into the Larynx

There can be several variations in the entry of the superior laryngeal artery into the larynx [1]. The superior laryngeal artery usually enters the larynx through the thyrohyoid membrane (Plate 1). If the superior cornu of the thyroid cartilage is missing, the superior laryngeal artery originates from the superior thyroid artery at a slightly lower level and enters the larynx through an opening in the thyroid lamina (Plate 2). The artery very rarely enters the larynx below the inferior border of the thyroid cartilage. In such cases (Plate 3), the artery passes through the cricothyroid ligament.

Surgical Importance

Vocal cord lateral fixation by external approach, coniotomy, partial laryngectomy.

Reference

1. Krmpotić-Nemanić J (1978)

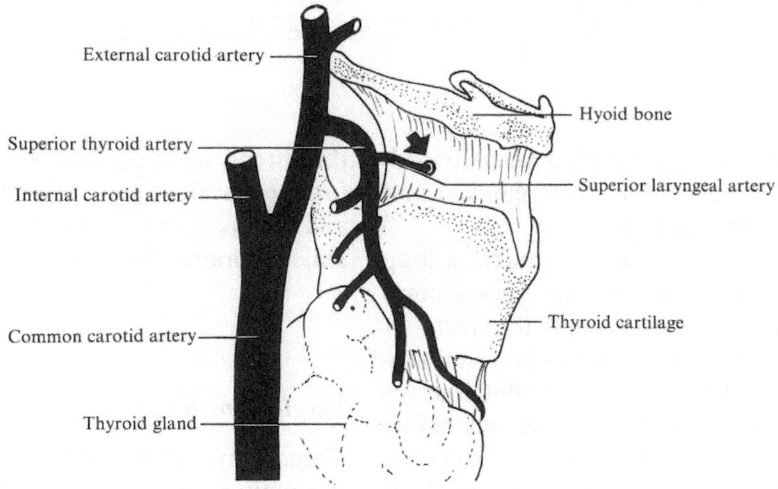
External carotid artery
Hyoid bone
Superior thyroid artery
Superior laryngeal artery
Internal carotid artery
Thyroid cartilage
Common carotid artery
Thyroid gland

1

2

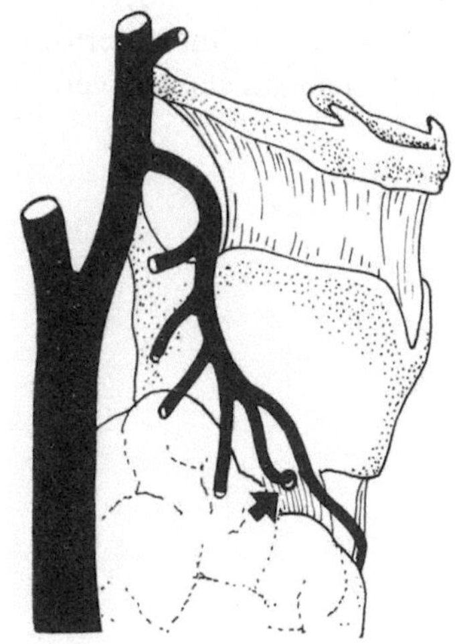

3

2.9 Relationship of the Thyroid Gland to the Larynx, Trachea, and Nerves of the Neck (Left Lateral View)

The isthmus of the thyroid gland has been divided. The left lobe of the gland has been removed and is schematically represented by a line.

The right recurrent laryngeal nerve makes a loop around the subclavian artery (not represented) while the left nerve passes around the aorta or the obliterated ductus arteriosus (*Botalli*) (not represented, see Sect. 2.17, Plates 1–2). Both nerves lie between the trachea and the esophagus before entering the larynx. The relationship of the recurrent nerve to the inferior thyroid artery can vary considerably (see Sect. 2.18).

The scalene foramina occur at the level of the first rib, and are situated between the sternocleidomastoid muscle (not represented) and the anterior and middle scalene muscles. The subclavian vein and the phrenic nerve pass through the anterior scalene foramen (scalene gap), while the subclavian artery and the brachial plexus pass through the posterior scalene foramen (see Sect. 1.15).

Surgical Importance

Coniotomy, tracheotomy, partial or total laryngectomy, surgery of the trachea, neck dissection, scalene node biopsy, mediastinoscopy, Zenker's diverticulectomy, lateral approach to the cervical spine, cervical rib resections, excisions of medial and lateral cervical fistulae or cysts, procedures involving the brachial plexus.

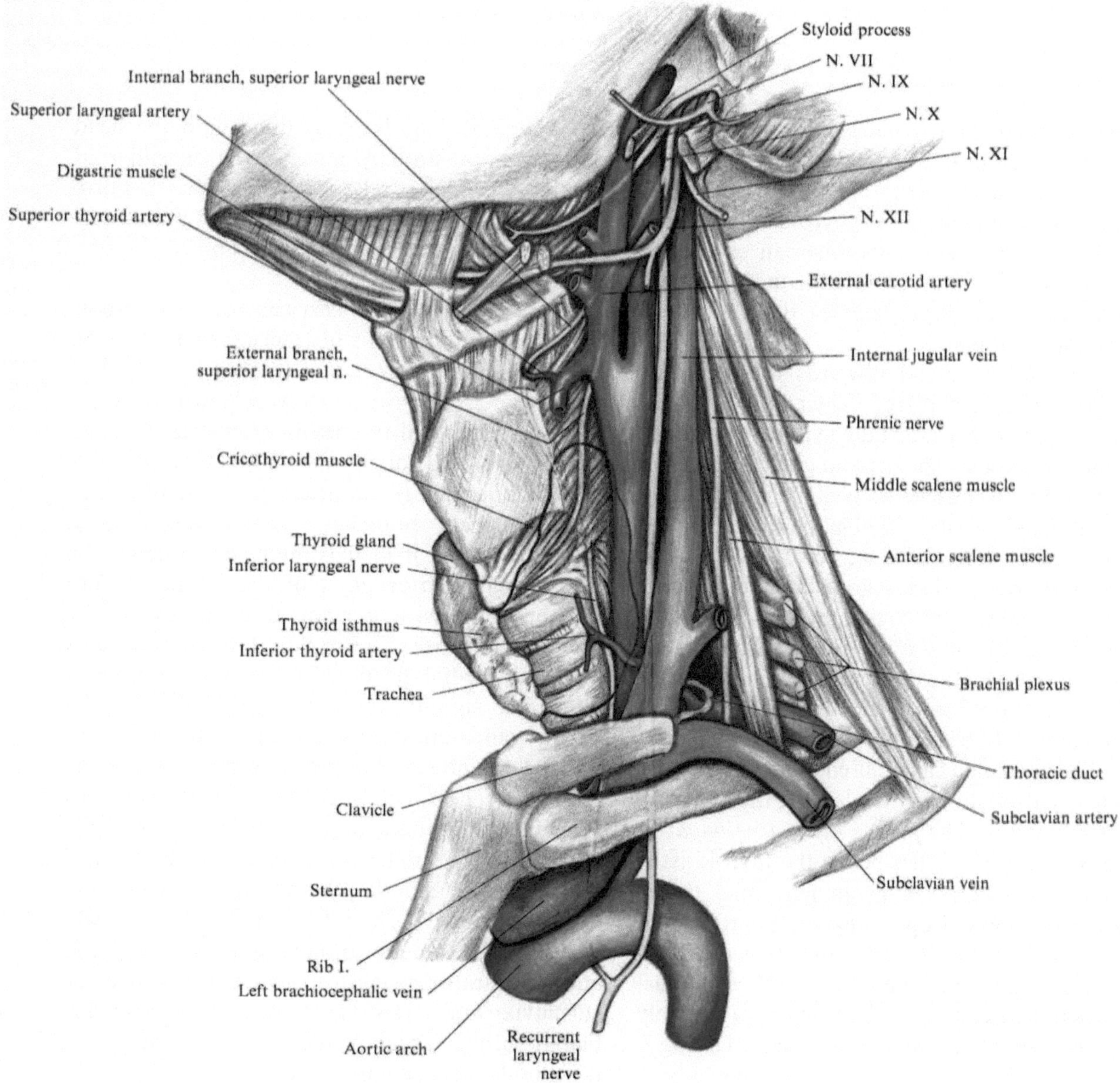

Styloid process
N. VII
N. IX
N. X
N. XI
N. XII
Internal branch, superior laryngeal nerve
Superior laryngeal artery
Digastric muscle
Superior thyroid artery
External carotid artery
Internal jugular vein
External branch, superior laryngeal n.
Phrenic nerve
Cricothyroid muscle
Middle scalene muscle
Thyroid gland
Inferior laryngeal nerve
Anterior scalene muscle
Thyroid isthmus
Inferior thyroid artery
Trachea
Brachial plexus
Thoracic duct
Clavicle
Subclavian artery
Sternum
Subclavian vein
Rib I.
Left brachiocephalic vein
Aortic arch
Recurrent laryngeal nerve

2.10 Topography of the Thyroid Gland, Neck Vessels and Nerves (Anterior View)

The two lobes of the thyroid gland are joined by an isthmus which usually extends from the second to the fourth tracheal rings. The pyramidal lobe is a cranial prolongation of thyroid tissue, and occurs in 50% of cases. This lobe can extend to the hyoid bone (not represented), and thus can interfere with soft tissue exposed during coniotomy or tracheotomy.

The thyroid lobes extend upwards to the lower part of the thyroid cartilage. Their lateral extensions can be variable toward the hypopharynx, trachea, and esophagus. The thyroid gland is covered by the infrahyoid muscles as well as by a median reduplication of the superficial and middle cervical fascia (*lamina pretrachealis*; see Sect. 1.9). The origins of the sternohyoid muscles occur laterally and in front of the sternothyroid muscles. The two groups of muscles delimit an elongated rhomboid which is covered by the cervical fasciae. These fascial layers are doubled vertically and are called the *linea alba colli* because of the white color produced. The linea alba can be used as an anatomical landmark in dissections of the larynx, thyroid isthmus, and trachea. The *plexus thyroideus impar* is a venous plexus which lies below the thyroid isthmus and may contain a prominent vein, the *v. thyroidea ima*. When operating in this region of the neck, the surgeon should realize that the innominate artery and even the aortic arch can be displaced here and can cross the trachea to possibly be injured during dissection (see Sects. 1.4, 5, 7, 11.10).

The topography of the thyroid isthmus and the variable arterial supply of the different segments of the cervical trachea previously resulted in tracheotomies being classified as "superior", "middle", or "inferior", depending on whether the incision made into the trachea was above, through, or below the isthmus. However, this classification also implied an emergency procedure that might involve the thyroid gland and result in additional bleeding from the thyroid vessels. Tracheotomy is *only* exceptionally done as a true emergency operation at the present time and has become a "planned" procedure in most cases following intubation of the patient for protection of the airway. If intubation is *not* possible, emergency coniotomy may be the safest procedure to establish an airway (see Sect. 2.6). The latter is done with the head in retroflexion and by incising the cricothyroid membrane, which is relatively avascular. This procedure is then followed by a properly placed tracheotomy which may require division of the thyroid isthmus.

The blood supply of the cervical trachea is important in the pathogenesis of tracheal stenosis [1–5]. The larynx and the cervical trachea are supplied by the superior thyroid artery, while the inferior part of the cervical trachea is vascularized by the inferior and occasionally the middle thyroid arteries. The inferior thyroid artery gives off three tracheoesophageal branches. The lowest branch enters the superior mediastinum and supplies the lateral and anterior parts of the trachea before anastomosing with peripheral branches of the bronchial arteries. The middle branch originates from the medial portion of the inferior thyroid artery and supplies the lateral tracheal wall. The upper branch anastomoses with the superior thyroid artery and supplies the uppermost part of the cervical trachea. The tracheoesophageal branches form intercommunicating arcades with both ipsilateral and contralateral vessels so that the cervical trachea has an abundant blood supply on occasion (not represented). The inferior thyroid artery will also give off a superior branch to the thyroid gland and a separate branch for the trachea. While the subclavian and vertebral arteries may likewise participate in the vascularization of the trachea, such is generally an uncommon occurrence.

Surgical Importance

Coniotomy, tracheotomy, thyroidectomy, surgery of the cervical trachea and esophagus, partial or total laryngectomy, neck dissection, surgery of the carotid sheath, scalene node (Daniel's) biopsy, mediastinoscopy.

References

1. Freeland AP (1974)
2. Miura T and Grille HC (1966)
3. Saito H (1976)
4. Salassa JR et al. (1977)
5. Staindl O (1984)

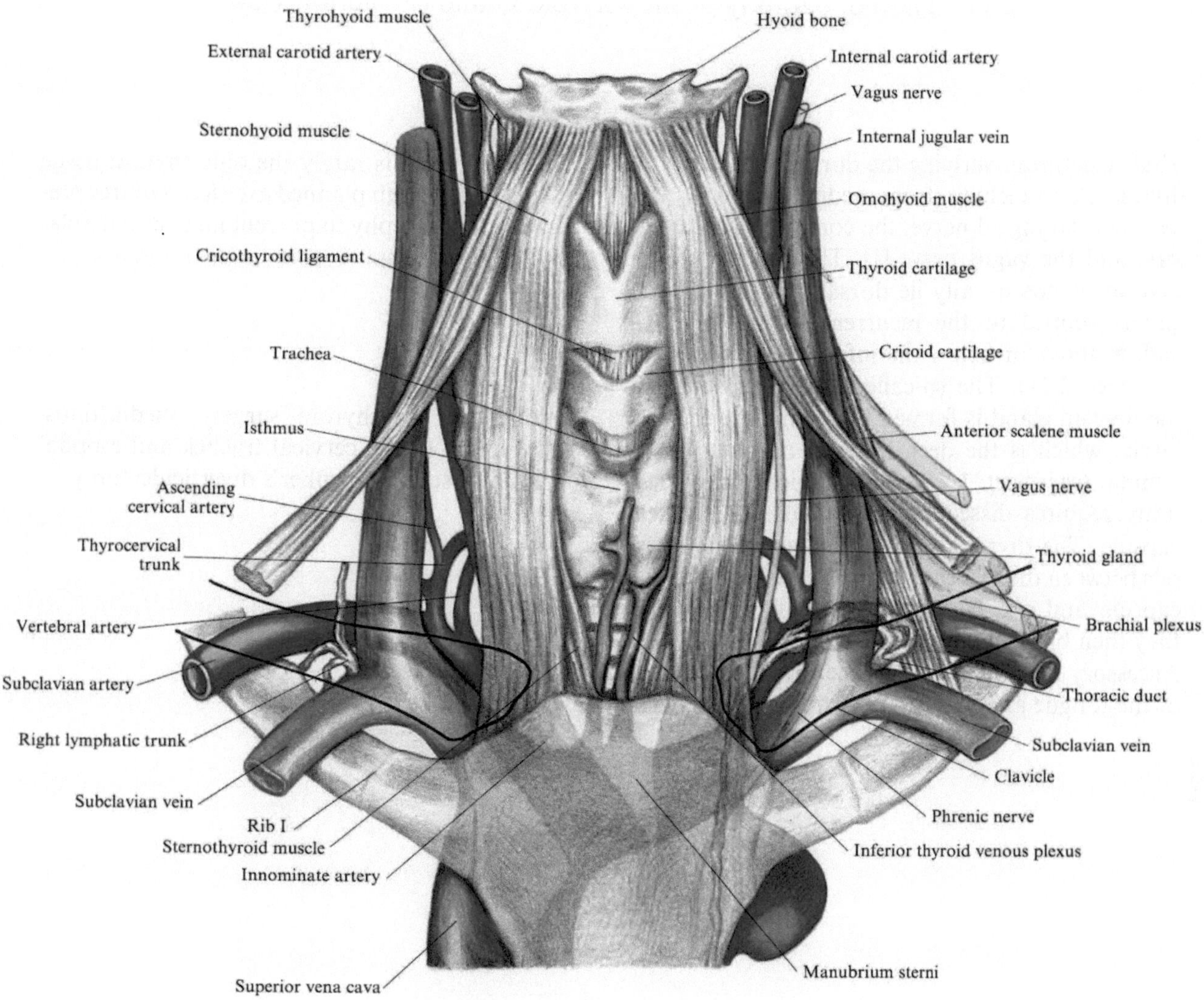
Thyrohyoid muscle
Hyoid bone
External carotid artery
Internal carotid artery
Vagus nerve
Sternohyoid muscle
Internal jugular vein
Omohyoid muscle
Cricothyroid ligament
Thyroid cartilage
Trachea
Cricoid cartilage
Isthmus
Anterior scalene muscle
Ascending cervical artery
Vagus nerve
Thyrocervical trunk
Thyroid gland
Vertebral artery
Brachial plexus
Subclavian artery
Thoracic duct
Right lymphatic trunk
Subclavian vein
Subclavian vein
Clavicle
Rib I
Phrenic nerve
Sternothyroid muscle
Inferior thyroid venous plexus
Innominate artery
Manubrium sterni
Superior vena cava

2.11 The Topography of the Thyroid Gland (Posterior View)

Vital structures overlying the dorsal surface of the thyroid gland include (from medial to lateral): the recurrent laryngeal nerve, the common carotid artery, and the vagus nerve [1]. The superior parathyroid glands usually lie dorsal and the inferior glands ventral to the recurrent laryngeal nerve and/or above or below the inferior thyroid artery (see Sect. 2.21). The so-called "false capsule" of the thyroid gland is formed by the middle cervical fascia, which is the deep layer of the pretracheal lamina (see Sect. 1.9, 2.16). Surgical thyroidectomy requires dissection in the false or surgical capsule. The thyroid arteries pass as isolated vessels between the external false and the internal true capsule and are best ligated in this plane since they then branch within the capsule.
Accessory thyroid glands can be found in the base of the tongue and at various locations in the neck. A lingual thyroid is rarely the only thyroid tissue present [2], although planned excision requires preoperative scintigraphy to prevent inadvertent ablation of all functional thyroid tissue.

Surgical Importance

Thyroid and parathyroid surgery, mediastinoscopy, surgery of the cervical trachea and esophagus, neck dissection, Zenker's diverticulectomy.

References

1. Guleke N (1953)
2. Helms J and Kimpel G (1978)

Plate from Guleke [1]

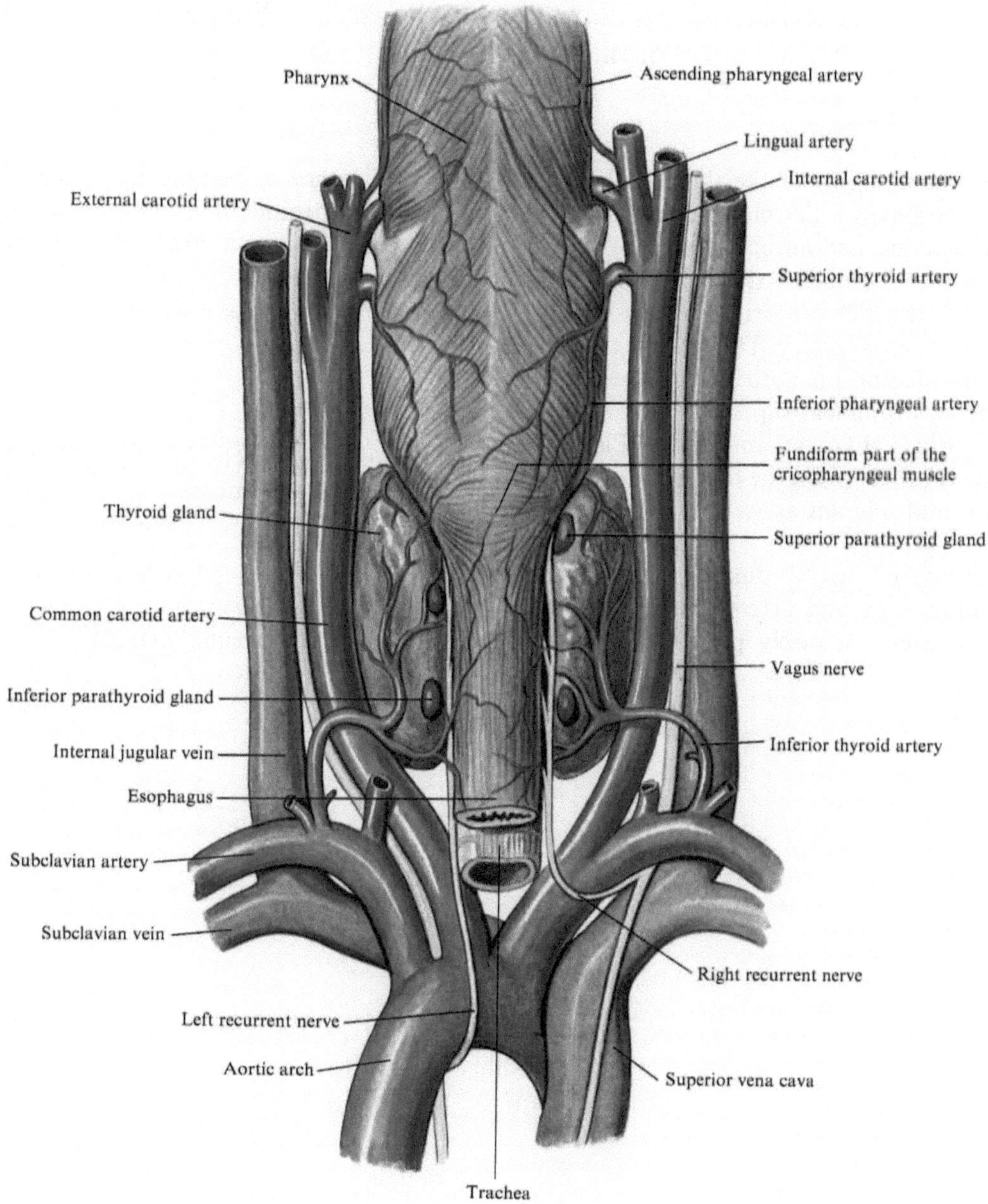
Pharynx
Ascending pharyngeal artery
Lingual artery
Internal carotid artery
External carotid artery
Superior thyroid artery
Inferior pharyngeal artery
Fundiform part of the
cricopharyngeal muscle
Thyroid gland
Superior parathyroid gland
Common carotid artery
Vagus nerve
Inferior parathyroid gland
Inferior thyroid artery
Internal jugular vein
Esophagus
Subclavian artery
Subclavian vein
Right recurrent nerve
Left recurrent nerve
Aortic arch
Superior vena cava
Trachea

2.12 (Plates 1–3) Variations in the Arterial Supply of the Thyroid Gland

The arterial blood supply of the thyroid gland is provided by the superior thyroid arteries, the inferior thyroid arteries, and an occasional thyroidea ima artery [1, 2]. The middle thyroid artery is an uncommon vessel which can originate as a branch of the superior thyroid artery. The inferior thyroid artery can be missing on both sides (Plates 1–2) and replaced by prominent superior thyroid arteries (Plate 1) or by a thyroidea ima artery (Plate 3). The arterial branches within the thyroid gland generally show multiple intraparenchymal anastomoses.

The inferior thyroid artery commonly passes behind the common carotid artery and in front of the vertebral artery. It rarely passes in front of the carotid artery or behind the vertebral artery. It usually lies behind the sympathetic trunk, and uncommonly occurs in front of or through the trunk. For the vessel's location in relationship to the recurrent nerve, see Sect. 2.18.

Surgical Importance

Thyroid surgery, tracheotomy, neck dissection, microvascular anastomoses

References

1. Krmpotić-Nemanić J (1978)
2. Lanz T v and Wachsmuth W (1955)

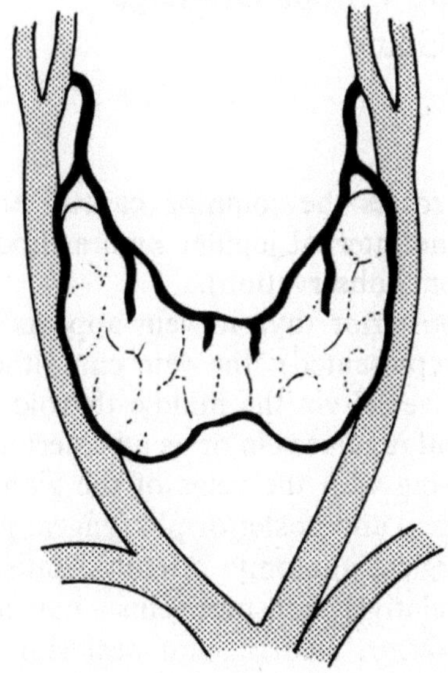

1

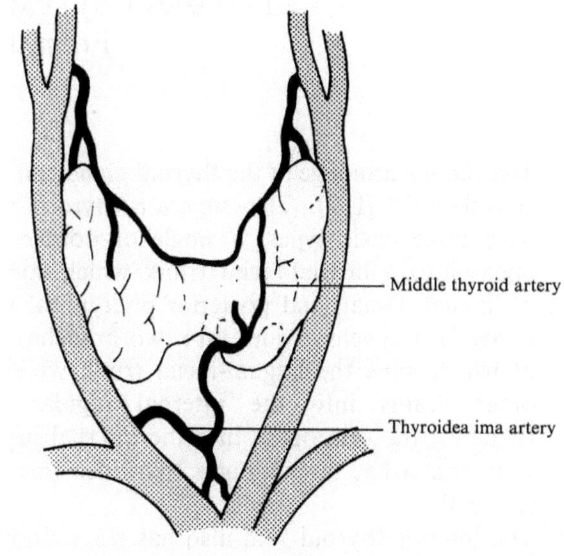
Middle thyroid artery
Thyroidea ima artery

2

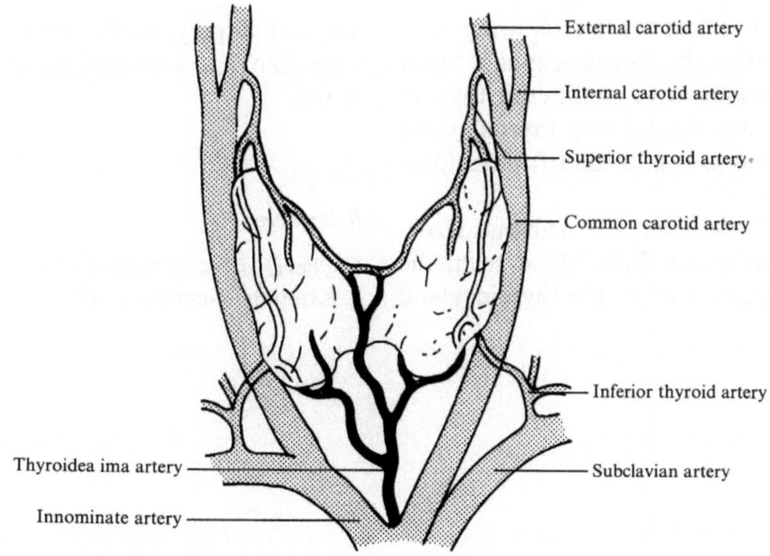
External carotid artery
Internal carotid artery
Superior thyroid artery
Common carotid artery
Inferior thyroid artery
Thyroidea ima artery
Subclavian artery
Innominate artery

3

2.13 (Plates 1–9) Variations of the Venous Drainage From the Thyroid Gland

The venous drainage of the thyroid gland can vary considerably [1, 2]. The superior thyroid veins have three basic types: a single or double vein opens into a lingual-facial trunk which consists of lingual, facial, and posterior pharyngeal veins (Plate 1); the vein divides into two branches, one of which joins the lingual-facial trunk while the other drains into the internal jugular vein (Plate 2); the vein opens into the internal jugular vein, somewhat resembling a thyroidea ima vein (Plate 3).

The inferior thyroid vein also has three drainage patterns: the vein opens on both sides into the brachiocephalic vein or middle thyroid veins (Plate 4); it opens into the brachiocephalic and internal jugular vein (Plate 5); it opens into the inferior part of the internal jugular vein, and corresponds to a middle thyroid vein (Plate 6).

The median thyroid vein shows the following variations: it opens into the brachiocephalic vein (Plate 7); it anastomoses with the inferior thyroid vein (Plate 8); it joins the inferior thyroid veins to drain with these vessels into the left brachiocephalic vein (Plate 9).

The middle thyroid vein is very often absent. However, when present, it can drain blood from the inferior and the lateral part of the thyroid gland and crosses the common carotid artery to open into the internal jugular or brachiocephalic veins (personal observation).

The posterior thyroid vein appears as two types (not represented): the vein can either drain as a single vessel via the middle thyroid vein into the internal jugular vein or as a posterior plexus anastomosing with the veins of the gland and with a laryngeal and posterior pharyngeal plexus.

In surgical anatomy, attention must be given to four relatively constant veins – namely, the superior, inferior, median, and posterior thyroid veins – as well as to the middle thyroid vein in one-third of cases [2].

Surgical Importance

Thyroid surgery, tracheotomy, partial or total laryngectomy, neck dissection, microvascular anastomoses.

References

1. Chevrel JP et al. (1965)
2. Krmpotić-Nemanić J (1978)

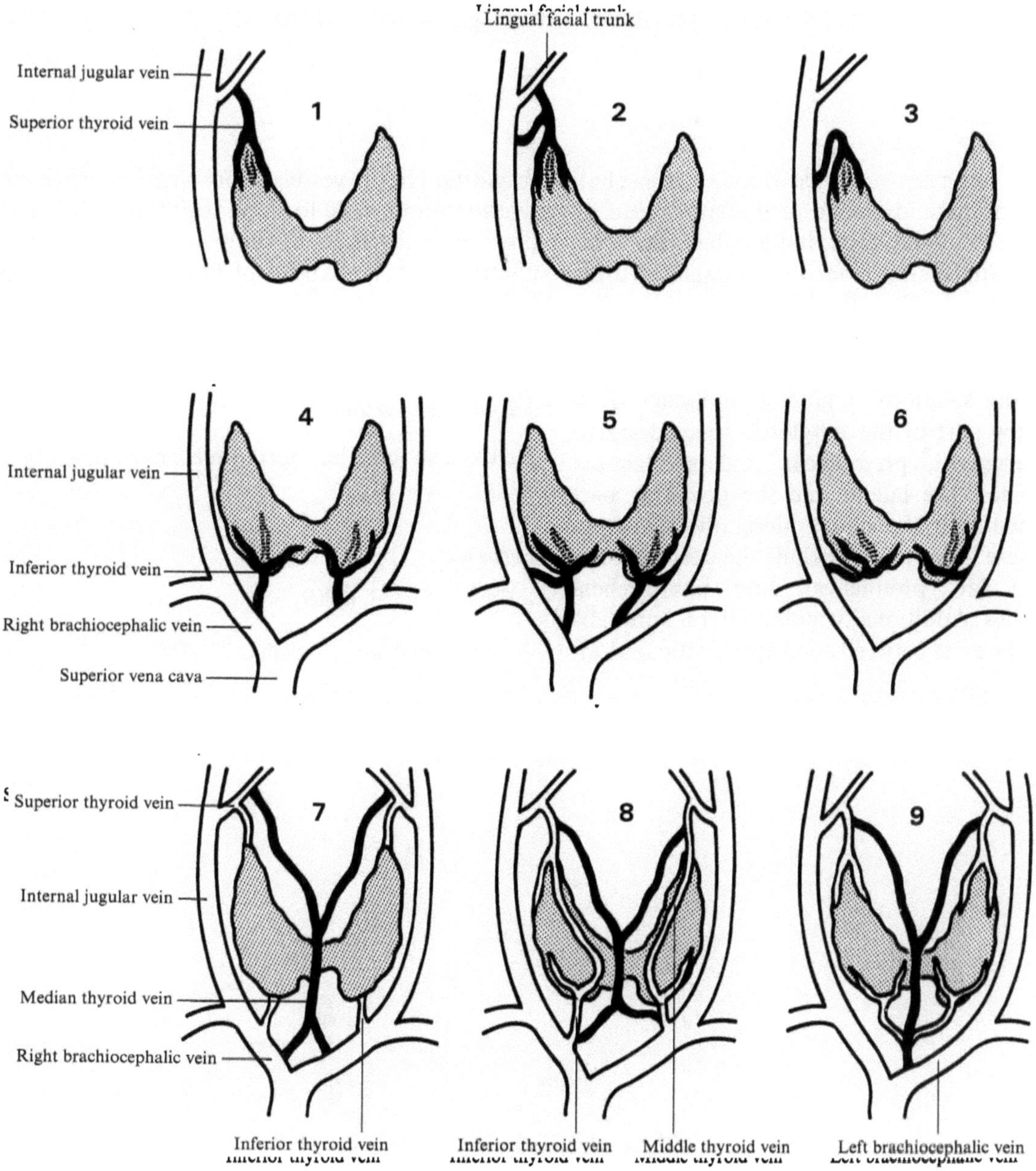
Lingual facial trunk
Internal jugular vein
Superior thyroid vein
1
2
3
4
5
6
Internal jugular vein
Inferior thyroid vein
Right brachiocephalic vein
Superior vena cava
7
8
9
Superior thyroid vein
Internal jugular vein
Median thyroid vein
Right brachiocephalic vein
Inferior thyroid vein
Inferior thyroid vein
Middle thyroid vein
Left brachiocephalic vein

The figure has been modified from Cachin et al. [1]. The lymphatic drainage of the larynx can be divided into a sugraglottic and a subglottic zone. The lymphatic vessels from the supraglottic zone drain through the thyrohyoid membrane into the deep cervical nodes. In contrast, the glottis has very few lymphatic vessels, and these generally drain to the subglottic region. Lymphatics from the anterior part of the subglottic space drain to the prelaryngeal, pretracheal, and mediastinal lymph nodes; the lateral and the posterior part are drained into the nodes along the recurrent nerve. These latter nodes include the deep jugular, supraclavicular, pretracheal, and paratracheal lymph nodes. Additionally, contralateral lymphatic pathways exist between the supraglottic and infraglottic lymph vessels. This rich lymphatic network is responsible for a poor prognosis when supra- and subglottic carcinomas occur, in distinct contrast to malignancies of the same size presenting in the glottic region.

Surgical Importance

Partial and total laryngectomies, neck dissection.

Reference

1. Cachin Y et al. (1969)

Plate modified from Cachin et al. [1]

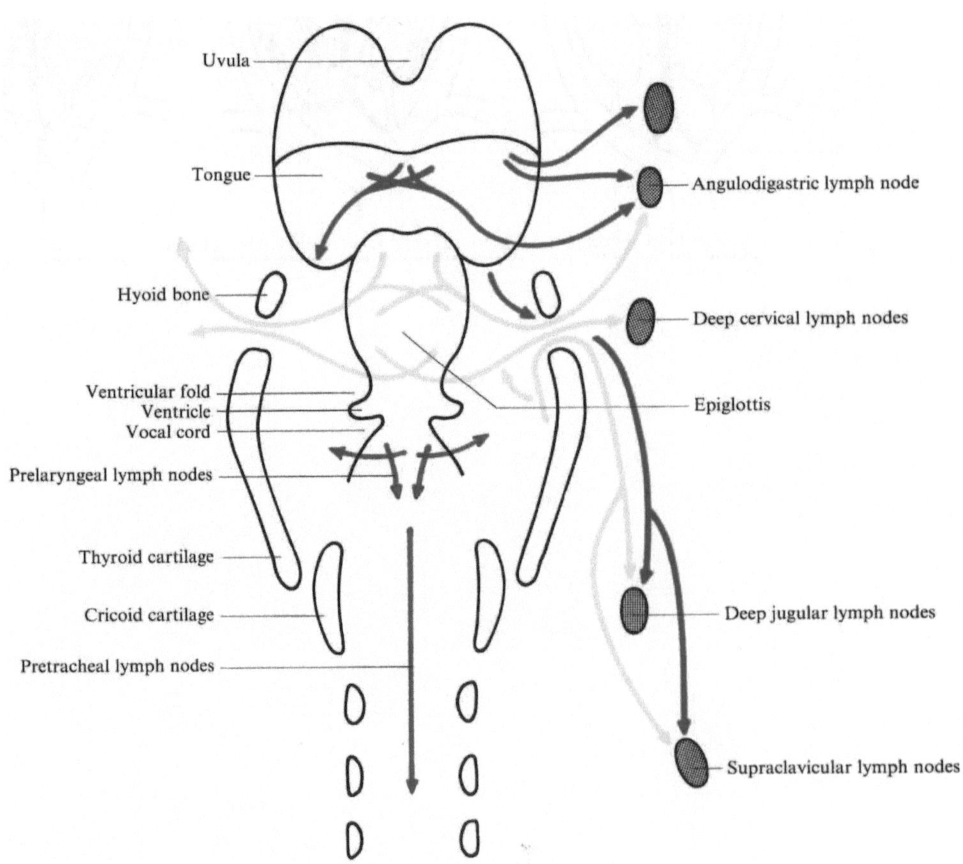

2.15 Lymphatic Drainage of the Thyroid Gland

The lymphatic vessels of the thyroid gland can pass in different directions to form "satellite" lymphatics [1, 2]. The vessels from the upper pole of the gland accompany the superior thyroid artery and vein, while those from the lateral side accompany the inferior thyroid artery. The lymphatics from the lower pole accompany the inferior thyroid vein.
The lymphatic vessels which do not follow the thyroid arteries and veins run along the margins of the gland to leave the gland at its poles. These vessels finally drain to the deep cervical, supraclavicular, prelaryngeal and pretracheal lymph nodes.

Surgical Importance

Thyroid tumors, partial or total extirpation of the thyroid gland, neck dissection.

References

1. Cachin Y et al. (1969)
2. Eickhoff W and Herberhold C (1968)

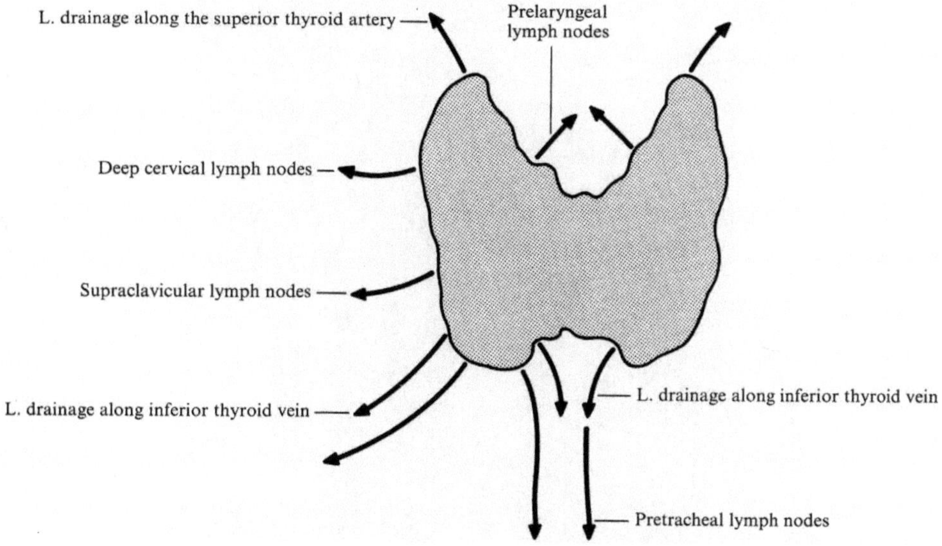

2.16 The Relationship of the Left Recurrent Laryngeal Nerve to the Thyroid Gland and the Inferior Thyroid Artery

The plate shows that the left lower lobe of the thyroid gland has been removed, and a hemostat has been placed on the false capsule of the gland for retraction. The dorsal relationship of the left recurrent laryngeal nerve to the inferior thyroid artery and sympathetic nerve fibers is shown.

Surgical Importance

Strumectomy, parathyroid surgery, tracheotomy, surgery of the cervical trachea, neck dissection, partial and total laryngectomy, resection of chemodectomas, Zenker's diverticulectomy, lateral approach to the cervical spine.

Reference

1. Guleke N (1953)

Plate adopted from Guleke [1]

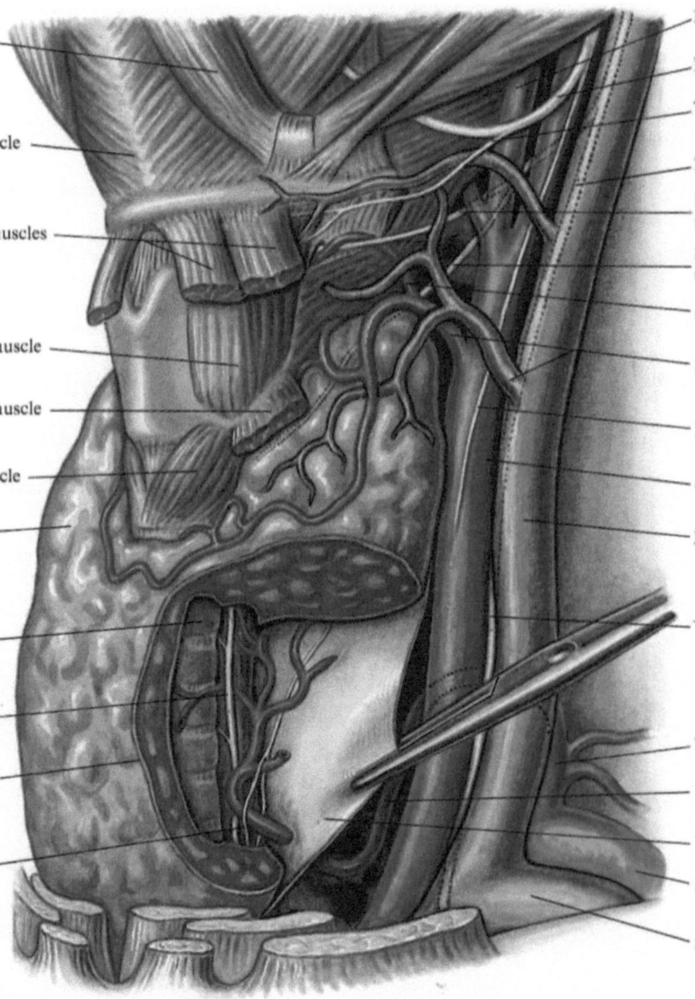

2.17 (Plates 1–2) The Course of the Recurrent Laryngeal Nerve in Relationship to a "Normal" Location of the Major Vessels and Due to an Anomaly of the Right Subclavian Artery

The right recurrent laryngeal nerve usually leaves the vagus nerve where it crosses the innominate artery or the subclavian artery (Plate 1). The recurrent nerve passes behind the subclavian artery to enter a groove between the trachea and the esophagus before reaching the larynx. The left recurrent nerve leaves the vagus at the level of the aortic arch, makes a loop around the fibrous remnant of the *ductus arteriosi Botalli* (*lig. arteriosum*) and then passes between the trachea and the esophagus to reach the larynx. If the right subclavian artery originates from the left part of the aortic arch (lateral to the left subclavian artery) and passes in front of or behind the trachea or behind the esophagus to the larynx (Plate 2), the right inferior laryngeal nerve leaves the vagus at a higher level (at approximately the first tracheal ring) and enters horizontally into the larynx [3]. This nerve is much shorter than normally, and can either be situated within a duplication of connective tissue (*ligament of Berry*) or forms a loop around the inferior thyroid artery [1, 5]. According to Freche and Boudin [2], in 35% of cases there is a deep branching of the recurrent laryngeal nerve within the larynx into abductor and adductor fibers. This division can also be replicated by a prominent posterior pharyngeal branch (*r. pharyngeus*).

A non-recurrent laryngeal nerve is not necessarily associated with an anomaly of the right subclavian artery [6]. In such cases, the non-recurrent laryngeal nerve can leave the vagus nerve at any level from C_5 to the upper pole of the thyroid gland [4].

Surgical Importance

Surgery of the larynx and trachea, thyroid and parathyroid surgery, surgery of lateral and median cervical cysts and fistulae, surgery of the great vessels of the neck, mediastinoscopy, surgery of the anterior upper mediastinum, neck dissection.

References

1. De Antoni E et al. (1971)
2. Freche C and Boudin G (1982)
3. Locatelli E et al. (1977)
4. Papadatos D (1978)
5. Runne J (1957)
6. Stewart GR et al. (1972)

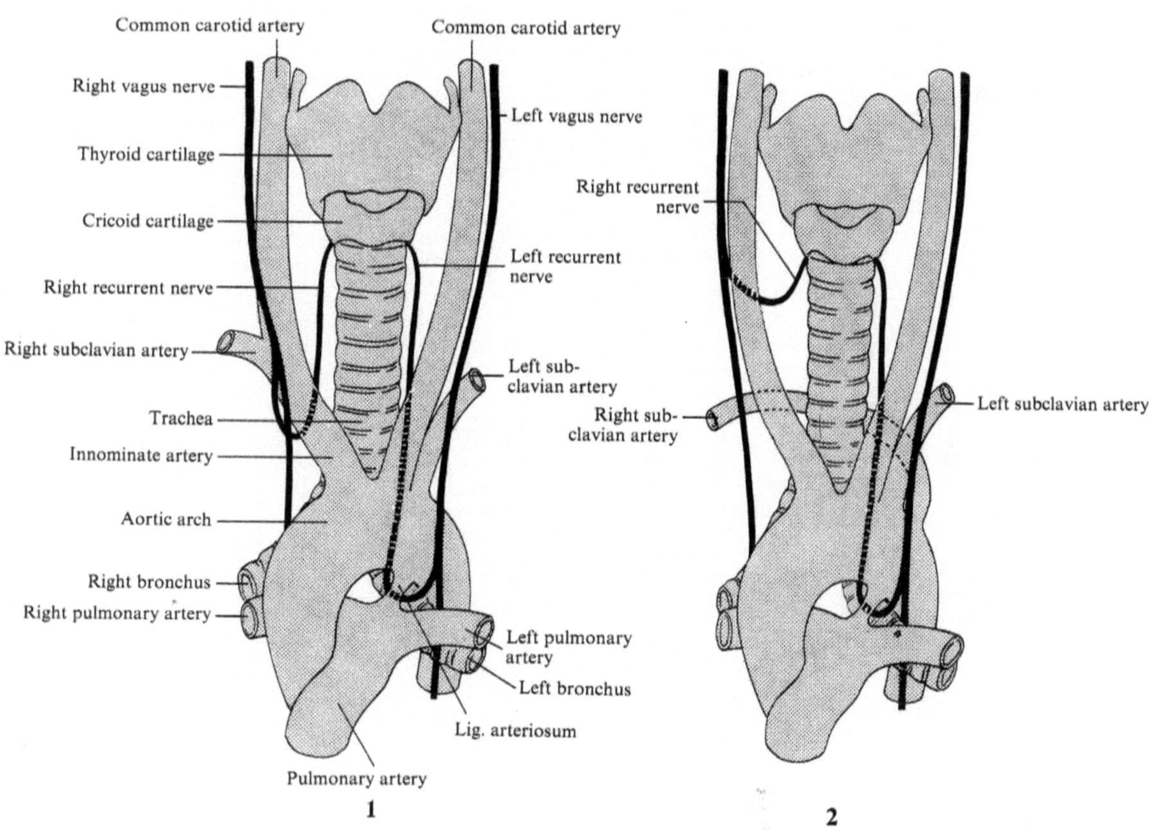

2.18 (Plates 1–10) Variations of the Recurrent Laryngeal Nerve in Relationship to the Inferior Thyroid Artery

An important anatomical relationship is that of the recurrent laryngeal nerve to the inferior thyroid artery [1–4]. The right recurrent laryngeal nerve usually runs in front of and lateral to the inferior thyroid artery. However, this relationship can vary, as shown by Keros and Nemanić [2]. The right nerve lies more anteriorly than the left nerve in 40.3% of cases (Plate 1). In 57.2% of cases, the left nerve lies medial and dorsal to the artery. In 13% of cases, there are atypical courses of the nerves. Additionally, the right nerve runs in front of the artery in 23% of cases, within the bifurcation of the artery in 50% of cases, and retrovascularly in 27% of cases. In contrast, the left nerve runs prevascularly in 10% of cases, within the bifurcation in 35%, and retrovascularly in 55%.

When the recurrent nerve divides at the level of the inferior thyroid artery or below (11% of cases), both branches of the nerve pass behind, between, or in front of the arterial branches (Plates 5–9). The posterior branch can also extend as far cranially as the level of the digastric muscle. According to Guerrier [1], this branch can innervate the pharynx as the *ramus pharyngeus*. This terminal nerve branch lies in the tracheoesophageal groove more often on the right side than on the left (see Sect. 2.19).

Surgical Importance

Partial resections of the larynx, thyroid and parathyroid surgery, tracheotomy, surgery of the cervical esophagus and trachea.

References

1. Guerrier Y (1953)
2. Keros P and Nemanić G (1967)
3. Krmpotić-Nemanić J (1978)
4. Pech A et al. (1976)

Plates 1–10 from Keros [2]

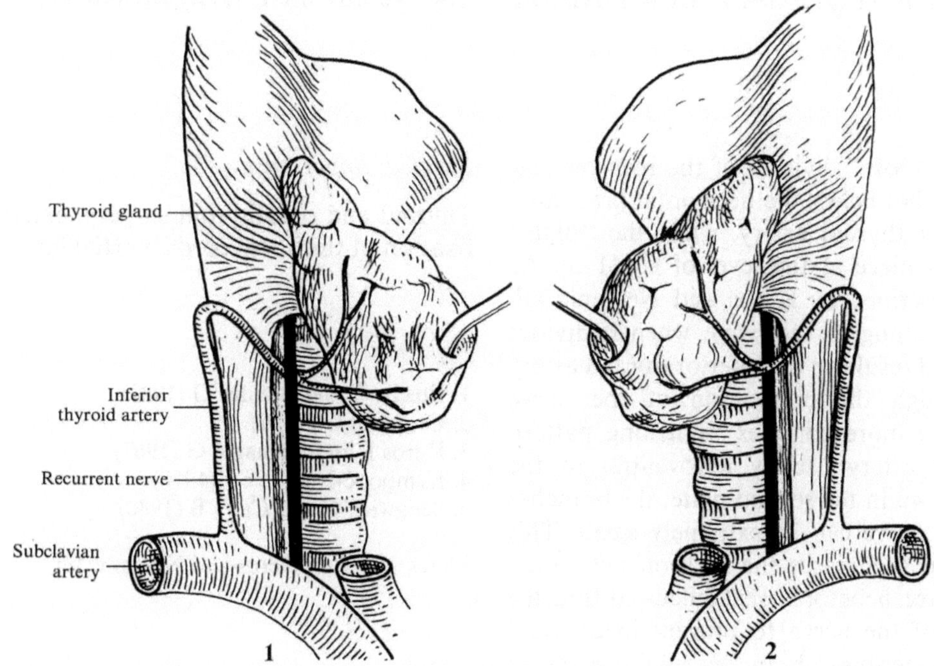
Thyroid gland
Inferior thyroid artery
Recurrent nerve
Subclavian artery
1
2

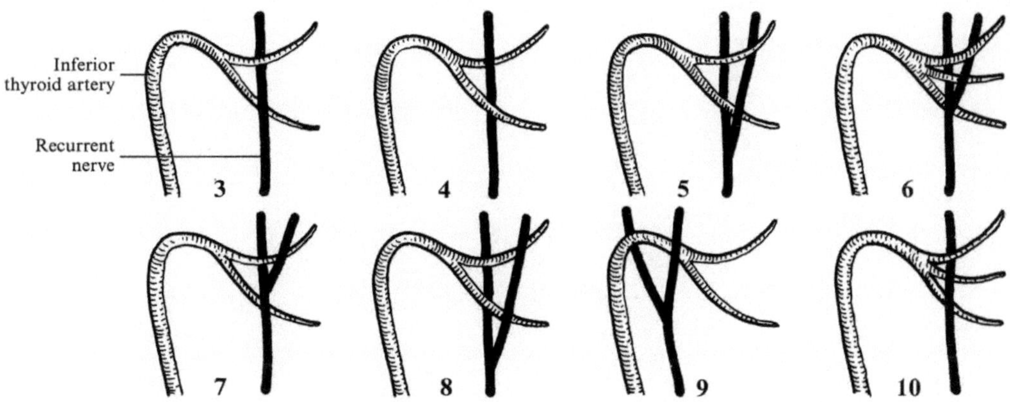
Inferior thyroid artery
Recurrent nerve
3
4
5
6
7
8
9
10

2.19 (Plates 1–2) Low Division of the Recurrent Laryngeal Nerve

The so-called "low" division of the recurrent laryngeal nerve lies below the horizontal branching of the inferior thyroid artery, while the "high" division takes place at the level of C_5 [1–5]. As previously described by Keros and Nemanić [3], a simple branching of the nerve when it divides at a lower level results in the inferior thyroid artery passing through the bifurcation of the nerve (Plate 1). In a more complex branching pattern (Plate 2), the artery usually lies ventral to the nerve. As shown in the second plate, the branches of this latter nerve can be extremely small. This means that in such cases the surgeon must look *lower* in the tracheoesophageal sulcus to find the main trunk of the nerve to prevent inadvertent injury to fine terminal branches of the nerve at the thyroid gland or in the vicinity of the inferior thyroid artery.

Surgical Importance

Thyroid and parathyroid surgery, external laterofixation of the vocal cords, partial laryngectomy.

References

1. Freche C and Boudin G (1982)
2. Grapulin G (1976)
3. Keros P and Nemanić G (1967)
4. Krmpotić-Nemanić J (1978)
5. Sedgwick C and Cady B (1980)

Plates 1–2 from Keros and Nemanić [3]

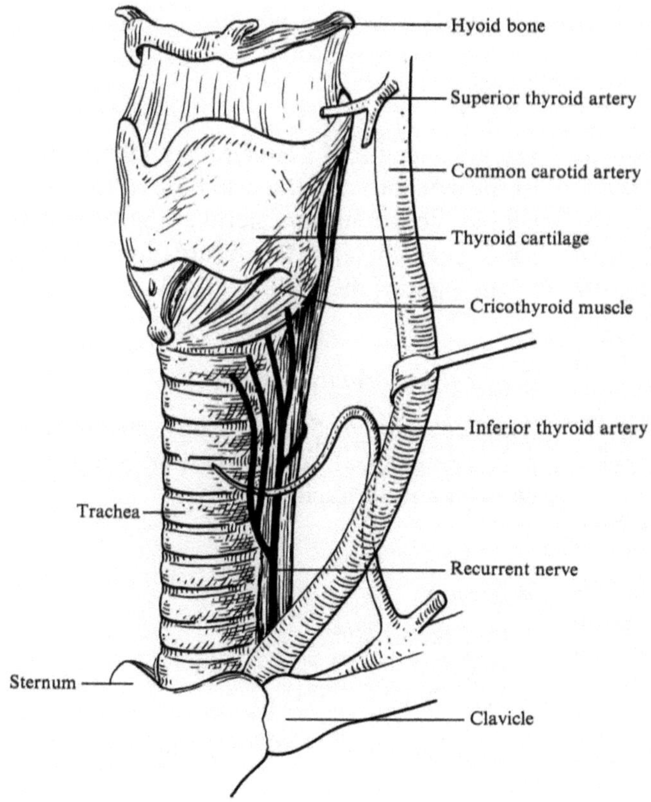
Hyoid bone
Superior thyroid artery
Common carotid artery
Thyroid cartilage
Cricothyroid muscle
Inferior thyroid artery
Trachea
Recurrent nerve
Sternum
Clavicle

1

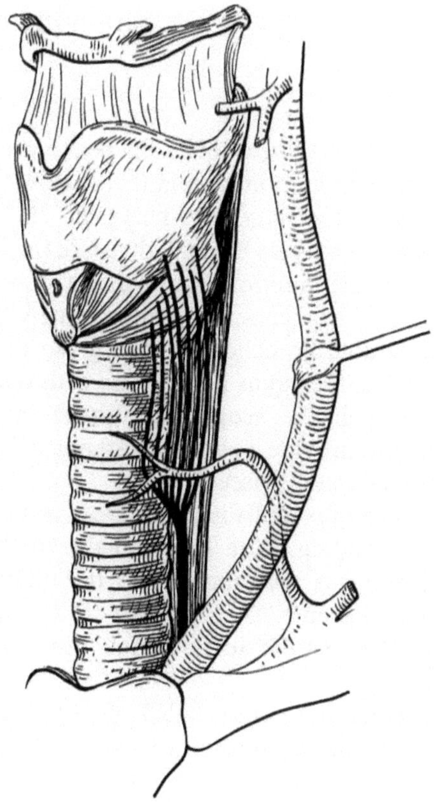

2

2.20 Topography of the Vagus Nerve and the Recurrent Laryngeal Nerve Within the Mediastinum (Anterior View)

The vagus nerves give off branches to the bronchi and bronchopulmonary segments below the levels of the recurrent laryngeal nerves [1–2]. The right vagus nerve courses posteriorly while the left nerve passes to the anterior esophageal wall, giving off esophageal branches. At approximately the same level, cardiac branches are also given off. At the level of T_9, the vagus nerves pass with the esophagus through the esophageal hiatus in the diaphragm to provide a parasympathetic nerve supply to the abdominal viscera.

The esophagus and the aorta cross in the thorax. The descending aorta lies cranially to the left of the esophagus and approaches the midline in its inferior descent. At the level of T_{12} the aorta passes through the aortic hiatus into the abdominal cavity.

The neonatal trachea is 5.5 cm long and has a 6 mm diameter. Within the next 24 months, it reaches a length of 8 cm and has a 9 mm diameter. In the newborn, the tracheal bifurcation is situated behind the manubrium sterni at the level of T_2–T_3. By 2 years of age, the bifurcation comes to lie to the right of the midline [3].

Surgical Importance

Tracheotomy, mediastinoscopy, mediastinal surgery, esophageal reconstruction, surgery of the cervical and thoracic trachea.

References

1. Denecke HJ (1980)
2. Hafferl A and Thiel W (1969)
3. Jacquemard C (1971)

Plate adopted from Hafferl and Thiel [2]

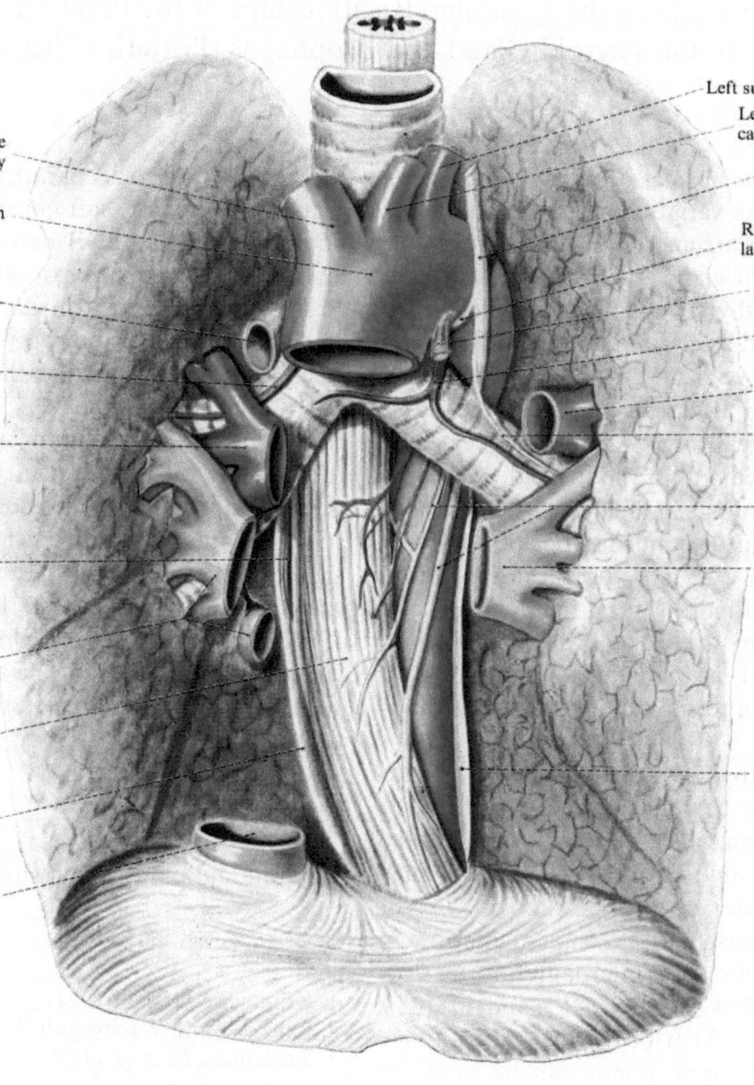
Left subclavian artery
Left common carotid artery
Innominate artery
Vagus nerve
Aortic arch
Recurrent laryngeal nerve
Lig. arteriosum
Azygos vein
Left bronchial artery
Right bronchial artery
Left pulmonary artery
Right pulmonary artery
Bronchial rami (X)
Esophageal branches (X)
Esophageal branch and vagus nerve
Pulmonary vein
Pulmonary veins
Esophagus
Mediastinal pleura (cut)
Mediastinal pleura (cut)
Inferior vena cava

2.21 Variations in the Anatomic Relationships of the Parathyroid Glands to the Thyroid Gland and Esophagus (Posterior View)

The plate shows the normal positions of the parathyroid glands. The variations in the locations of the superior glands are marked by dark circles and those of the inferior glands by light circles.

The superior parathyroid glands usually lie above the inferior thyroid artery and dorsal to the recurrent laryngeal nerve near the upper pole of the thyroid gland, and are located between the gland and the paratracheal fascia [2–4]. However, they can also be situated within the thyroid tissue (see Sect. 2.11). According to Wang [5], the upper parathyroids are in their normal position in 74% of cases, higher in 9%, more dorsal in 7%, more caudal in 6%, and more ventral in 2%. They can also be situated in the carotid sheath, near the carotid bifurcation, in the isthmus of the thyroid gland, as well as in retropharyngeal and retroesophageal locations.

The inferior parathyroid glands normally lie caudal to the inferior thyroid artery and ventral to the recurrent laryngeal nerve at the level of the lower pole of the thyroid gland (54%) [5]. They are situated anterior to this position in 7% of cases, in 6% posterior, in 5% superoanterior, in 3% superoposterior and inferoanterior in 2%. They lie 1 cm below the thyroid gland in 13%, 2 cm below in 7%, 3 cm below in 1%, and 4 cm below in 1%. Aberrant glands can be found at the carotid bifurcation, above the upper pole of the thyroid, on the aortic arch, or in the tracheoesophageal or laryngopharyngeal sulcus.

Macroscopically, the parathyroid glands display a lightbrown color, soft consistency, and are supplied by a single vessel derived from one of the branches of the superior or inferior thyroid artery. A significant problem in doing surgical parathyroidectomy involves not only the variable number of the glands (3–6) but also their variable locations [2].

In 6% of cases of primary hyperparathyroidism, the parathyroid glands can be found within the thyroid capsule [1]. When looking for a parathyroid adenoma, the surgeon should expose not only the posterior surface of the thyroid gland but also each carotid sheath up to the base of the skull. If necessary, total thyroidectomy may be required, as well as exploration of the anterior mediastinum with thymectomy (and possibly include a sternotomy).

Surgical Importance

Thyroid and parathyroid surgery.

References

1. Kümmerle F et al. (1974)
2. Lanz T v and Wachsmuth W (1955)
3. Rothmund M et al. (1980)
4. Sedgwick C and Cady B (1980)
5. Wang CA (1976, 1977)

Plate modified from Lanz-Wachsmuth [2]

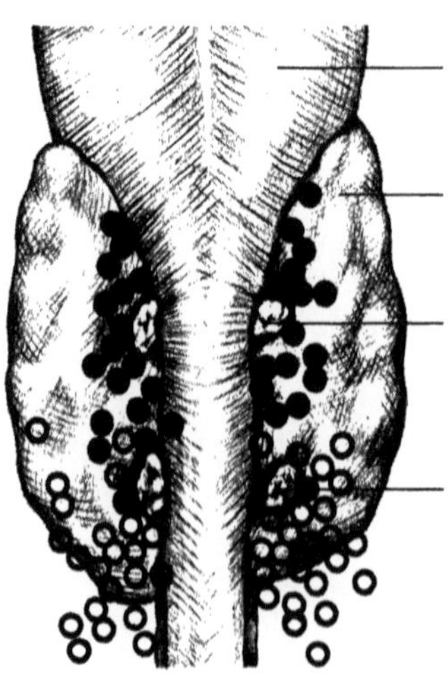

3 The Face

3.1 (Plates 1–3) Cutaneous Sensory Innervation of the Head and Neck

Sensory branches of the trigeminal nerve (V) and the ventral branches of C_1 and C_2 supply the face, with the peripheral branches of the opthalmic, maxillary and mandibular nerves extending as far as the corners of the eyelids and the mouth [1, 2]. The neck is supplied by C_2–C_6. Since the spinal accessory nerve (XI) receives sensory and motor fibers from C_3 and C_4, division of the nerve above the cervical roots will produce variable postoperative muscular dysfunctions, the degree of which depends on this motor contribution.

The cervical branch of the facial nerve is motor to the platysma muscle and passes beneath this muscle. Since the mandibular branch of the facial nerve is important to the motor function of the lower face, this nerve can be preserved if any surgical incision in the upper neck is made 2 cm below the mandible (see Sects. 1.13, 3.5).

Infiltration anesthesia of the cervical plexus can be performed by placing the anesthetic solution at the mid-point of the posterior border of the sternocleidomastoid muscle between its insertion onto the mastoid process and its origin from the clavicle. In this way (Plates 1, 2), anesthesia of the anterior portion of the neck and of the skin and soft tissue supplied by the supraclavicular nerves and the greater auricular nerve can be achieved [3].

The various clinically important areas of sensory innervation in the head and neck for inducing local anesthesia are shown in Plate 3.

Surgical Importance

Facial or neck surgery requiring local anesthesia, limited or total rhytidectomy, neural transplantations, extirpations of the submandibular gland, dissections of the floor of the mouth.

References

1. Clara M (1942)
2. Lang J (1979)
3. Lanz T v and Wachsmuth W (1955)

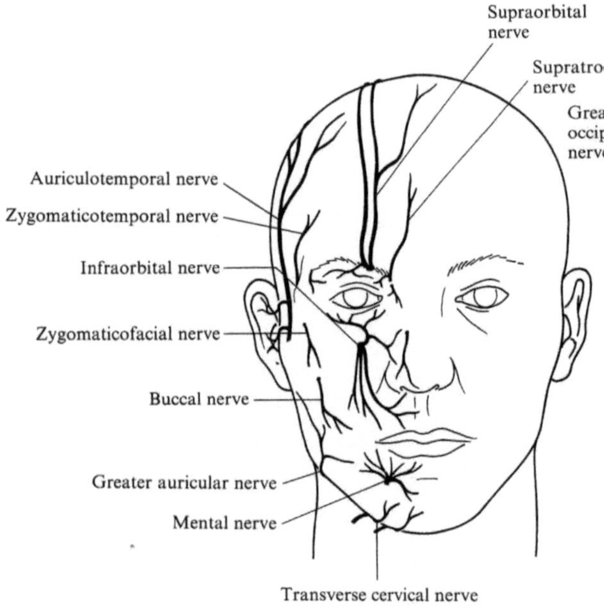

1

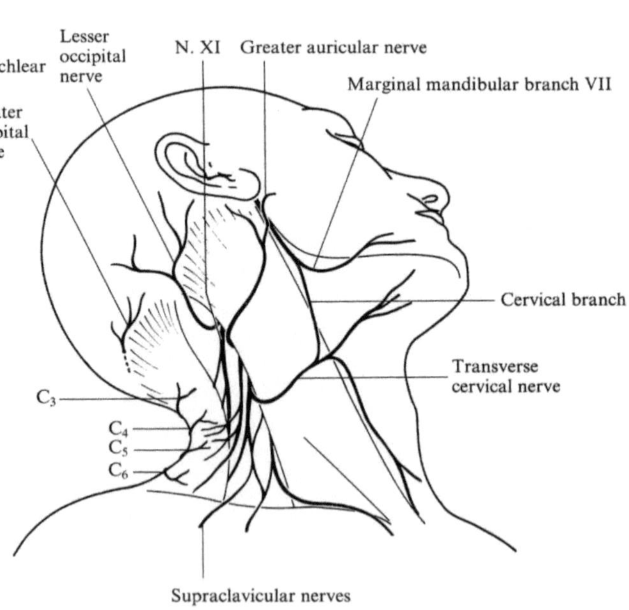

2

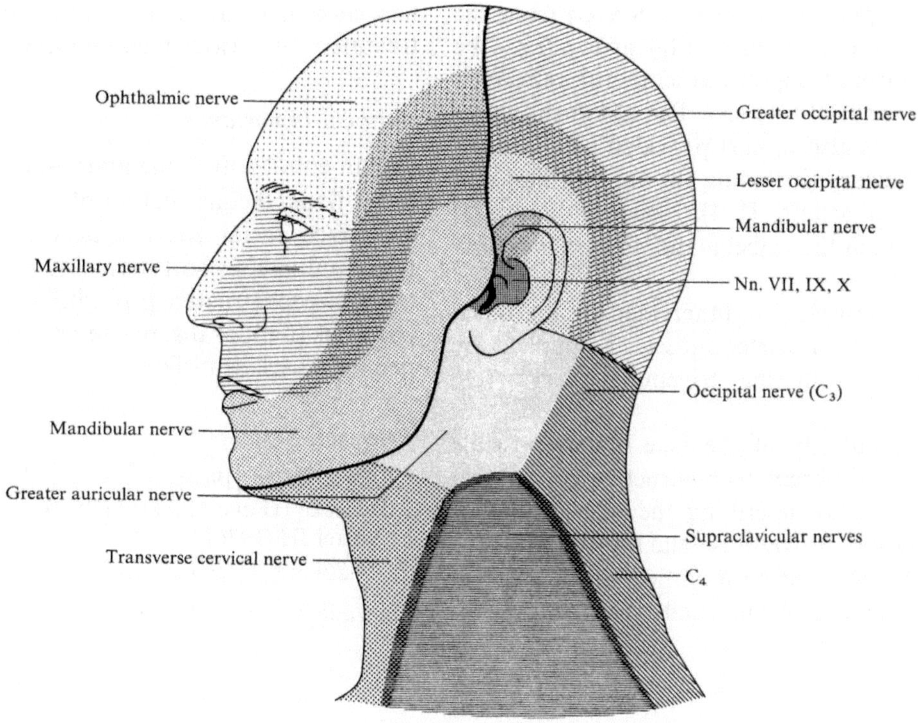
Ophthalmic nerve
Greater occipital nerve
Lesser occipital nerve
Mandibular nerve
Maxillary nerve
Nn. VII, IX, X
Occipital nerve (C_3)
Mandibular nerve
Greater auricular nerve
Supraclavicular nerves
Transverse cervical nerve
C_4

3

3.2 The Mimetic Muscles with their Arteries and Nerves (Anterior View)

The blood supply of the face is derived from the branches of the external carotid artery (i.e., the facial, infraorbital, superficial temporal, retroauricular, and occipital arteries). The blood supply of the forehead is also in part provided by the internal carotid artery (including the supraorbital and supratrochlear arteries [1–4]. However, the nasal dorsum between the glabella and its apex is poorly vascularized.

The mimetic muscles are innervated by the facial nerve, while the masseter muscle is supplied by motor fibers from the trigeminal nerve (see Sect. 3.8).

The rich vascularity of the face makes possible a number of different reconstructive procedures [1–2]. A 2:1 ratio describing the relationship of a flap's length to its width is valid for other regions of the body, but can be modified on the face in favor of its length. Additionally, as Converse [1] has shown, esthetic units are important in the treatment of various facial injuries.

Surgical Importance

Treatment of soft tissue injuries, cosmetic surgery, facial flaps, reconstruction of the lips and eyelids, cross-face facial nerve anastomoses, parotidectomy, cutaneous/soft tissue tumor surgery, secondary reconstructive procedures after facial nerve paralysis (temporalis, masseter and corrugator supercilii muscle transfer).

References

1. Converse JM (1964)
2. Denecke HJ and Gabka J (1972)
3. Warfel JH (1976)
4. Wright WK (1967)

Plate adapted from Warfel [3]

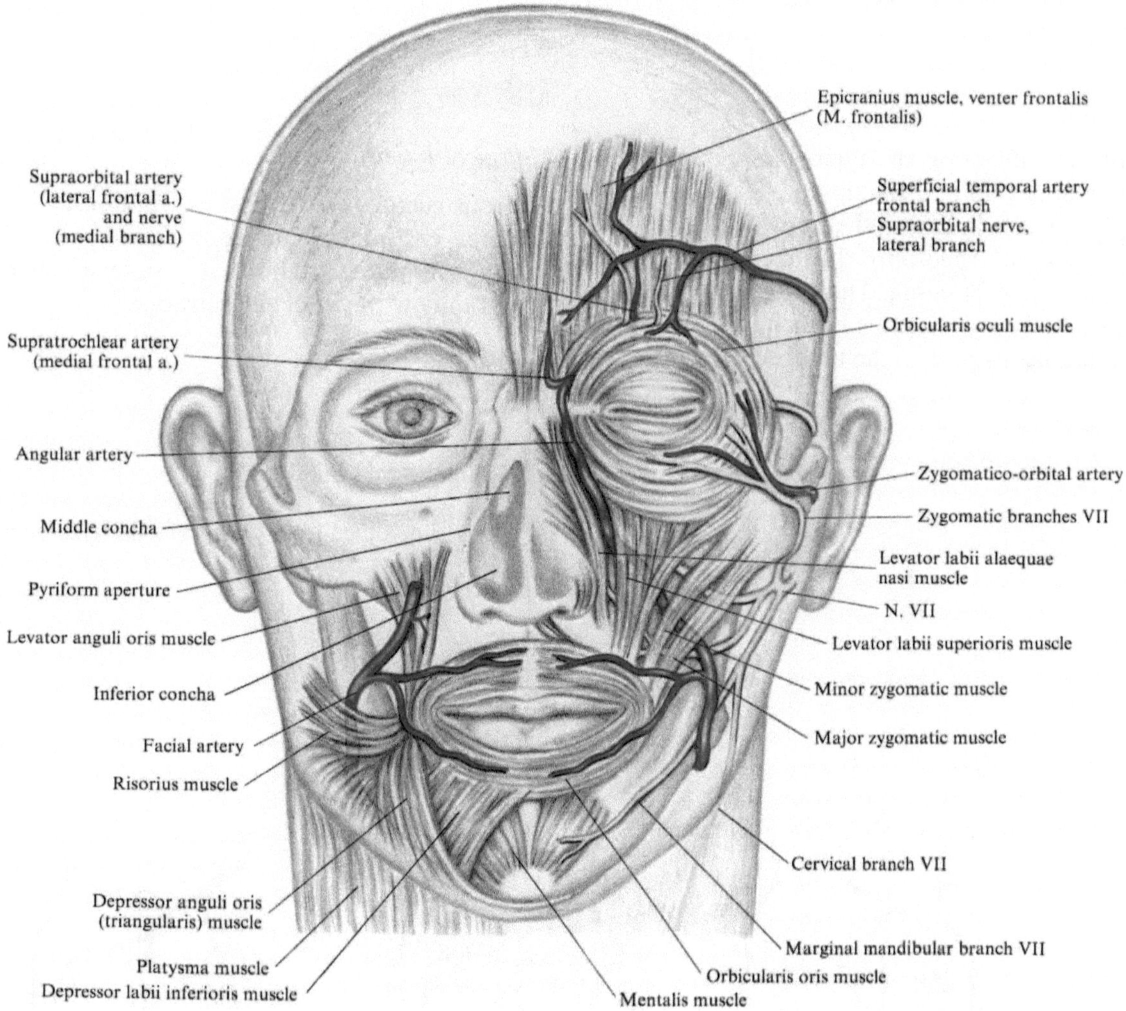
Epicranius muscle, venter frontalis (M. frontalis)
Supraorbital artery (lateral frontal a.) and nerve (medial branch)
Superficial temporal artery frontal branch
Supraorbital nerve, lateral branch
Orbicularis oculi muscle
Supratrochlear artery (medial frontal a.)
Angular artery
Zygomatico-orbital artery
Middle concha
Zygomatic branches VII
Levator labii alaequae nasi muscle
Pyriform aperture
N. VII
Levator anguli oris muscle
Levator labii superioris muscle
Minor zygomatic muscle
Inferior concha
Major zygomatic muscle
Facial artery
Risorius muscle
Cervical branch VII
Depressor anguli oris (triangularis) muscle
Marginal mandibular branch VII
Platysma muscle
Orbicularis oris muscle
Depressor labii inferioris muscle
Mentalis muscle

3.3 (Plates 1–3) Arterial Supply of the Upper Lip and the Base of the Nose

Plates 1–3 illustrate the typical vascular patterns of the upper lip and the nose [1–2]. Injury to the angular artery lateral to the base of the nose when doing lateral osteotomy during rhinoplasty can lead to severe bleeding. The superior labial artery usually also gives rise to a branch to the columella (the cutaneous part of the nasal septum).

Surgical Importance

Reconstruction of the lips, extirpation of hemangiomas, excision of cutaneous lesions, microvascular anastomoses, rhinoplasty, superselective angiography with vessel embolization.

References

1. Combelles RG et al. (1974)
2. Krmpotić-Nemanić J (1978)

Plates 1–3 modified from Combelles et al. [1]

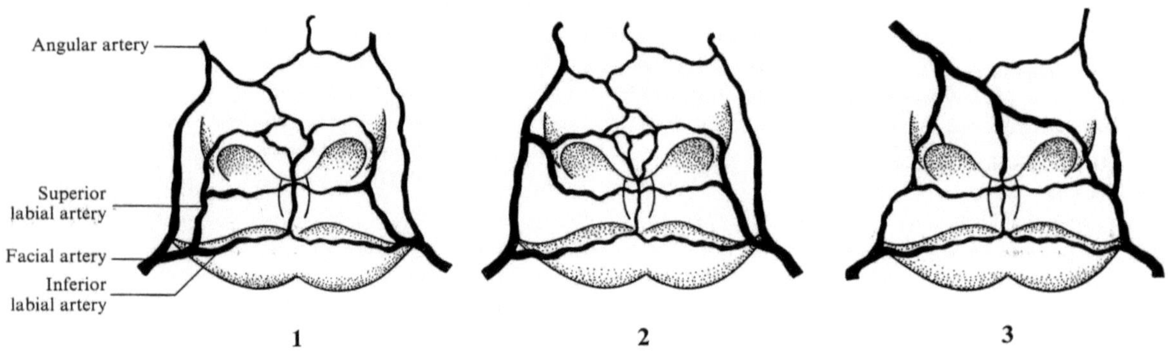

The typical vascular patterns of the upper and lower lips are illustrated in Plate 1. However, several variations do exist [1–4]. If a blood supply is present that is similar to the patterns shown in Plates 2–4, cone-shaped pedicle flaps from either lip to reconstruct defects in the other lip (as described by Abbé [1] and Estlander [3]) can cause disturbances in wound healing on the contralateral side of the supplying artery.

Surgical Importance

Reconstruction of lip defects, excision of hemangioma.

References

1. Abbé R (1898)
2. Combelles RG et al. (1974)
3. Estlander JA (1872)
4. Krmpotić-Nemanić J (1978)

Plates 1–4 modified from Combelles et al. [2]

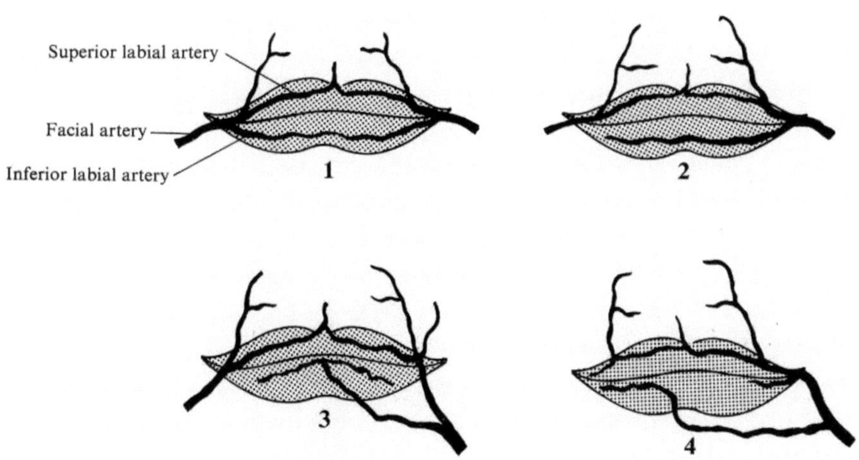

3.5 The Mimetic Muscles with their Arteries and Nerves (Lateral View)

The Plate depicts the facial or mimetic muscles and their major arteries and nerves. The branches of the external carotid artery include the facial, deep temporal, superficial temporal, retroauricular, and occipital arteries. The nerve branches are the temporal (temporo-frontal), zygomatic, buccal, marginal mandibular, and cervical (*R. colli*) nerves. Knowledge of this region is pertinent to angiography and tomographic studies [2].

According to Godlewski et al. [1], the temporomandibular joint is supplied by the superficial temporal, anterior tympanic, and deep temporal arteries. The deep auricular (*a. auricularis profunda*), transverse facial (*a. transversa faciei*), and the middle meningeal arteries are additional sources of vascularization (not represented).

Surgical Importance

Facial flaps, microvascular anastomoses, surgery of the parotid and the submandibular glands, surgery of the upper and lower jaw, rhytidectomy, facial nerve reconstructions, cross-face facial nerve anastomoses.

References

1. Godlewski G et al. (1979)
2. Warfel JH (1976)

Plate adapted from Warfel [2]

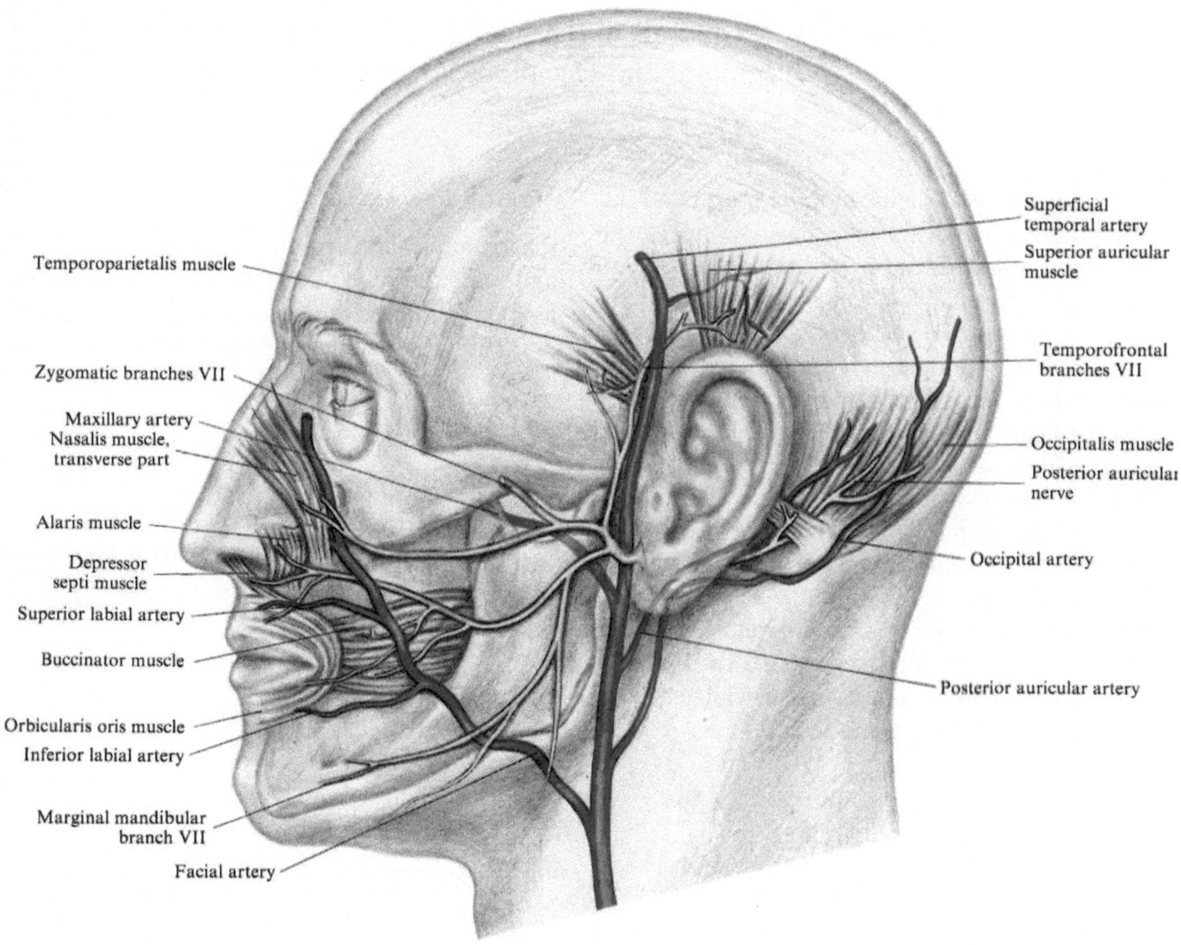
Temporoparietalis muscle
Zygomatic branches VII
Maxillary artery
Nasalis muscle, transverse part
Alaris muscle
Depressor septi muscle
Superior labial artery
Buccinator muscle
Orbicularis oris muscle
Inferior labial artery
Marginal mandibular branch VII
Facial artery
Superficial temporal artery
Superior auricular muscle
Temporofrontal branches VII
Occipitalis muscle
Posterior auricular nerve
Occipital artery
Posterior auricular artery

3.6 (Plates 1–2) Variations in the Locations of the Superficial Temporal and Occipital Arteries

The blood supply of the frontal, temporal and lateral regions of the scalp may have several surgically important variations. In the preparation of long and narrow forehead and scalp flaps [1–11, 13], the courses of these vessels should be determined by sonography and angiography in order to best incorporate the appropriate vessel in the pedicle flap. Two typical examples are depicted in Plates 1 and 2.

When extracranial-intracranial microvascular anastomoses are planned for managing patients with cerebral ischemia, the diameters of the superficial temporal artery (0.68–1.89 mm) as well as its frontal (0.40–1.38 mm) and parietal branches (0.50–1.29 mm) are important [12]. In 15 of 29 cases, the bifurcation of the superficial temporal artery lies above the zygomatic arch.

Surgical Importance

Plastic and reconstructive surgery of the forehead and scalp, extra- or intracranial microvascular anastomoses, biopsies of the superficial temporal artery (for diagnosing temporal arteriitis).

References

1. Bardenheuer B (1931)
2. Converse JM (1942, 1977)
3. Conway H and Murray JE (1953)
4. Donaghy R M P and Yasargil MG (1967)
5. Juri J (1975)
6. Juri J et al. (1978)
7. Lang J (1979)
8. McGregor JA (1963)
9. Ricbourg B et al. (1975)
10. Schaefer SD et al. (1980)
11. Schimmelbusch C (1931)
12. Stock AL et al. (1980)
13. Yasargil MG (1969)

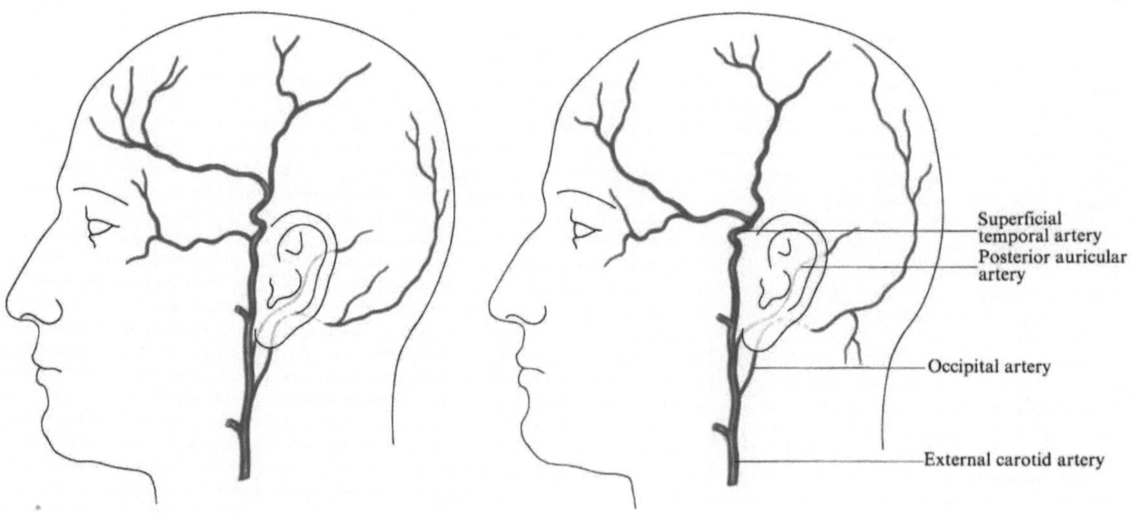

1 2

Anastomoses of the superficial temporal arteries form a semicircular vascular arcade on both sides of the forehead [1–3]. By placing radial lines from the orifice of the supratrochlear artery above the medial palpebral ligament, fan-shaped sectors will be delimited that are supplied by the supraorbital and supratrochlear branches of the ophthalmic artery, the facial artery and the superficial temporal artery.
The numerous variations of the forehead arteries make Doppler's sonography very useful in determining the courses of the major vessels when developing pedicle flaps.
Since the nasal dorsum is poorly vascularized (see Sect. 3.2), reconstructions of defects are best done by creating a flap that incorporates a named artery.
The vascularity of regional flaps can be determined by the intravenous injection of such dyes as fluorescein and observing tissues under an ultraviolet light for their fluorescence [1].

Surgical Importance

Pedicle flaps of the scalp, forehead, and eye lids; microvascular anastomoses; cosmetic surgery of the forehead.

References

1. Deutschmann H and Scharnagel E (1980)
2. Domarus Hv et al. (1984)
3. Mangold U et al. (1980)

Plate adapted from Mangold et al. [3]

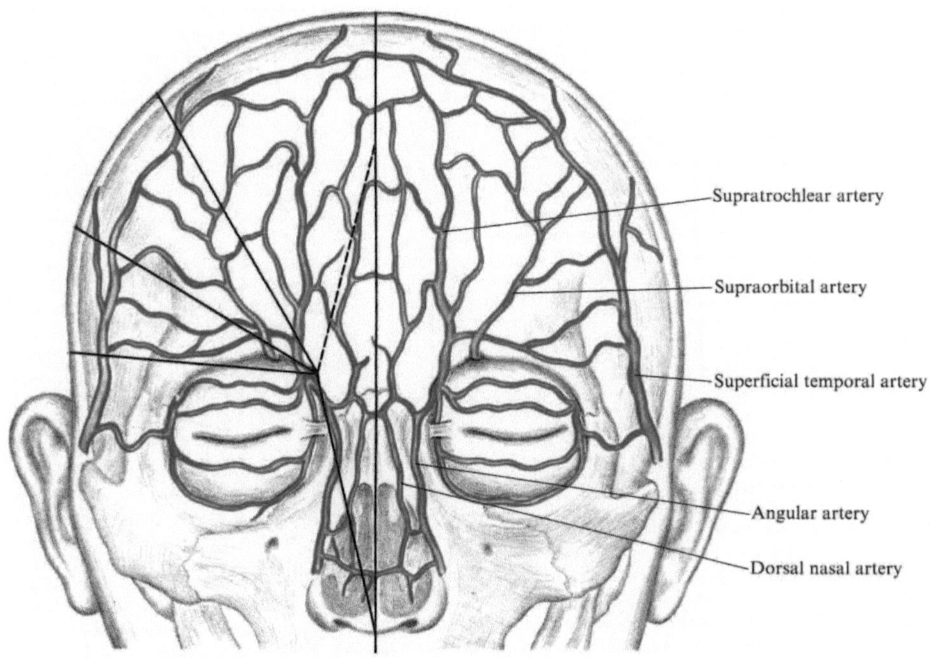

3.8 Arteries and Nerves of the Masseter, Temporalis, and Sternocleidomastoid Muscles

The arteries to the temporalis and masseter muscles are supplied by branches of the external carotid artery (the maxillary, anterior and posterior deep temporal, and masseteric arteries) [1–3, 6].

The blood supply of the sternocleidomastoid muscle is divided into three segments (see Sect. 1.27): the superior region is supplied by the sternocleidomastoid artery or possibly a branch of the occipital artery if the former vessel is missing [4]; the medial region is supplied by branches of the superior thyroid artery, while the inferior region is connected with the thyrocervical trunk by the inferior thyroid and transverse cervical arteries.

The temporalis and masseter muscles are innervated by the mandibular division of the trigeminal nerve via the deep temporal and the masseteric nerves. Both muscles can be used as dynamic transposition flaps to correct segmental facial paralyses [5]. In so doing, flaps of muscle are best prepared so as to run parallel to and encompass the long axis of the primary nerves and arteries present.

In creating myocutaneous flaps from the sternocleidomastoid muscle, the path of the accessory nerve should be carefully noted.

Surgical Importance

Muscular neurotization in facial paralysis [5], Denecke's [3] sternocleidomastoid flap (as used following surgery for glomus tumors).

References

1. Ariyan S (1979)
2. Conley JJ (1975)
3. Denecke HJ (1966)
4. Lanz T v and Wachsmuth W (1955)
5. Lexer E (1908, 1911)
6. Nakajima H et al. (1981)

Plate adapted from Ariyan [1], Conley [2]

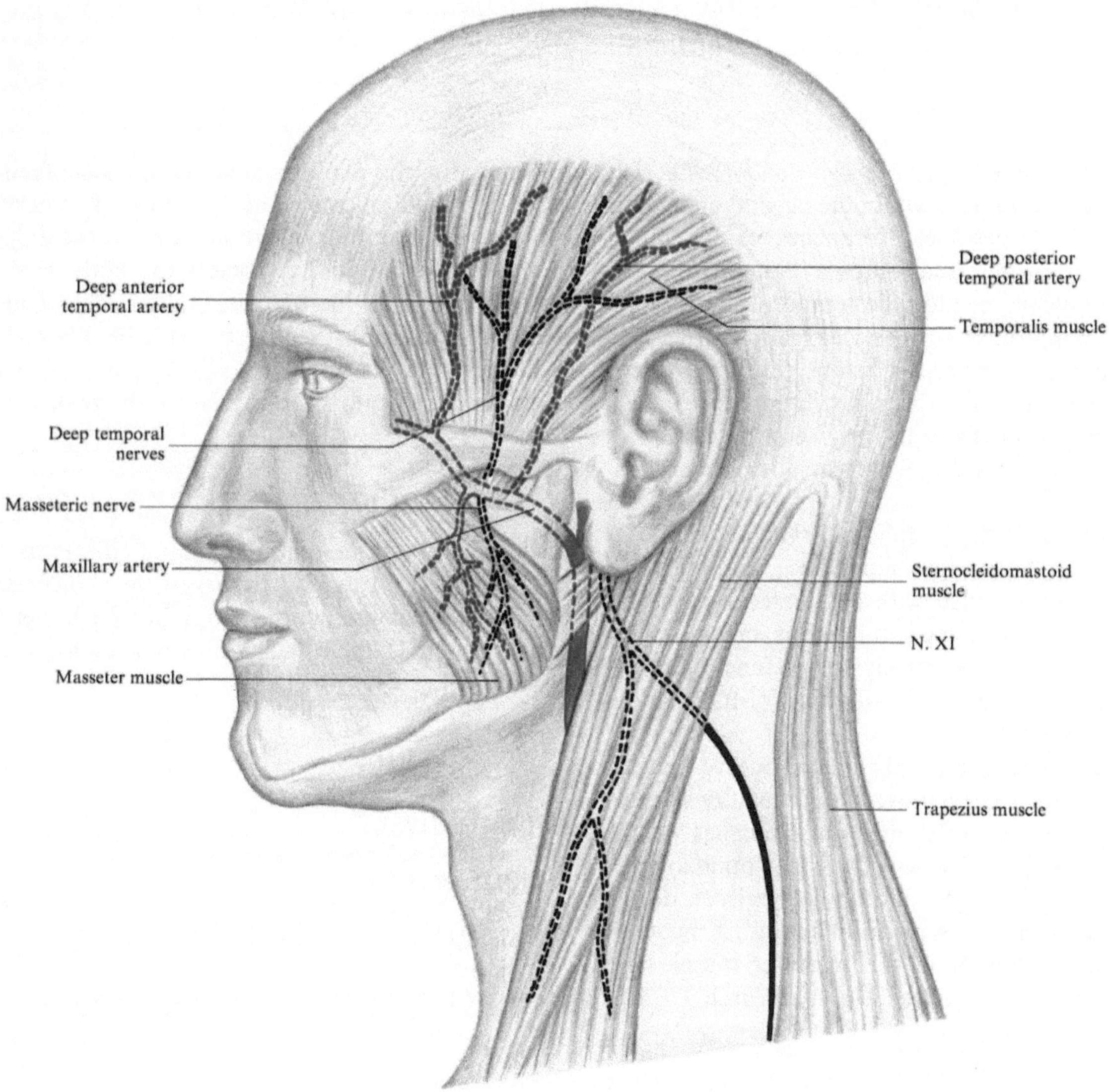

Deep anterior temporal artery
Deep temporal nerves
Masseteric nerve
Maxillary artery
Masseter muscle
Deep posterior temporal artery
Temporalis muscle
Sternocleidomastoid muscle
N. XI
Trapezius muscle

3.9 (Plates 1–6) Variations in the Extratemporal Branching Patterns of the Facial Nerve

The parotid plexus (*pes anserinus major*) is a rich neural plexus situated within the parotid gland and is formed by peripheral branches of the facial nerve with variable anastomoses (Plates 1–6) [1–3]. These branches include the temporal, zygomatic, buccal, mandibular and cervical branches. The terminal posterior auricular ramus runs in a retrograde fashion, while the digastric and stylohyoid branches run medially to their respective muscles as well as anastomose with the glossopharyngeal nerve.

The position of the peripheral branches of the facial nerve is of surgical importance, and injury to them can be avoided with adequate preparation. In general, the nerve's numerous cross-anastomoses explain the relatively minor functional disturbances that occur after division of its terminal small branches.

In doing cross-face (or face-facial) nerve anastomoses between the paralyzed and healthy sides of the face, one or two sufficient branches of the nerve to be joined can usually be found on the anterior border of the parotid gland. In so doing, a vertical incision is made in the angle between the anterior border of the masseter muscle and the inferior margin of the zygomatic arch.

Infrequently, the mandibular branch of the facial nerve can be duplicate or even trebled. The nerve is most prone to injury 2 cm anterior to the angle of the mandible and 2 cm below the body of the mandible, inferior to the location of the second premolar tooth [5]. (See Sects. 1.13, 30, Plates 1–3). The variations in the branching patterns seen clinically are given in percentages in the plates.

Surgical Importance

Parotid and submandibular gland dissections, aural and facial trauma, tumors of the skull base, facial nerve reconstruction, cross-face facial nerve anastomoses [1], facio-facial anastomoses [4], rhytidectomy.

References

1. Anderl H (1973)
2. Davis RA et al. (1956)
3. Miehlke A (1973)
4. Samii M (1975)
5. Ziarah HA and Atkinson ME (1981 a, b)

Plates 1–6 adapted from Davis [2] and Miehlke [3]

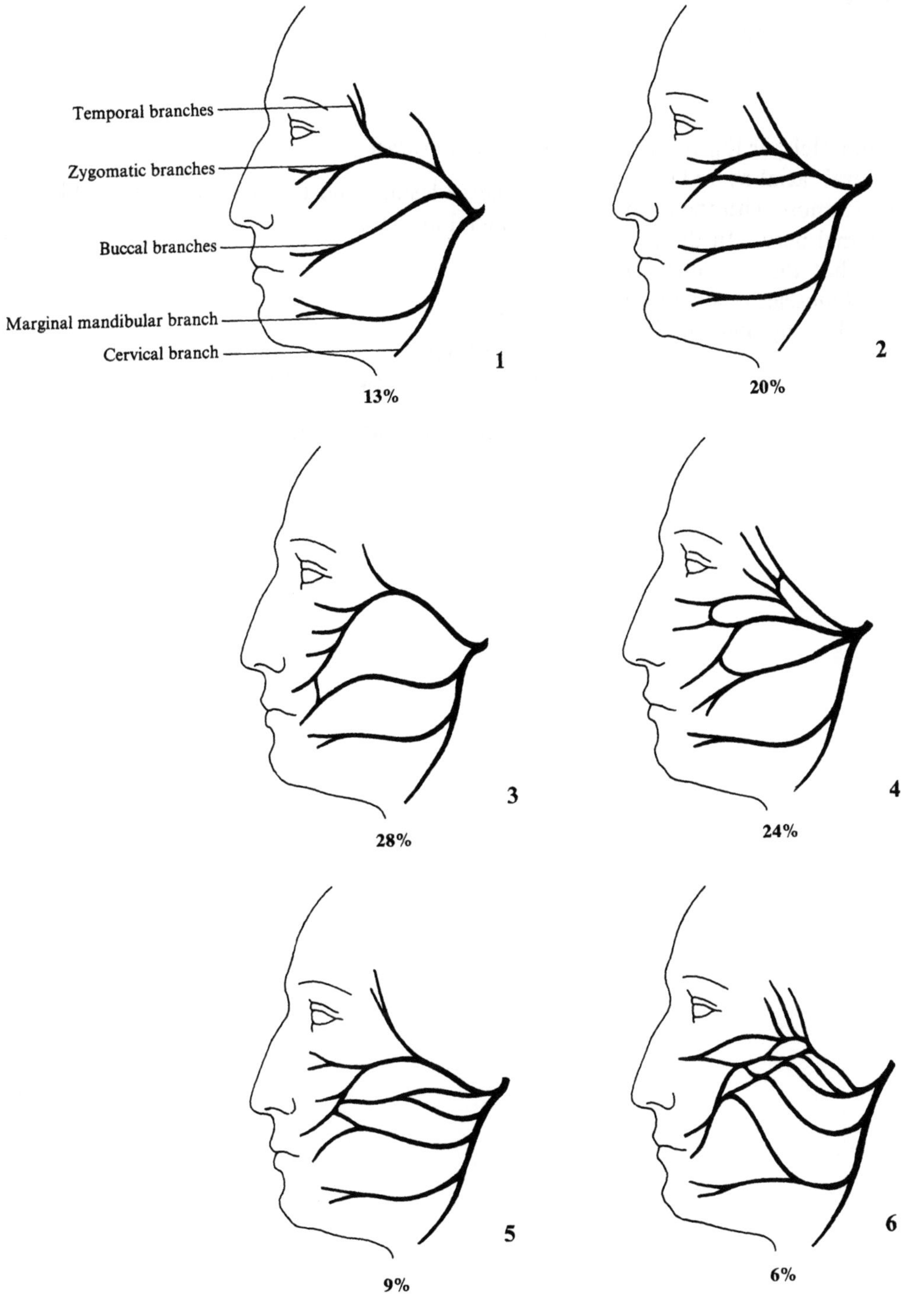
Temporal branches
Zygomatic branches
Buccal branches
Marginal mandibular branch
Cervical branch
1
13%
2
20%
3
28%
4
24%
5
9%
6
6%

3.10 Possible Sites for Peripheral Anastomoses Between the Facial and the Trigeminal Nerves

As described by Martin and Helsper [2], the end-branches of the extratemporal facial nerve lie very close to and may even anastomose with the terminal branches of the trigeminal nerve. In the past, Conley [1] has suspected that these connections might explain why the facial muscles retain a minimal degree of function following known facial nerve paralyses.

Surgical Importance

Facial nerve lesions and regeneration, myography of the facial muscles.

References

1. Conley JJ (1975)
2. Martin H and Helsper JP (1957)

Plate adapted from Martin and Helsper [2]

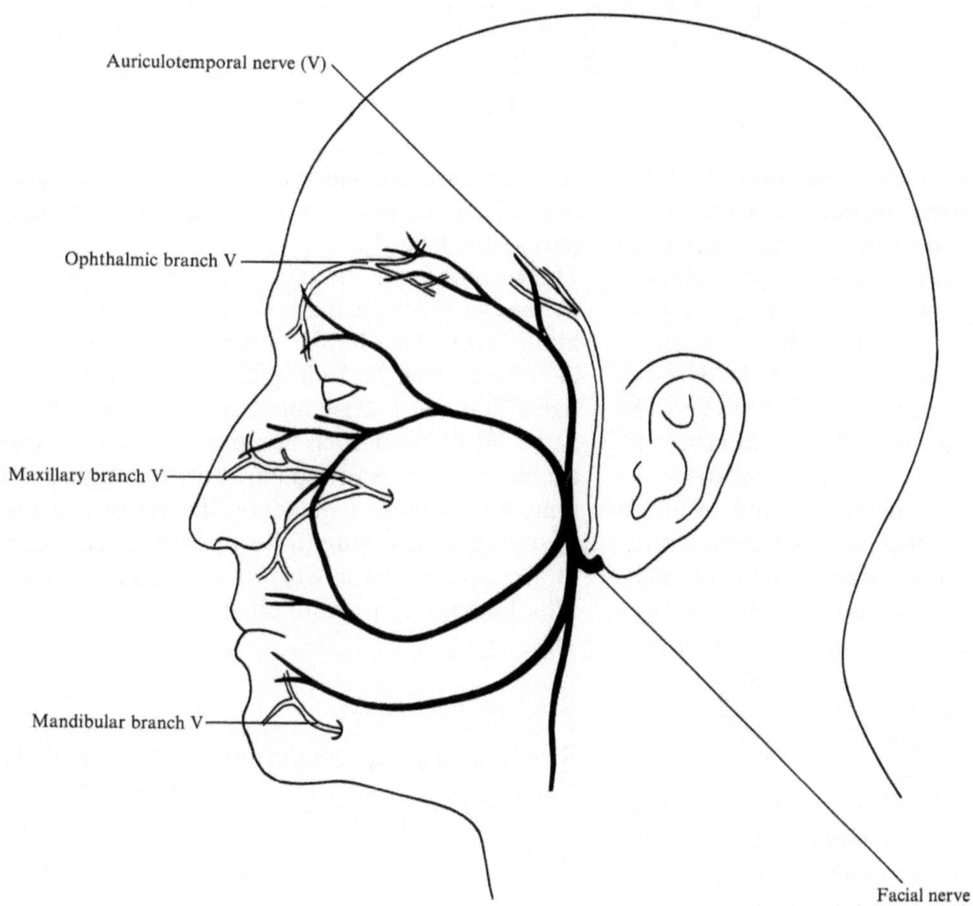

Auriculotemporal nerve (V)
Ophthalmic branch V
Maxillary branch V
Mandibular branch V
Facial nerve

3.11 Topographic Relationships of the Facial Nerve and the Parotid Gland

The parotid gland develops in the vestibule of the mouth, originating from the papilla at the orifice of its duct [2]. From this site, the anlage migrates to the retromandibular fossa (and is later reflected by the duct's location, which can be palpated over the clenched masseter muscle). The facial nerve lies medial to the anlage of the gland. As development proceeds, the glandular tissue grows between the extratemporal branches of the facial nerve medialy, such that the nerve becomes incorporated into the gland. As the dimensions and locations of various parts of the gland can differ individually, their topographical relationships to the facial nerve are also subject to variation. The gland can extend in varying degrees as far as the lateral pharyngeal wall and to the infra-auricular region. However, a strict division of the parotid gland into a lateral and medial lobe in relationship to the facial nerve does not exist.

The main branches of the facial nerve outside of the stylomastoid foramen include the posterior auricular nerve, the digastric and stylohyoid branches, the communicant branch to the glossopharyngeal nerve, and the parotid plexus. This latter plexus (see Sect. 3.9) includes the temporal, zygomatic, buccal, one or more marginal mandibular, and the cervical branches. The superficial temporal artery is usually located medial to the branching of the zygomatico-temporal or temporal ramus.

It appears that there is a functional differentiation clinically between the innervation of the mimetic muscles of the lower face and that of the upper part of the face [3].

The secretions of the paroid gland are regulated by the tympanic plexus and by an eventual contribution from the chorda tympani nerve [1]. There have been attempts to use neurectomy of the tympanic nerve and resection of the chorda tympani nerve [4] to treat such conditions as Frey's syndrome, chronic recurrent parotitis, neuralgia of the tympanic plexus, fistulae of the parotid gland, post-surgical drooling (following tumor resections of the lip or oral cavity) or excessive salivation after bulbar and pseudobulbar paralyses.

Surgical Importance

Rhytidectomy, parotidectomy, submandibular gland excision, facial nerve reconstruction, facio-facial anastomoses, Frey's syndrome, parotid fistulae, stenosis of the parotid duct (*ductus Stenonis*).

References

1. Diamant H and Wiberg A (1965)
2. Healy JE and Seybold WD (1969)
3. Stennert E (1979)
4. Thomas JR (1980)

Plate adapted from Healy and Seybold [2]

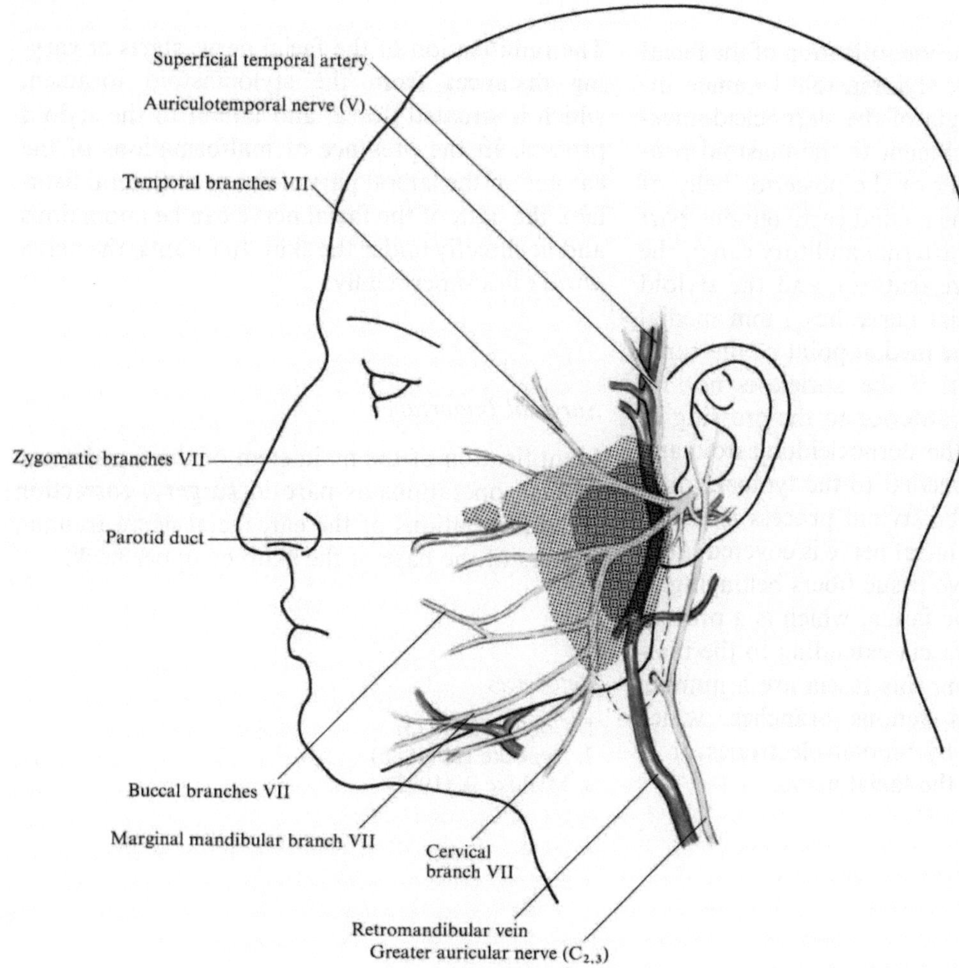
Superficial temporal artery
Auriculotemporal nerve (V)
Temporal branches VII
Zygomatic branches VII
Parotid duct
Buccal branches VII
Marginal mandibular branch VII
Cervical branch VII
Retromandibular vein
Greater auricular nerve ($C_{2,3}$)

3.12 Landmarks for Identifying the Facial Nerve in Parotid Surgery

Landmarks aiding in the identification of the facial nerve after it leaves the stylomastoid foramen include the anterior margin of the sternocleidomastoid muscle at its attachment to the mastoid process, the anterior border of the posterior belly of the digastric muscle, the medial cartilaginous part ("the pointer") of the external auditory canal, the tympanomastoid fissure (suture), and the styloid process [1–3]. The facial nerve lies 7 mm medial and 2 mm caudal to the medial point of the cartilaginous external canal if the auricle is not retracted. It lies slightly anterior to the crossing of the anterior border of the sternocleidomastoid and digastric muscles and medial to the tympanomastoid suture, crossing the styloid process laterally. The main trunk of the facial nerve is covered laterally by dense connective tissue fibers belonging to the parotideomasseteric fascia, which is a prolongation of the parotid fascia extending to the mastoid process. Traversing this fascia are a number of small arterial and venous branches, which should be coagulated by bipolar electrodes or ligated before exposing the facial nerve.

The ramification of the facial nerve starts at varying distances from the stylomastoid foramen, which is situated dorsal and lateral to the styloid process. In the presence of malformations of the ear and of the lateral part of the neck (lateral fistulae), the path of the facial nerve can be anomalous and lie directly under the skin. In infants, the nerve *always* lies superficially.

Surgical Importance

Identification of the main stem of the facial nerve in such operations as parotid surgery, correction of malformations of the ear, facial nerve trauma, tumors of the base of the skull or upper neck.

References

1. Conley JJ (1975)
2. Denecke HJ (1980)
3. Miehlke A (1973)

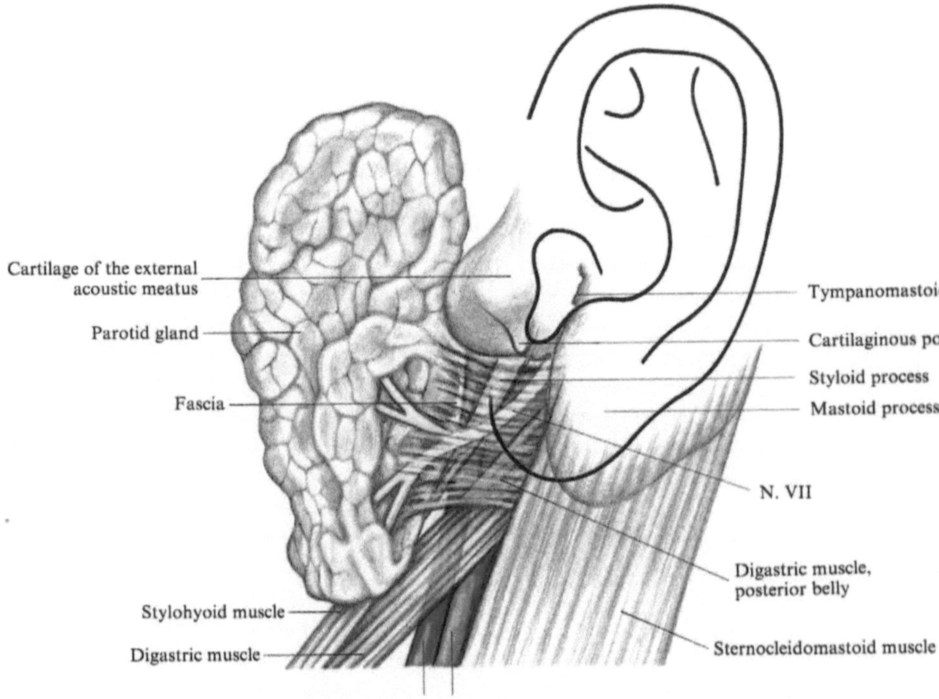

4 The Oral Cavity and the Oropharynx

4.1 Projections of the Atlas and the Axis (*Epistropheus*) in the Facial Region

The spatial relationships of the first and second cervical vertebrae, especially of the dens axis, to the facial region is of great help in posterior transoral surgery [1–2]. The first two cervical vertebrae are projected into the oral cavity when the mouth is opened. With the mouth closed, the body of the second vertebra corresponds to the level of the lower lip, the first vertebra to the upper lip, and the dens axis to the philtrum. The dens axis is normally inclined slightly backward at a 10° angle in relation to its body [1]. The relationship of the dens axis to the skull base is clinically important. The position of the dens is determined by different lines (such as the *bimastoid* or *Chamberlain*'s line). The transverse process of the first vertebra can be asymmetrical and may be mistaken clinically for a tumor beneath the mastoid process (see Sects. 11.11, 1.15.2).

Surgical Importance

Transoral surgery of the cervical spine (and dens axis), incision and drainage of retropharyngeal abscesses, removal of posterior nasopharyngeal tumors.

References

1. Helms J (1963)
2. Lanz T v and Wachsmuth W (1955)

Plate modified from Lanz and Wachsmuth [2]

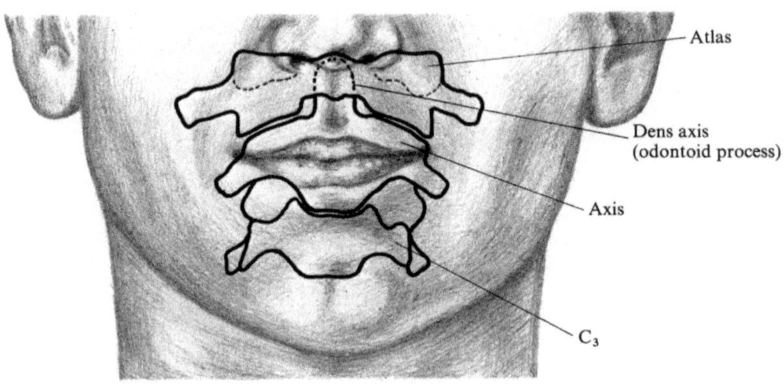

4.2 Topography of the Floor of the Mouth (Anterior View)

The surgically most important anatomic structures in this region are represented in the Plate, which has been modified from Lanz [1]. The lingual nerve crosses underneath the submandibular duct (*ductus Whartoni*), and passes from dorsolateral to ventromedial toward the tip of the tongue [2, 4]. The nerve and the submandibular ganglion lie near the mandible at the level of the posterior molar teeth. As a result, both structures can be injured in extraction of the last molar tooth, (or the *dens sapientiae*). The accessory ducts of the sublingual glands open along the sublingual folds (see Sect. 4.3). The main sublingual duct and the submandibular duct drain into the anterior floor of the mouth at the sublingual papilla (*caruncula sublingualis*).

The close relationship of the lingual nerve to the lower pole of the tonsil is important clinically. The hypoglossal nerve lies more caudally. Firm pressure placed against the tongue during such surgery as microlaryngoscopy may result in temporary paralysis of the lingual and hypoglossal nerves. This should be particularly kept in mind in patients with short thick necks who require such surgery.

The lingual nerve normally consists of one root; in 4% of cases it has two roots and in 8.3% of cases it is formed by a neural plexus. In 25% of cases, it anastomoses with the inferior alveolar nerve and in 33% with the mylohyoid nerve (*ansa mylohyoidea*). In 2.5% of cases, it supplies the lateral and medial pterygoid muscles and the palatoglossus muscle [3], which may explain different functional disturbances after its injury (not represented).

Surgical Importance

Incision of the submandibular duct, extirpation of the submandibular gland, dental extractions, neurolysis of the lingual nerve, extirpation or marsupialization of ranula, abscesses of the base of the tongue and the floor of the mouth, oral trauma, dissections of the floor of the mouth.

References

1. Lang J (1979)
2. Lanz T v and Wachsmuth W (1955)
3. Rácz L et al. (1981)
4. Sadeghi EM and Bell WA (1980)

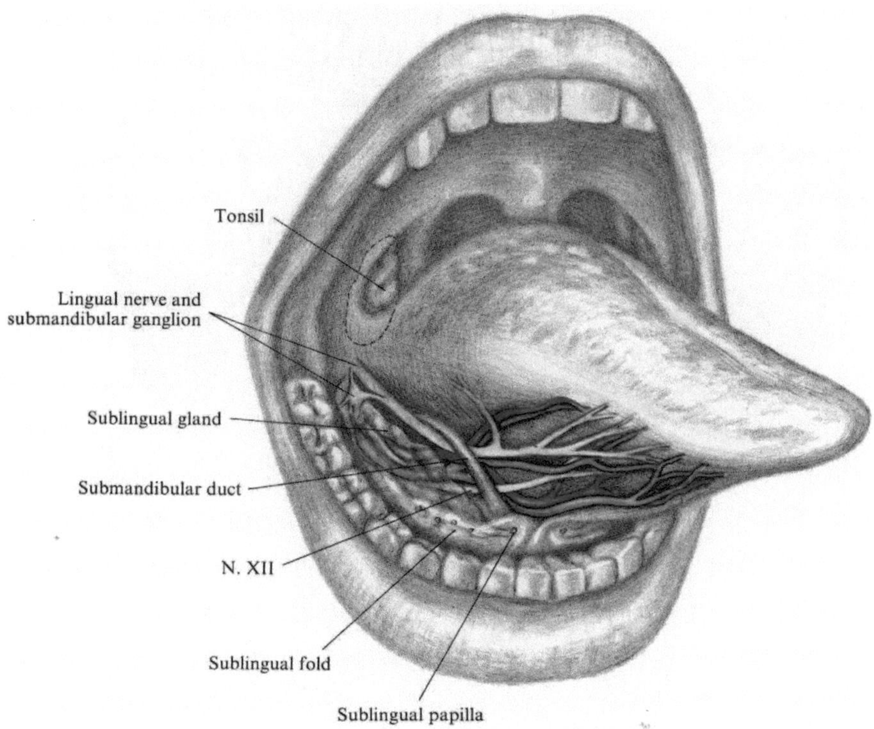

4.3 Topographic Relationships of the Floor of the Mouth (Inferior View)

The surgically most important structures are represented in the Plate. The hypoglossal nerve and the accompanying veins disappear under the mylohyoid muscle [1]. These structures are covered by the submandibular gland and can be injured in lateral transoral dissections of the floor of the mouth. The lingual artery normally lies medial and the hypoglossal nerve lateral to the hyoglossus muscle.

Surgical Importance

Extirpation of the submandibular gland or ranulae, dissection of the floor of the mouth, excision of medial cervical cysts.

Reference

1. Clemente CD (1975)

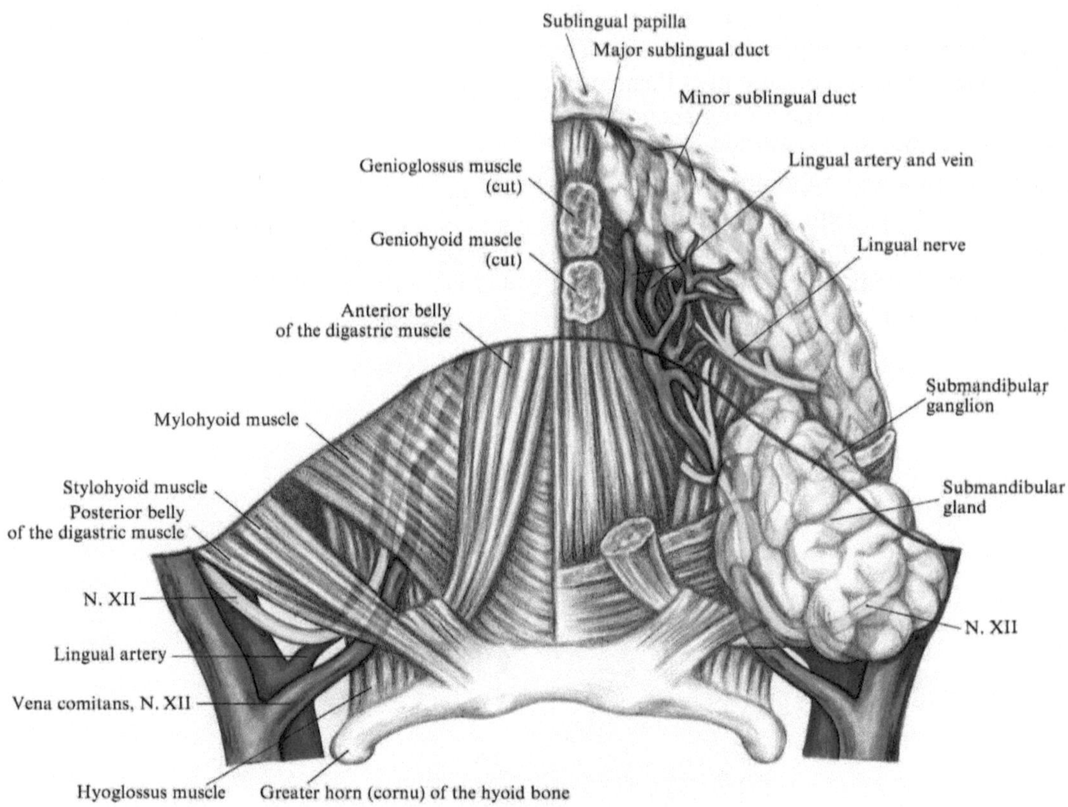

4.4 The Blood and Nerve Supply of the Hard and Soft Palates

The palatine artery and the greater palatine nerve pass through the greater palatine foramen to supply the hard palate and also communicate with the corresponding structures of the nasal cavity through the incisive foramen (see Sects. 5.9, 8). The lesser palatine arteries and nerves supply the soft palate. The midline of the hard palate is poorly vascularized. After removal of the mucosa, the muscles of the soft palate are readily visible in the plate.

When surgically developing flaps on the hard or soft palate, the major vessels must be protected in order to avoid soft tissue necroses.

Surgical Importance

Closure of alveolar fistulae, cleft palate repair, velopharyngoplasty, transpalatal surgery, correction of choanal atresia, resection of nasopharyngeal tumors, etc.

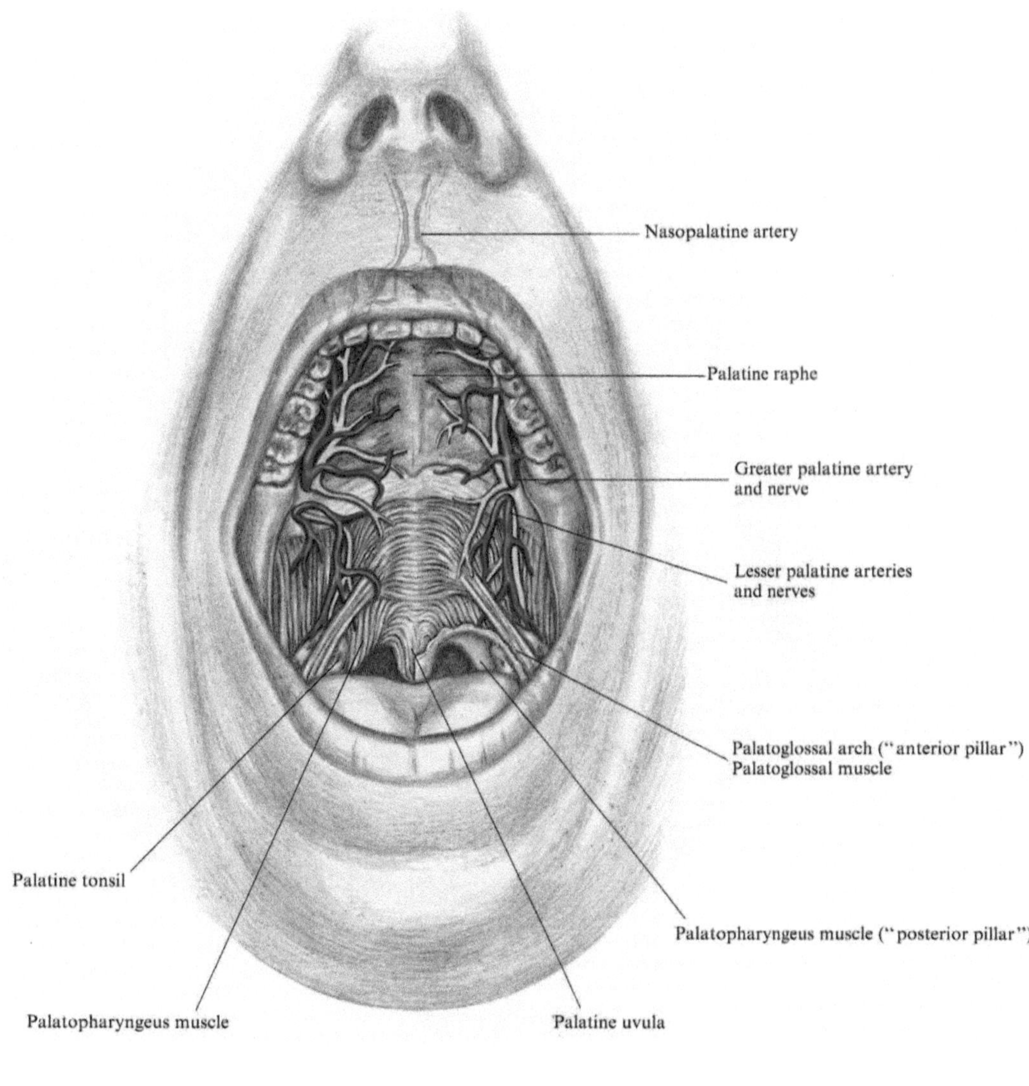

4.5 Horizontal Section Showing Topography of the Skull Base (Inferior View)

The topographically important structures of the skull base are depicted in the Plate so that they correspond to their locations in a horizontal CT section. Some side differences present are due to the asymmetry of the outline of the nasopharyngeal space.

Surgical Importance

Radiological diagnosis of the skull base, surgery of the skull base and the nasopharynx, retropharyngeal and parapharyngeal diseases.

Reference

1. Hafferl A and Thiel W (1969)

Plate adopted from Hafferl and Thiel [1]

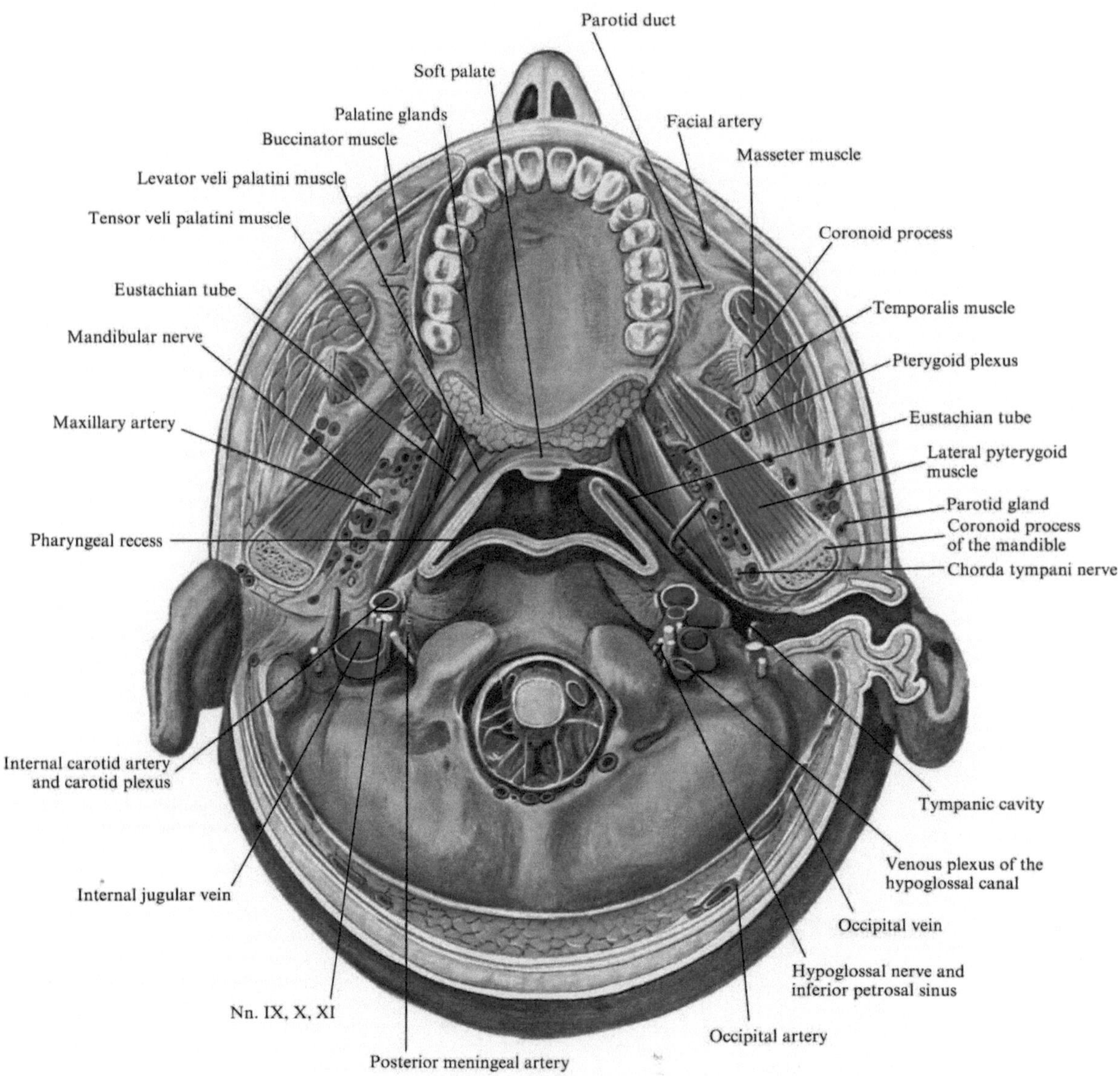

The palatine tonsil lies within the tonsillar fossa at a variable depth between the facial pillars, i.e., the palatoglossus and the palatopharyngeus muscles. The anterior surface of the tonsil can be covered by the *plica triangularis*, an enlargement of the anterior pillar at the base of the tongue. The supratonsillar fossa is situated above the tonsil and provides ready access to the tonsil's "capsule". The tonsil is not always clearly delimited from the base of the tongue, so that difficulties may occur during tonsillectomy when removing the tonsil from this site. When delivering the tonsil from its fossa, the surgeon should keep in mind the proximity of the glossopharyngeal, lingual, and hypoglossal nerves (see Sect. 4.7).
A thin muscular layer from the superior constrictor muscle is located lateral and posterior to the tonsillar capsule and joins the pharyngeal wall dorsally to protect the nearby neural and vascular structures [1].

Surgical Importance

Tonsillectomy; surgery of the lingual tonsil, tongue, and soft palate.

References

1. Denecke HJ (1980)
2. Lautenschläger A (1934)

Plate from Lautenschläger [2]

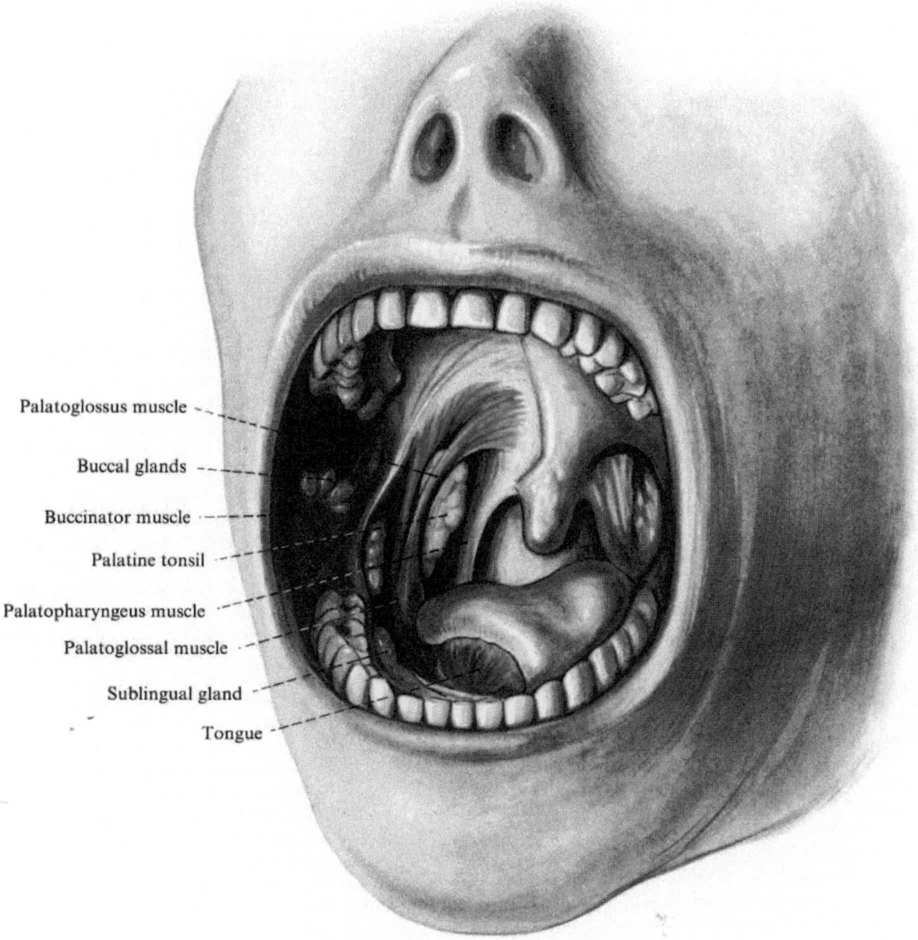

4.7 The Relationship of the Palatine Tonsil to the Major Adjacent Arteries and Nerves

The arterial supply of the tonsil is variable [1]. The primary tonsillar artery is usually a branch of the ascending palatine artery, but it may originate directly as a branch of the facial artery. Other tonsillar vessels originate as branches of the lesser palatine artery, the dorsal lingual artery, and the ascending pharyngeal artery. The facial or lingual artery can form a loop within the tonsillar fossa, and can be injured during tonsillectomy.

The possible variations in the tonsil's blood supply are represented in the Plate. In this illustration, the distance of a medially displaced tortuous internal carotid artery from the tonsil is most important (see Sects. 11.1, 2). Before doing a tonsillectomy, the surgeon must assure himself by careful inspection that no such pulsatile vessel is present in the submucosal soft tissues in order to prevent inadvertent injury. The normal distance between the tonsil and the internal carotid artery is 1–1.5 cm.

Surgical Importance

Palatine or lingual tonsillectomy, surgery of the soft palate and tongue, transoral approach to the cervical spine.

Reference

1. Hafferl A and Thiel W (1969)

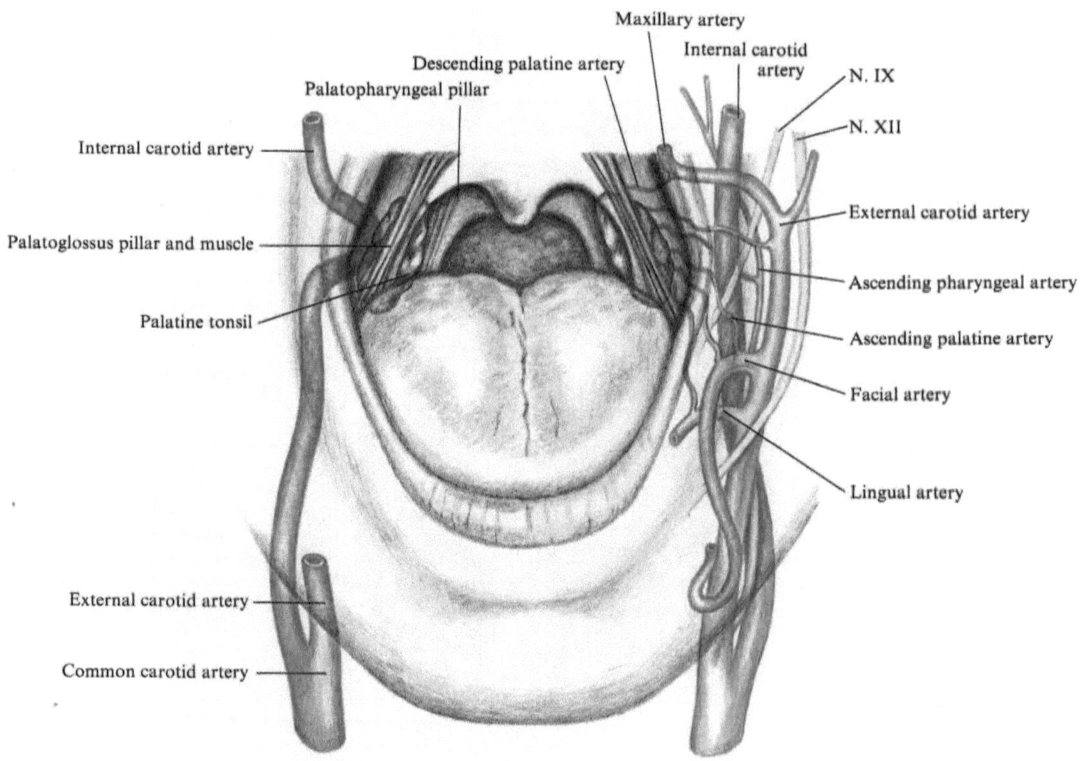

4.8 The Lymphatic Drainage of the Tongue (Lateral View)

The tongue has a rich ipsi- and contralateral lymphatic drainage to the lymph nodes of the neck. In general, the more anteriorly the lymphatic vessels of the tongue are situated, the more caudally any lymph is drained to the corresponding cervical nodes. The juguloomohyoid nodes serve as the first filtering system for drainage from the anterior one-third of the tongue. Lymphatics from the middle third drain to the submandibular nodes and along the internal jugular chain of nodes. Finally, lymphatics from the posterior third of the tongue drain to the jugulodigastric nodes and then to the nodes along the internal jugular vein. However, the superficial lymphatic network also provides anastomoses to the lymph nodes of both sides of the neck.

Surgical Importance

Inflammations and tumors of the tongue and the floor of the mouth; pull-through operations of the tongue and the floor of the mouth, dissection of the floor of the mouth, suprahyoid lymph node extirpations, radical and functional neck dissections.

References

1. Denecke HJ (1980)
2. Droulias C and Whitehurst JO (1976)

Plate from Denecke [1], Droulias and Whitehurst [2]

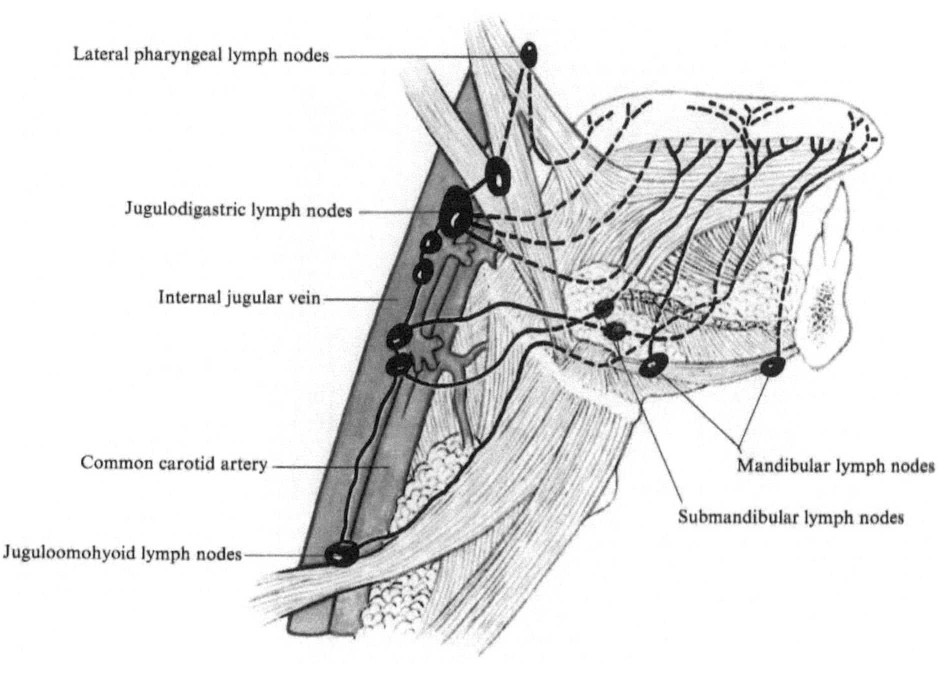

4.9 (Plates 1–2) Lymphatic Drainage of the Tongue (Anterior View)

The deep lymphatic vessels of the tongue are represented in Plate 1. The interrupted lines indicate the lymphatic drainage from the anterior third of the tongue, the dotted lines that from the middle third, and the unbroken lines that from the posterior third. The lymphatic vessels cross in the midline so that lymph is drained to both sides of the neck [1, 2]. This is also true for the superficial lymphatic vessels from the dorsum of the tongue (Plate 2).

Surgical Importance

Inflammatory and neoplastic diseases of the tongue, partial resections of the tongue, pull-through operations of the tongue and the floor of the mouth, dissection of the floor of the mouth, suprahyoid lymph node extirpations, radical and functional neck dissections.

References

1. Denecke HJ (1980)
2. Droulias C and Whitehurst JO (1976)

Plates 1 and 2 from Denecke [1], Droulias and Whitehurst [2]

1

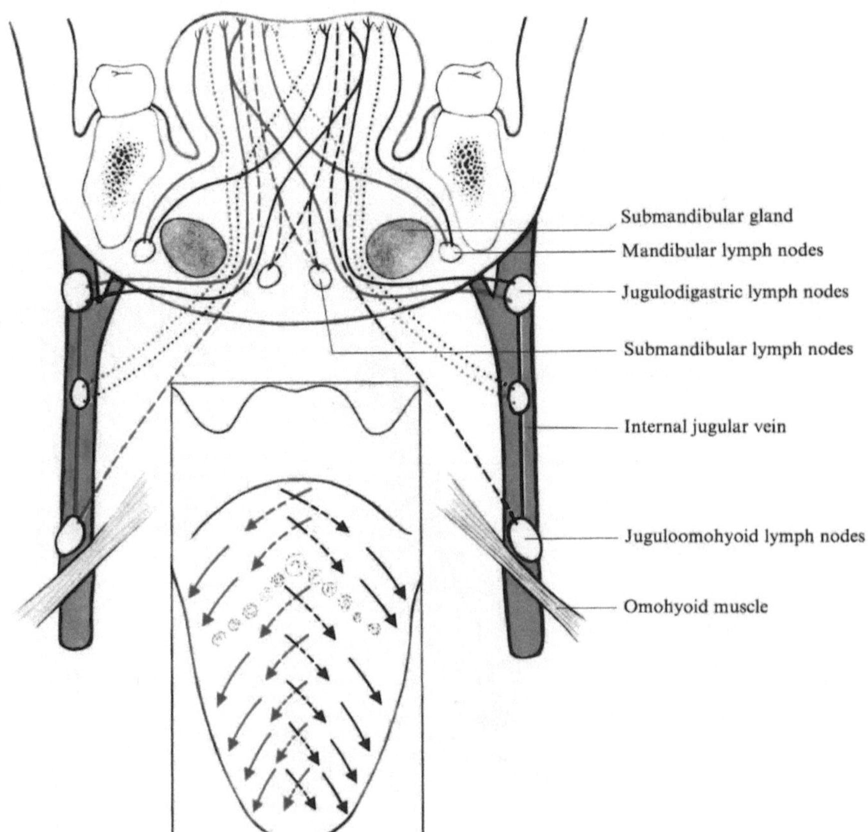

2

5 The Nose

The bony infrastructure of the nose consists of the nasal bones and the frontal process of the maxilla. The cartilage is derived from the upper lateral (triangular) cartilages as the semimobile part, and the alar or apical cartilages as the mobile part. This latter part also includes duplication of the adjacent skin. All of these structures are age, sex, and race dependent. The comparative ratio of the three nasal parts in Indo-European adults is more or less 1:1:1. In children, the osseous part (i.e., the nasal bones) is much shorter, whereas the semimobile and mobile parts are longer in old age [1, 2, 4].

The nasal bones can be hypo- or hyperplastic [4, 8]. Their length varies from 8 to 33 mm, while the width superiorly varies from 2 to 17 mm and inferiorly from 7 to 24 mm. Because of the trapezoid shape of the (upper) lateral nasal cartilage, the term "triangular cartilage" is not completely correct [6]. The cartilage penetrates a variable distance beneath the sharp margin of the nasal bones and is connected in its upper part with the cartilaginous septum [5]. Its inferior border corresponds to the limen nasi and tends to "return" in its shape (as described by Cottle [3]). This cartilage is covered to a variable extent by the lateral crus of the apical or lower lateral cartilage (lobular cartilage, greater alar cartilage). The apical cartilage has a lateral and a medial crus. The zone between the upper lateral cartilages nasal bone and septum is called the "keystone area" by Cottle. The cleft between the septal and upper lateral cartilage is rarely missing. The zone between the upper lateral cartilage, the dorsal septum, and the apical cartilage is called the "weak triangle" by Converse [2]. The doubling of skin between the lateral and medial crus of the apical cartilage is termed the "soft triangle." The variations in the shape of the lateral crus have been classified by Zelnik and Gingrass [9] into five groups: (1) smooth and convex; (2) convex in front, concave behind; (3) concave in front, convex behind; (4) concave in front and behind, and centrally convex; (5) concave (not represented).

The height of the upper lateral crus varies from 0.7 to 1.5 cm (average 1.1 cm) and the width from 1.7 to 3 cm (average 2.2 cm). The distance of the lower margin of the alar cartilage from the lower margin of the ala nasi is 0.4–0.8 cm in front, 0.3–0.7 cm in the middle, and 0.9–1.9 cm posteriorly. The medial crus can be seen in one of three forms [7]: (1) medial convex; (2) S-shaped symmetrical; (3) S-shaped asymmetrical.

The external shape of the nose also depends on the thickness of the skin and the underlying subcutaneous tissue. This configuration is also important in surgery to reshape the alae nasi.

Surgical Importance

Excisions of nasal tumors and fistulae, rhinoplasty.

References

1. Berendes J et al. (1977)
2. Converse JM (1955)
3. Cottle MW (1955)
4. Krmpotić-Nemanić J (1977)
5. Masing H (1964a,b)
6. Milhalkovics V v (1899)
7. Natvig P et al. (1971)
8. Pech A and Cannoni M (1976)
9. Zelnik J and Gingrass GR (1979)

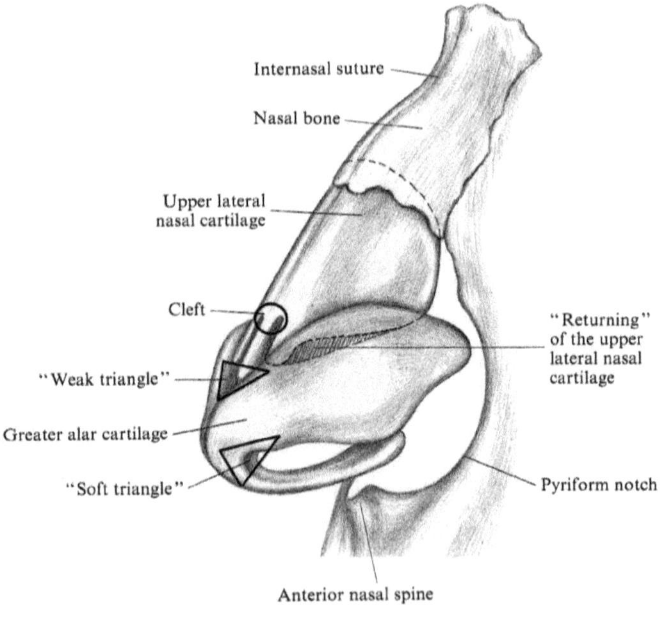

5.2 The Cartilaginous Skeleton of the Nasal Tip

The inferior view of the tip of the nose shows the transition of the medial crus into the lateral crus of the alar cartilage. This is characterized by a medial and lateral angulation which should be kept in mind when doing tip rhinoplasty on the alar cartilages. An external deviation of the inferior extremity of the medial crus can produce a widening of the columella. The anterior (caudal) border of the cartilaginous septum is an additional support to the nasal tip.

Surgical Importance

Nasal trauma, rhinoplasty, reconstructive nasal surgery of the cleft nose.

Reference

1. Denecke HJ and Meyer R (1964)

Plate from Denecke and Meyer [1]

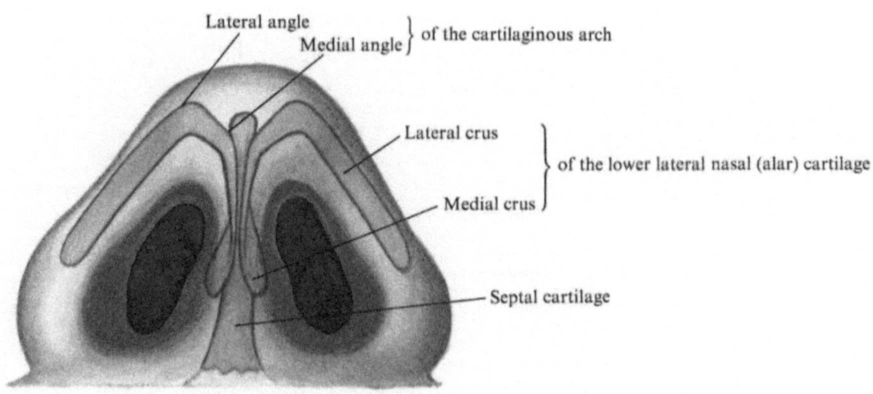

5.3 Age-Dependent Segmentation of the Lateral and Tip Cartilages of the Nose

The lateral crus of the alar cartilage is convex and elastic in the younger individual. With advancing age, this cartilage generally becomes thin, soft, and straight. The inferior margin of the upper lateral cartilage and the superior margin of the lateral crus of the alar (lower lateral) cartilage can display segmentation, resulting in the presence of sesamoid cartilages [2].

The lateral crus of the alar cartilage can extend to the pyriform aperture and be fixed to it by connective tissue. A lateral process exists in some cases that can be separated into the minor alar cartilages. Such small cartilages must be taken into consideration in cosmetic rhinoplasty [1].

Surgical Importance

Cosmetic rhinoplasty.

References

1. Jost G (1975)
2. Krmpotić-Nemanić J et al. (1972)

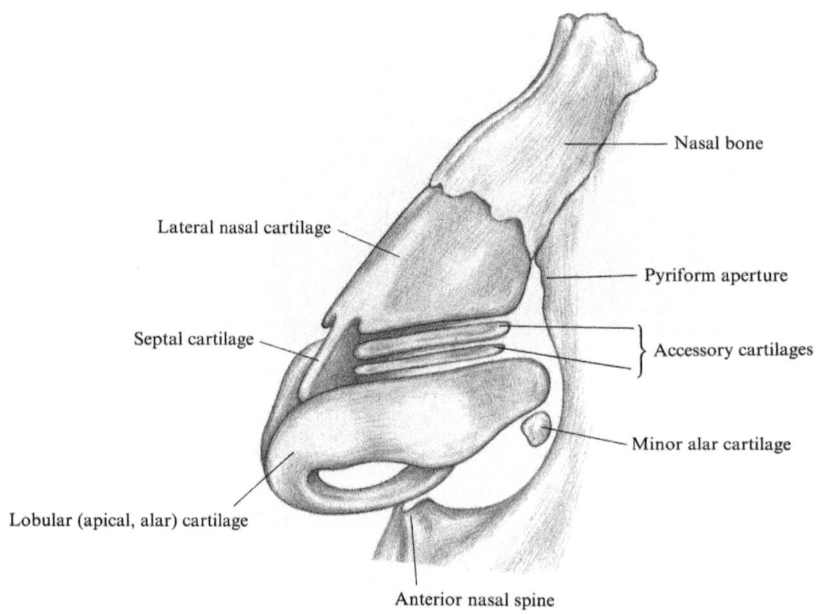

The pyriform aperture is delimited by the frontal process of the maxilla and the nasal bones. As described by Hovorka [3], its shape is variable (see Sect. 5.5). Its lower two-thirds are important for respiration and have more constant dimensions. The anterior nasal spine lies between the two nasal cavities and can protrude considerably or be absent, as well as show various transitional forms [1, 2, 4].

The anterior pole of the inferior nasal concha can extend into the region of the *os incisivum* and return medially to form the so-called *pars evoluta* [5]. This latter structure forms a 3 mm or less cleft with the inferolateral margin of the pyriform aperture and can serve as a guide to placement of a needle for puncture of the maxillary sinus through the inferior nasal meatus. If the needle is guided through this cleft, it passes through the thinnest part of the medial wall of the sinus.

Surgical Importance

Septoplasty, rhinoplasty, turbinectomy (conchotomy), sinus endoscopy, maxillary sinus puncture, antral fenestration (antrostomy).

References

1. Berendes J et al. (1977)
2. Denecke HJ and Meyer R (1964)
3. Hovorka cit. Denecke HJ and Meyer R (1964)
4. Krmpotić-Nemanić J (1977)
5. Perović D (1954)

Plate from Krmpotić-Nemanić in Berendes et al. [1]

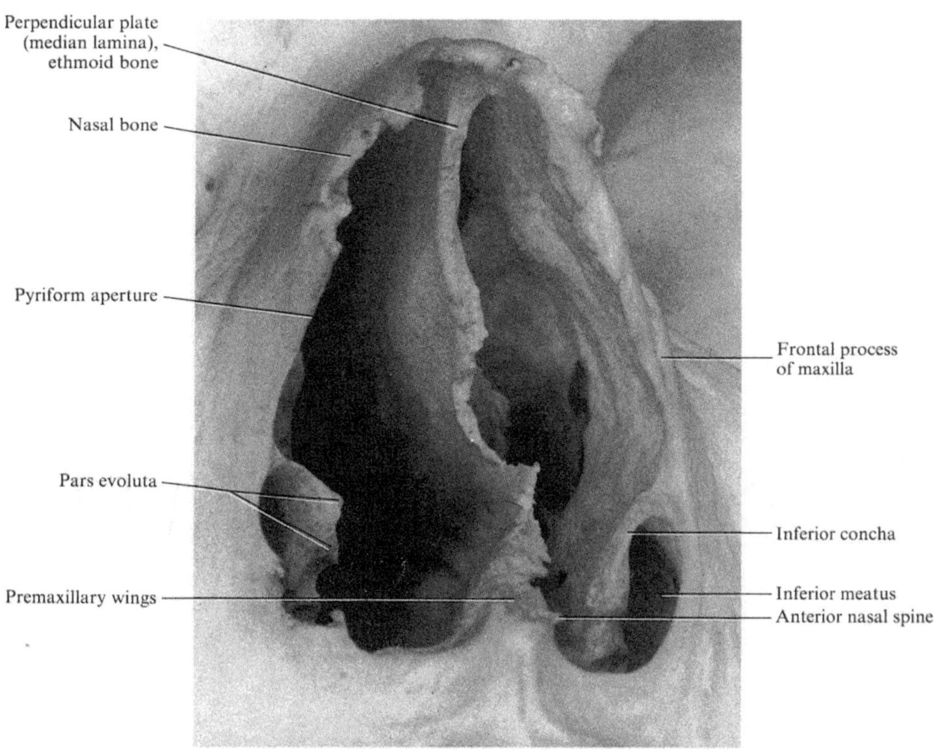

5.5 (Plates 1–3) Variations of the Pyriform Aperture

Plates 1–3 show:

1. Different shapes of the inferior pyriform aperture;
2. Different shapes of the entire pyriform aperture;
3. Variations of the upper part of the pyriform aperture.

Familiarity with variations in the shapes of the pyriform apertures is important in surgical elevation of the mucosa from the nasal floor during septoplasty and in freeing the periosteum from the underlying bone when doing lateral osteotomies in cosmetic rhinoplasty.

Surgical Importance

Nasal trauma, nasal malformations, septoplasty, cosmetic rhinoplasty.

References

1. Denecke HJ and Meyer R (1964)
2. Hovorka cit. Denecke HJ and Meyer R (1964)

Plates 1–3 from Hovorka [2]

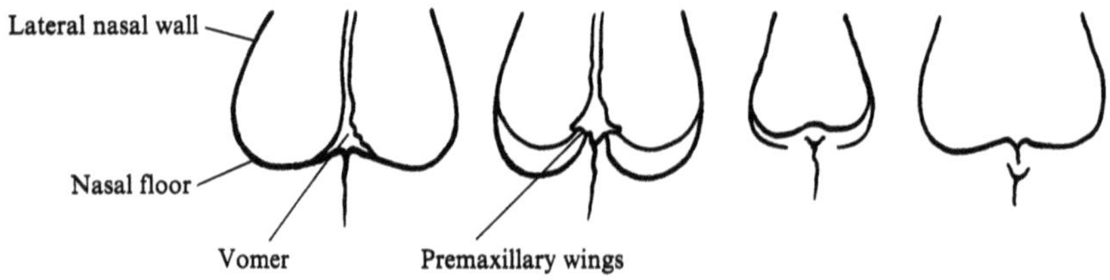

1

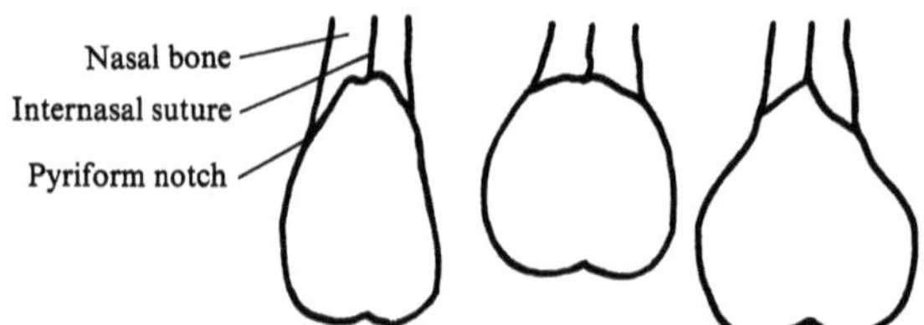

2

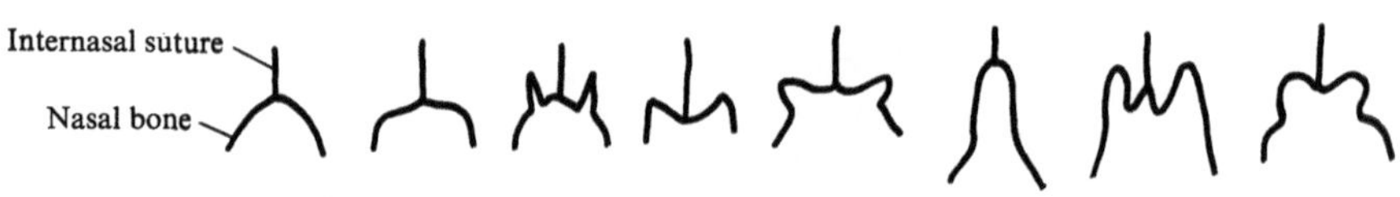

3

5.6 (Plates 1–4) Variations in the Shapes of the Nasal Bones and Dorsal Nasal Sutures

Plates 1–4 show the following:

1. Variations of the nasofrontal sutures;
2. Variations of the internasal suture;
3. The so-called sutural bones;
4. Variations of the sutures.

The periosteum is fixed on the dorsum of the nose in the region of its bony sutures so that its complete elevation subperiosteally is not possible without being torn [1, 2].
Injuries to the sutural bones can lead to unfavorable fractures of the nasal dorsum following nasal trauma or during medial or lateral osteotomies.

Surgical Importance

Treatment of nasal trauma, cosmetic rhinoplasty, excisions of nasal fistulae, correction of nasal malformations.

References

1. Denecke HJ and Meyer R (1964)
2. Hovorka cit. Denecke HJ and Meyer R (1964)

Plates 1–4 from Hovorka [1]

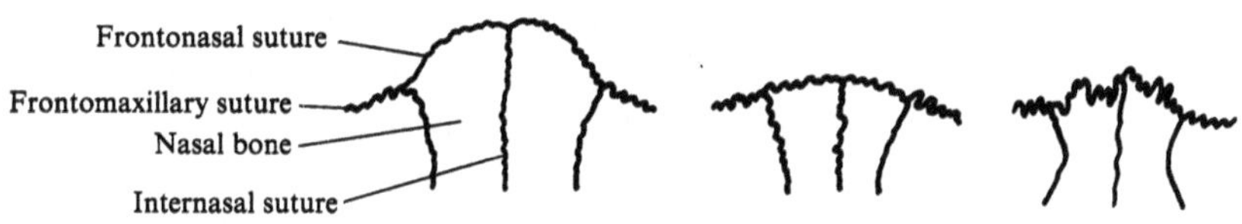

1

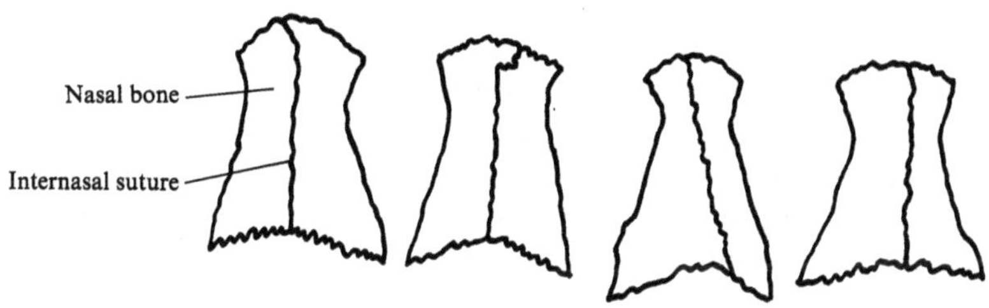

2

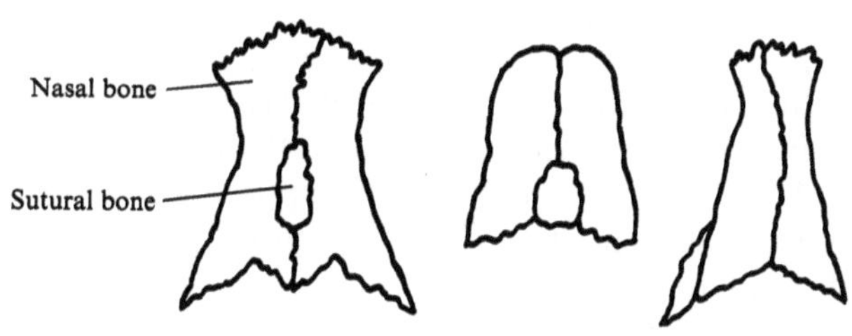

3

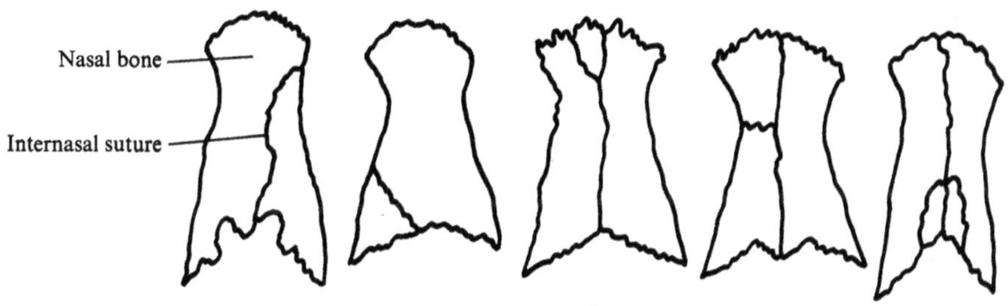

4

5.7 The Nasal Septum as the Medial Wall of the Nasal Cavity

Cartilaginous and bony septa form the medial wall of the nasal cavity. The osseous part consists of the perpendicular lamina of the ethmoid bone, the vomer, the crista galli and the sphenoid rostrum, the maxillary nasal crest and the palatine bone. The cartilaginous part (or *cartilago quadrangularis*) is situated between the vomer and the median lamina of the ethmoid bone. Additionally, a septal process (*processus sphenoidalis*) can extend to the sphenoid bone. Anteriorly, the osteocartilaginous part of the septum is joined by a cutaneous part, or membraneous septum, to the columella. This latter structure is formed by the anterior border of the cartilaginous septum, the medial crura of the alar cartilages, connective tissue and skin. The basal portion of the cartilaginous septum lies in a groove of the vomer and ends anteriorly in the region of the anterior nasal spine, which is formed by fusion of bony continuations from the maxillary crest (the *cristae incisivae*) [1–3].

The nasal opening of the incisive canal lies between the crista incisiva and the maxillary crest of each side and is filled with connective tissue, nerves, and vessels. This aperture lies 8–15 mm behind the anterior nasal spine and should be anesthetised prior to septal or rhinoplastic surgery. The inferior or osseous part of the nasal septum consists of a part of the premaxilla and its wings, the anterior nasal spine, the nasal crest of the maxilla (which is related anteriorly to the premaxillary wings and superiorly to the vomer), and finally the vomer, which is related to the premaxillary processes (see Sects. 5.12, 13). The premaxillary wings are small at birth and remain essentially unchanged until the sixth year of life. Thereafter, their growth is relatively rapid, although fusion with the vomer occurs at approximately age 15. Since the septal periostea and perichondria come together at the sites of fusion of the septal cartilage and bones to penetrate the sutures, a subperichondral dissection of the septum is possible *only* to the sutures. If the connective tissue present is very dense, sharp division of the penetrating fibers is necessary at the time of surgery.

An articulation is somewhat formed at the point of contact between the septal cartilage and the anterior nasal spine. Perichondrium ensheathes the inferior margin of the cartilage, as does periosteum on the upper margin of the nasal spine and the premaxillary wings.

The anterior-superior region of the bony septum is formed by the perpendicular lamina of the ethmoid bone, and is occasionally pneumatized by one or both frontal sinuses to form the *sinus septi nasi*. In the same way, the vomer can be pneumatized from the sphenoid bone and its sinuses. In such cases, X-ray or CT pictures may give the impression of a "double" septum.

The height of the nasal cavity can show sex-dependent variations. The average is 48.15 mm in males and 47.84 mm in females, with the right nasal cavity in general being higher [4].

The vomeronasal organ or *accessory olfactory zone* is a phylogenetically interesting organ and lies on the septum in the region of the incisive canal. Numerous capillaries can be found at this site, particularly in the adult, and can cause bleeding in the same way as does Kiesselbach's vascular plexus (see Sect. 5.8).

In transseptal surgery on the sphenoid sinus, a part of the osseous septum must be removed. In so doing the perpendicular lamina of the ethmoid (with its mucosa attached on one side) can be temporarily displaced laterally.

Surgical Importance

Treatment of epistaxis, endoscopy of the nose and nasopharynx, septoplasty, cosmetic rhinoplasty, endonasal surgery of the paranasal sinuses, transseptal approaches to the sphenoid sinus and the hypophysis.

References

1. Berendes J et al. (1977)
2. Denecke HJ and Meyer R (1964)
3. Krmpotić-Nemanić J (1977)
4. Lang J and Sakals LE (1981)

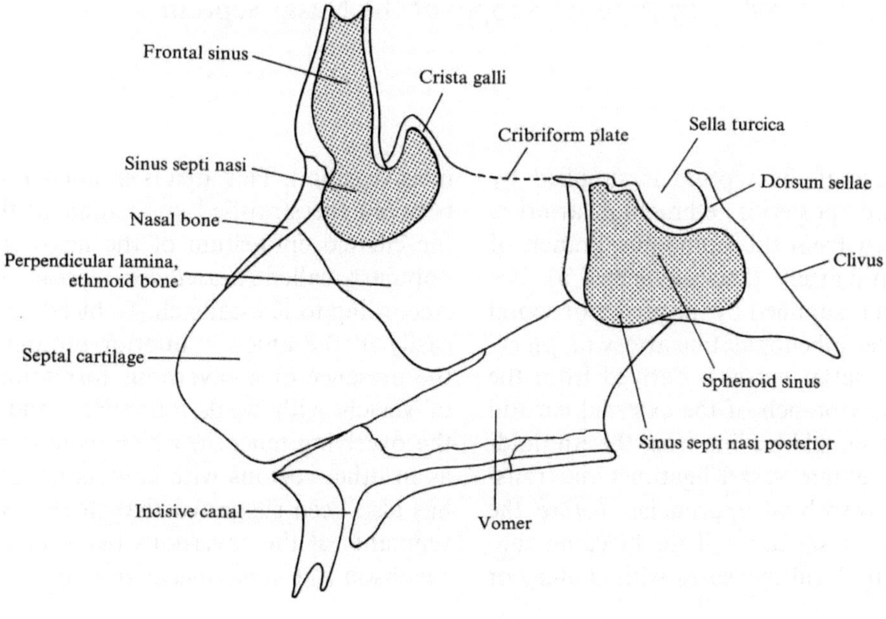

Frontal sinus
Crista galli
Cribriform plate
Sella turcica
Sinus septi nasi
Dorsum sellae
Nasal bone
Perpendicular lamina, ethmoid bone
Clivus
Septal cartilage
Sphenoid sinus
Sinus septi nasi posterior
Incisive canal
Vomer

The superior part of the septum is supplied by the anterior and posterior ethmoidal arteries, which are derived from the opthalmic branch of the internal carotid artery. (See also Sect. 7.5). The posterior septum is supplied by the posterior septal branches from the sphenopalatine artery (*a. pterygopalatina*). This latter vessel is derived from the maxillary artery, a branch of the external carotid artery [2]. Severe septal bleeding from the ethmoidal arteries may require vessel ligations via transethmoidal or transorbital approaches *before* the arteries reach the nasal cavity if the bleeding cannot be stopped by local measures with cautery or packing (see Sects. 11.3, 4). Persistent bleeding from the sphenopalatine artery may require ligature of the maxillary artery in the sphenopalatine fossa (see Sects. 6.21, 22).

The deep capillaries of the nasal mucosa have a continuous relatively thick endothelial layer, while the superficial capillaries – especially in the anterior part of the turbinates or conchae – display an open endothelial structure with signs of increased permeability [3].

An arterial vascular network lies in the anterior inferior part of the septum and consists of branches from the maxillary artery (the posterior septal artery), the ethmoidal arteries, the facial artery, and the greater palatine artery (within the incisive canal). This area is situated at the junction between the stratified epithelium of the skin and the ciliated epithelium of the nasal cavity and is popularly called *Kiesselbach's locus*.

According to Kiesselbach [1], bleeding occurs very easily in the anterior inferior septum because of the presence of a cavernous formation consisting of vessels with weak muscularis and fixation of the overlying mucosa, which cannot be distended as in other regions with cavernous formations. It has also been suggested that these vessels may be remnants of the cavernous tissue of the organ of Jacobson (the *vomeronasal organ*).

Surgical Importance

Septoplasty, cosmetic rhinoplasty, trans-septal sphenoidectomy, endonasal and external operations on the paranasal sinuses, treatment of frontobasal injuries, cleft surgery.

References

1. Kiesselbach W (1884, 1885)
2. Lautenschläger A (1934)
3. Watanabe K and Watanabe J (1980)

Plate from Lautenschläger [2]

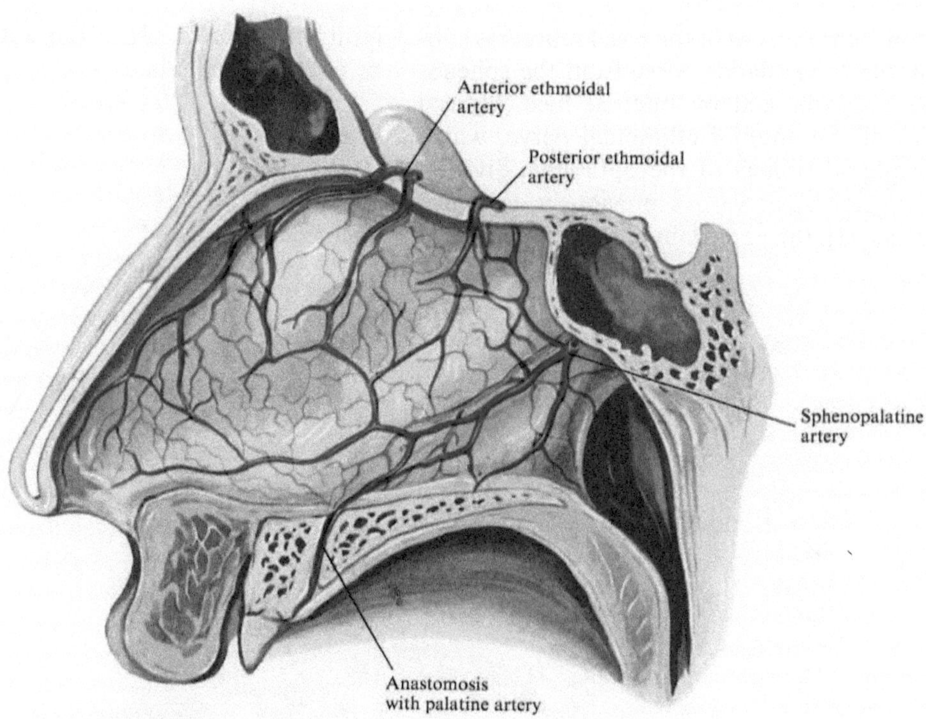
Anterior ethmoidal artery
Posterior ethmoidal artery
Sphenopalatine artery
Anastomosis with palatine artery

5.9 The Nerve Supply of the Nasal Septum

The respiratory mucosa of the nasal septum is supplied by the nasopalatine nerve from the sphenopalatine ganglion and by internal medial nasal branches of the anterior ethmoidal nerve, which is a peripheral branch of the opthalmic division of the trigeminal nerve. As previously described by Denecke [1], the important sites for topical and infiltration anesthesia of the septum are indicated by circles on the Plate.

Reference

1. Denecke HJ (1953)

Plate from Denecke [1]

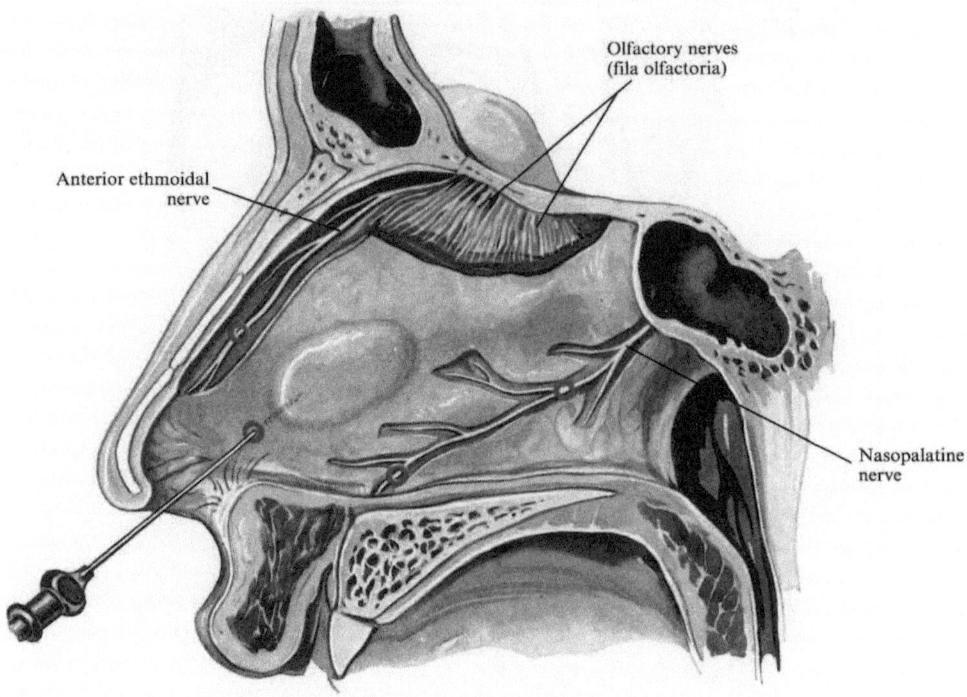

5.10 The Lateral Nasal Wall

The specimen illustrated in the Plate was taken from the 1954 Perović collection in the Department of Anatomy at the University of Zagreb. This details the lateral nasal wall, which is composed of the following bones: the maxilla, the palatine bone, the ethmoidal bone with the superior and middle conchae, possibly an agger nasi cell (or in some cases, a so-called *agger nasi concha* [1]), the inferior concha, the lacrimal bone, the nasal bone, part of the frontal bone, as well as the pterygoid process and part of the body of the sphenoid bone.

Surgical Importance

Nasal and paranasal sinus endoscopy; sinus surgery, antral puncture, cosmetic and reconstructive (cleft) rhinoplasty.

Reference

1. Perović (1961)

Plate from Perović [1]

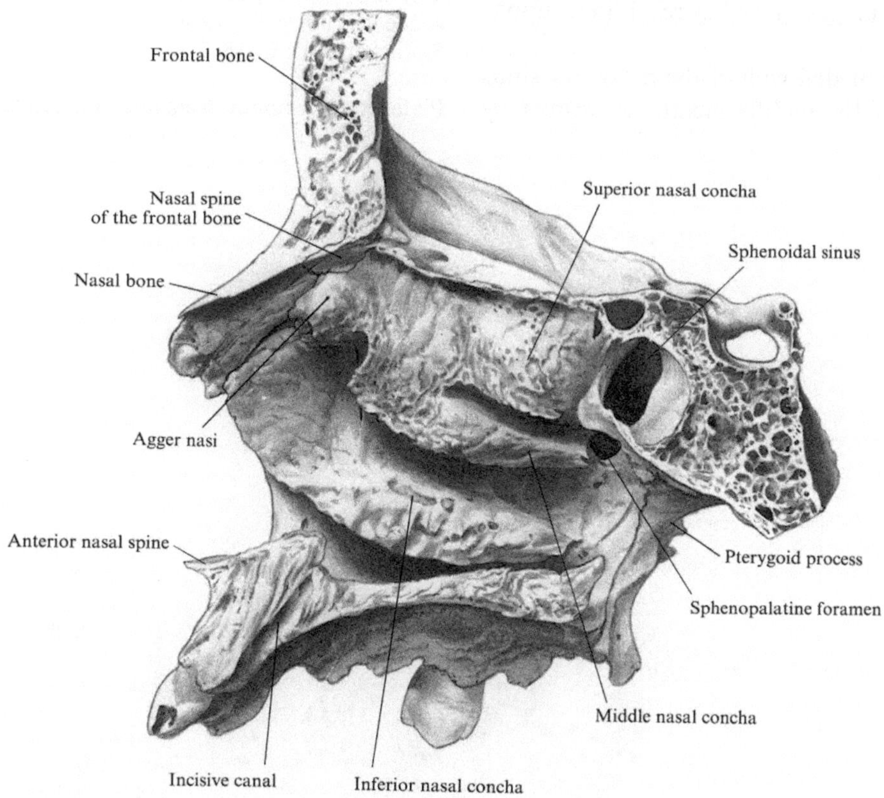

5.11 The Lateral Wall of the Nasal Cavity After Removal of the Superior Concha and Part of the Middle Concha

The specimen shown in the Plate was further prepared in the Perović collection. The superior concha and part of the middle concha have been removed to show the ostia of the posterior ethmoid air cells into the superior meatus and the opening of the hiatus semilunaris into the middle meatus. The ostia of the frontal sinus, the anterior ethmoid cells, and the maxillary sinus open into the middle meatus [1, 3]. The so-called *agger nasi* is the anterior part of the uncinate process and must be removed in endonasal approaches to the frontal sinus. The agger nasi can form an additional concha, but usually consists of solid bone or contains one or more pneumatized cells lying medial to the lacrimal sac and/or duct. As such, these structures can impede access to the sac (see Sect. 5.10). These pneumatized cells can also be affected by isolated inflammatory changes as well as result in a suppurative sinusitis.

One part of the medial wall of the maxillary sinus in the region of the middle meatus is formed by fontanellae [5], which consist of a doubling of the adjacent mucosa (making possible direct cannulation of the sinus).

Surgical Importance

Endoscopy of the nose, maxillary sinus and sphenoid sinus [2]; endonasal operations on the paranasal sinuses; surgery of the lacrimal apparatus and the nasal conchae; endoscopic surgery of the middle meatus [4].

References

1. Berendes J et al. (1977)
2. Draf W (1978, 1983)
3. Krmpotić-Nemanić J (1977)
4. Messerklinger W (1978)
5. Zuckerkandl E (1882)

Plate from Krmpotić-Nemanić in Berendes et al. [1]

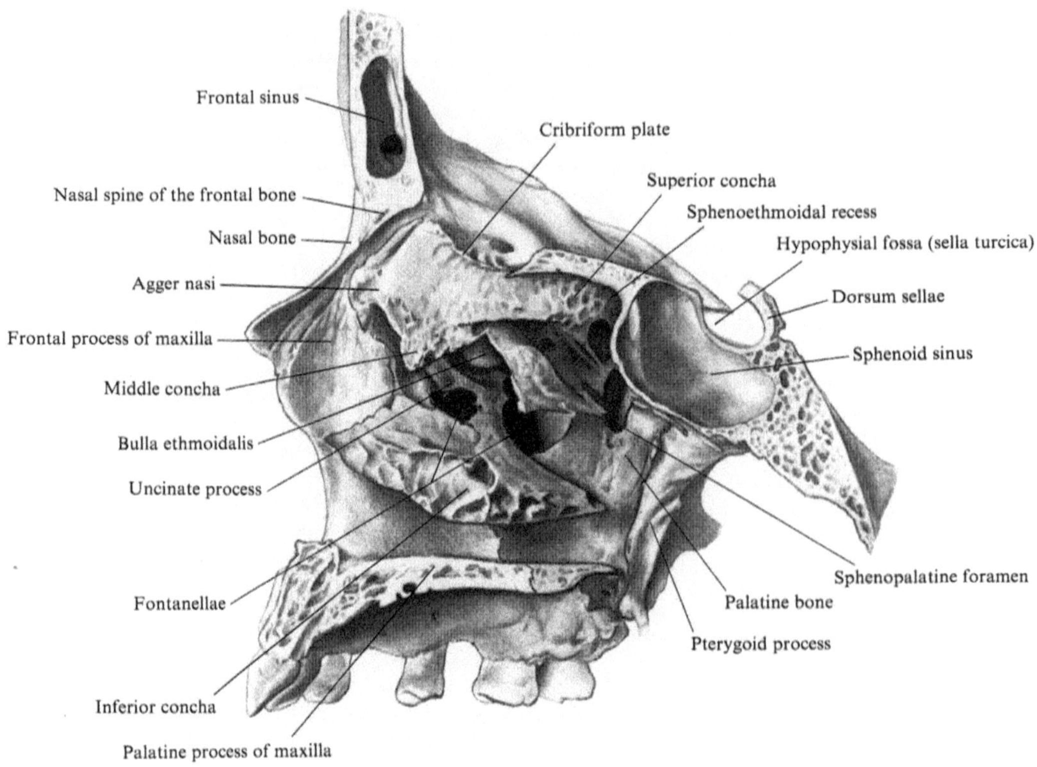

5.12 The Bony Lateral Wall of the Nasal Cavity (Frontal View)

The nasolacrimal duct is schematized in the Plate and its orifice is shown in the inferior meatus behind the anterior attachment of the inferior concha. The bulla ethmoidalis and the uncinate process can be seen beneath the anterior part of the middle concha. The semilunar hiatus lies between the bulla and the uncinate process, and contains the opening of the nasofrontal duct, represented by the dotted lines as the infundibulum of Killian [1, 2].

Differentiation should be made between the frontal infundibulum of Killian (the orifice of the frontal duct within the frontal sinus) and the ethmoid infundibulum (not represented), lying lateral to the semilunar hiatus and connecting the maxillary sinus with the frontal sinus.

Surgical Importance

Nasal endoscopy, endonasal surgery of the lacrimal apparatus and the paranasal sinuses, fenestration of the maxillary sinus (antrostomy), submucous resections of the conchae or partial turbinectomies.

References

1. Hafferl A and Thiel W (1969)
2. Killian G (1903)

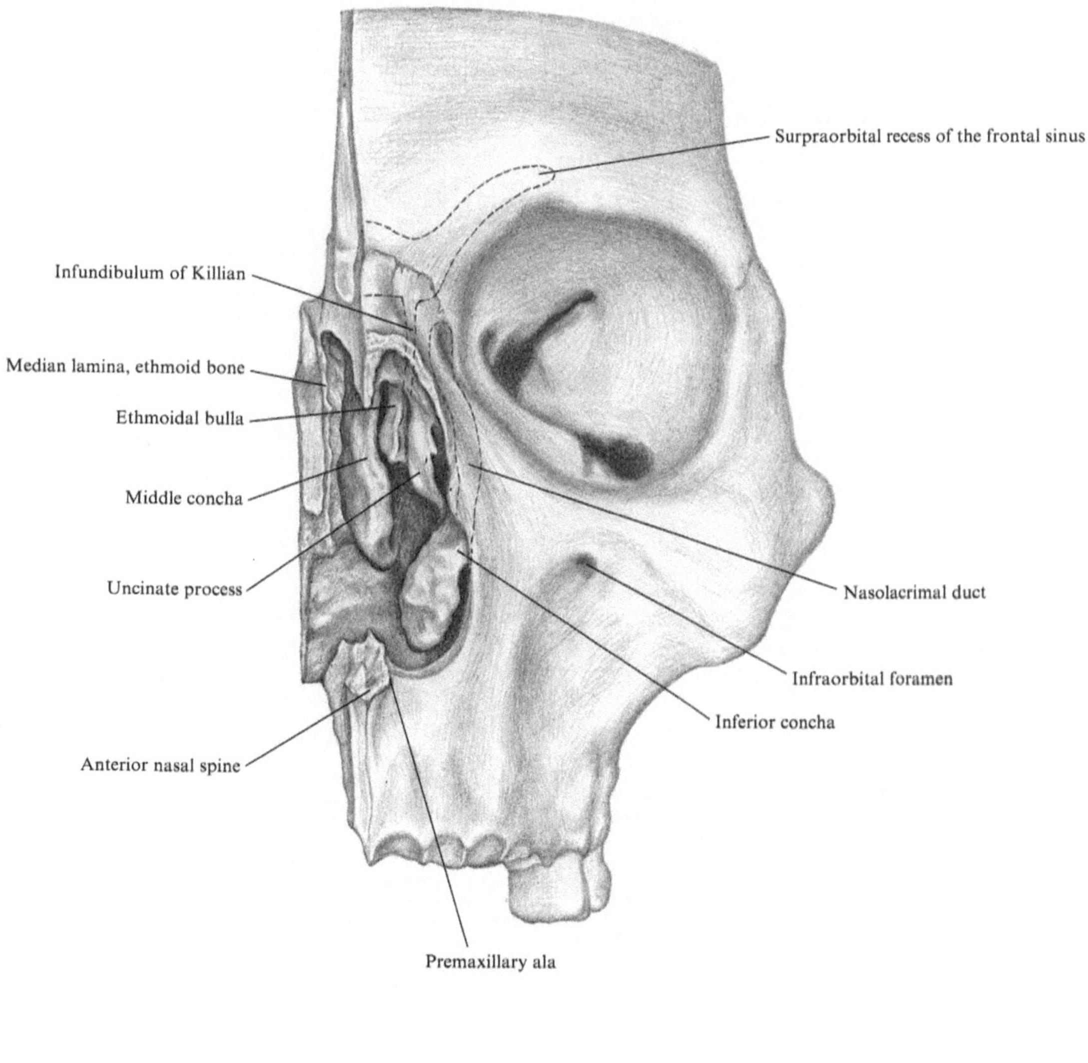

5.13 The Bony Wall of the Nasal Cavity (Oblique Anterior View)

The oblique anterior view of the nasal cavity is useful for its orientation of the bony structures of the lateral wall. The anterior and posterior fontanelles of the hiatus sinus maxillaris are located in front of and behind the uncinate process and its contact with the ethmoidal process of the inferior nasal concha. These fontanelles are covered with a double layer of mucosa (see Sect. 5.11).

Surgical Importance

Nasal endoscopy, pansinus surgery, fenestration of the maxillary sinus, endonasal surgery of the lacrimal apparatus, surgery of the conchae (turbinates).

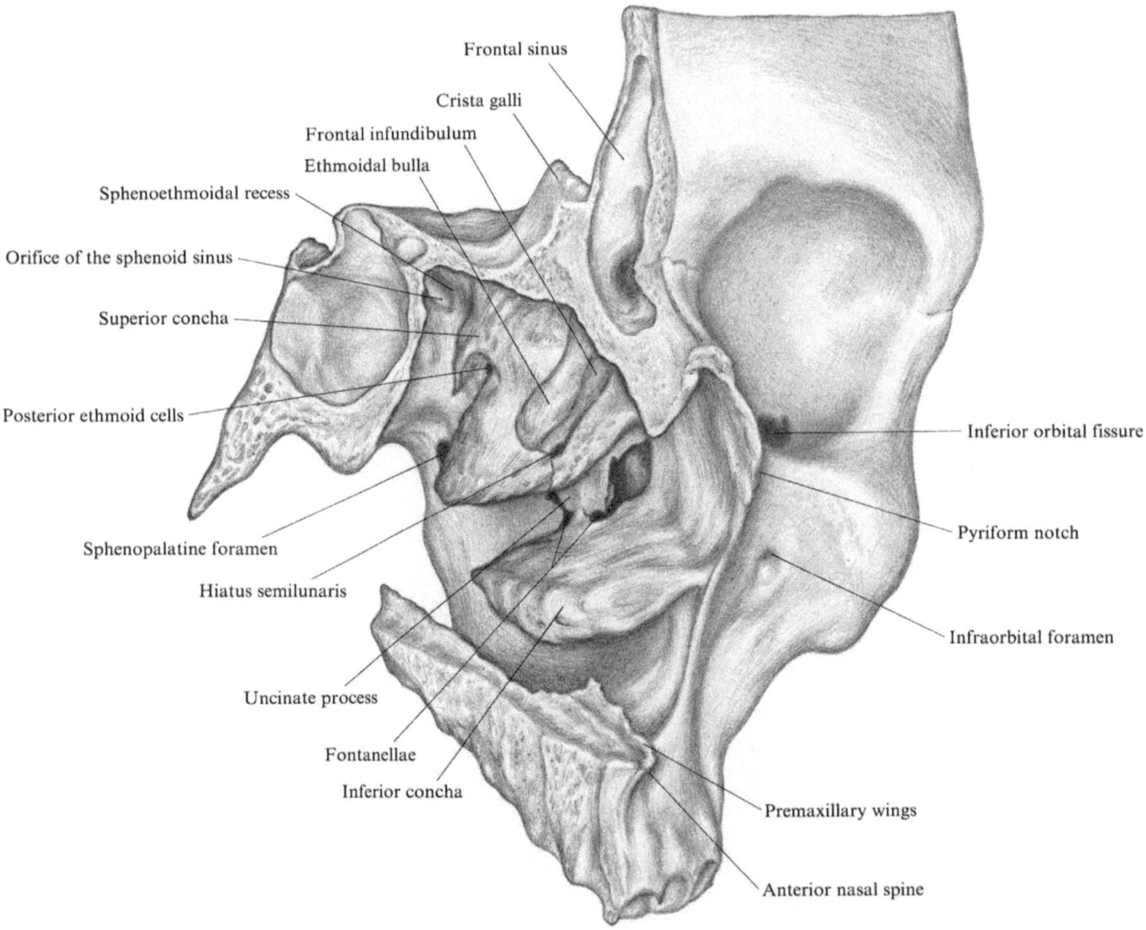

5.14 The Blood Supply of the Lateral Wall of the Nasal Cavity and the Palate

As shown by Lautenschläger [26], the lateral wall of the nasal cavity is supplied by vessels from the internal and external carotid systems [1, 2]. These include the anterior and posterior ethmoidal arteries, which branch from the ophthalmic artery. This latter vessel is in turn normally a branch of the internal carotid artery (see Sects. 11.3, 4). The external carotid artery gives rise to the sphenopalatine or pterygopalatine artery and the posterior lateral nasal arteries via the maxillary artery. The greater palatine artery anastomoses through the incisive canal with the arteries of the nasal cavity.

Surgical Importance

Conchal (turbinate) surgery, endonasal surgery of the lacrimal apparatus, extranasal and endonasal surgery of the paranasal sinuses, skull base surgery, transsphenoidal hypophysectomy, retromaxillary dissections.

References

1. Denecke HJ (1953)
2. Lautenschläger A (1934)

Plate adapted from Lautenschläger [2]

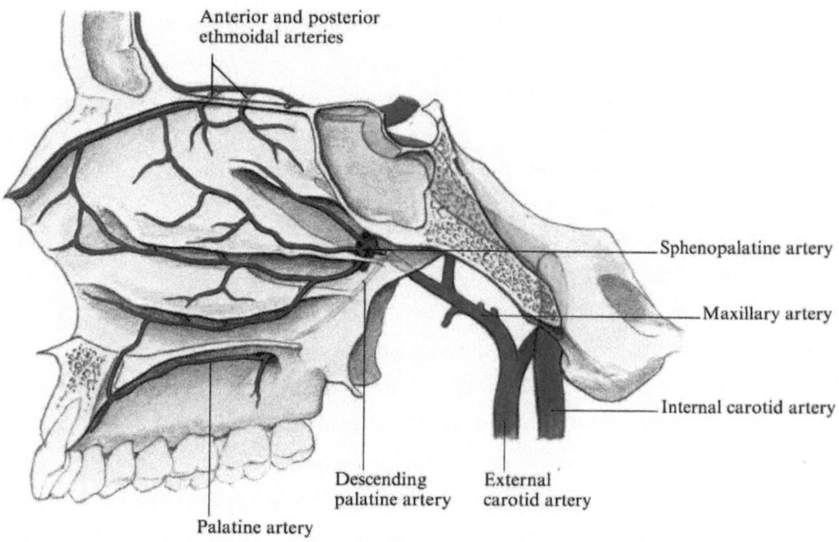

6 The Paranasal Sinuses and the Retromaxillary Space

6.1 Nasal Cavity of the Newborn child

The anlage of the nasal cavity is formed by the olfactory placodes and pits [4]. These structures are blind sacs, which later open into the oral cavity to form the primitive choanae. The palate forms as a horizontal plate between the anlagen of the nasal and oral cavities, and subsequently becomes the floor of the nose. The secondary palate develops from the laterally situated maxillary processes. The anterior portion of the palate is formed by the intermaxillary bone (*os incisivum*) [1, 3]. The nasal conchae, meati, and choanae develop following fusion of the palatine plate with the nasal septum, as shown in the Plate.

The middle and superior chonchae form the first and second ethmoturbinal bones from the medial septal border of the nasal cavity, while the nasoturbinal bone and the inferior concha as the maxilloturbinal evolve on the lateral wall of the nasal cavity. Later, the middle concha and the septum can – although uncommonly – become pneumatized (see Sects. 5.12, 13).

The superior meatus is formed between the superior and middle conchae, while the middle meatus develops between the middle and inferior conchae. Accessory conchae can be found in the superior and middle nasal meati [2]. These include the agger nasi, bulla ethmoidalis, uncinate process, and the supreme nasal concha (which is located above the superior nasal concha).

The bones forming the walls of the nasal cavity that ossify by membranous ossification include the maxilla, palatine bone, nasal bones, and vomer, and develop in the second and third gestational month. In contrast, those bones ossifying by chondral ossification begin to take form in the fifth month of gestation, with the inferior concha forming in the seventh gestational month. The ethmoidal labyrinth together with the ethmoidal cells take form in the primitive cartilaginous nasal capsule. Consequently, this bone already has well-developed pneumatic spaces by birth, while the other sinuses appear only as small mucosal evaginations at this time.

The average dimensions of the frontal recess in the newborn infant are 4.5×3.2 mm and that of the maxillary sinus are $5 \times 10 \times 3.5$ mm. Other sinus dimensions are: the sphenoid sinus, $4 \times 2 \times 2$ mm; the anterior ethmoid cells, $1-3 \times 1-5 \times 1-3$ mm; and the posterior ethmoid cells, $4.5-5 \times 2.5-5 \times 1.5-2$ mm.

Surgical Importance

Orbital complications following paranasal sinus inflammations (e.g. as in nursing).

References

1. Goethe J W v (1784, 1831)
2. Grünwald L (1925)
3. Haecker V (1927)
4. Peter K (1925)

Plate adapted from Peter [4]

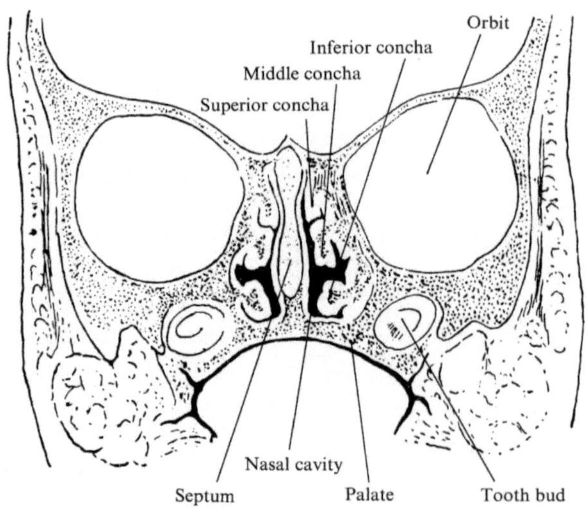

6.2 Development of the Paranasal Sinuses in a 4-Year Old (Right on the Plate) and an 11-Year Old Child (Left on the Plate)

The developmental stages of the maxillary and frontal sinuses start in the first year of life. During this time, pneumatization of the sphenoid sinus reaches *Bertin*'s *ossicle* (Sect. 6.5) and then proceeds until the body of the sphenoid bone is pneumatized by age 5 years [1].

In contrast, the ethmoid air cells are already well-developed by the first year of life. This is clinically very important, since inflammation or infections of the sinus (as ethmoiditis) can cause orbital complications, even in a one-year old child.

Pneumatization of the sinuses develops slowly from birth until age 7 years and then becomes more intensive but irregular. By the age of 12–14 years, the paranasal sinuses have nearly completed taking their final shapes and dimensions, but may continue to increase slightly in size to age 25. In old age, however, the maxillary sinus can penetrate more deeply into the edentulous alveolar process.

The nasolacrimal duct is 7.5 mm long in the newborn and is 20 mm long in adults. Depending on the development of the nose, the paranasal sinuses and the teeth, the orifice of the nasolacrimal duct will change its position over time, being situated first at the anterior pole of the inferior concha and then later on becomes somewhat more posterior or within the inferior nasal meatus.

Surgical Importance

Endoscopy of the nasal cavity and maxillary sinus, external ethmoidectomy, endonasal surgery of the paranasal sinuses and lacrimal apparatus, surgical exploration of the orbit (and management of orbital complications).

Reference

1. Peter K (1925)

Plate adapted from Peter [1]

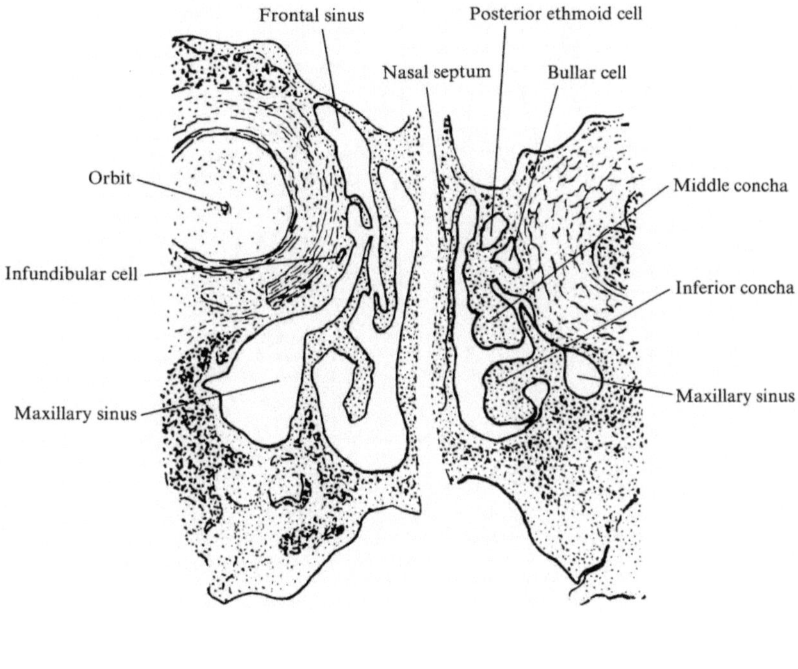

6.3 Development of the Frontal Sinus from Infancy to its Adult Size

The frontal sinus is shown in the figure in relationship to the middle concha and the septum. The development of the sinus begins in infancy as a small evagination from the middle nasal meatus. The changing dimensions of the sinus are indicated chronologically in the figure (see *newborn* and *adult*). The numbers 1–12 indicate the extension of the sinus by the corresponding year of life [1–3].

Surgical Importance

Beck's puncture of the frontal sinus [1], endoscopy and endonasal or extranasal surgery of the frontal sinus.

References

1. Beck K (1933)
2. Krmpotić-Nemanić J (1977)
3. Peter K (1925)

Plate modified from Peter [3]

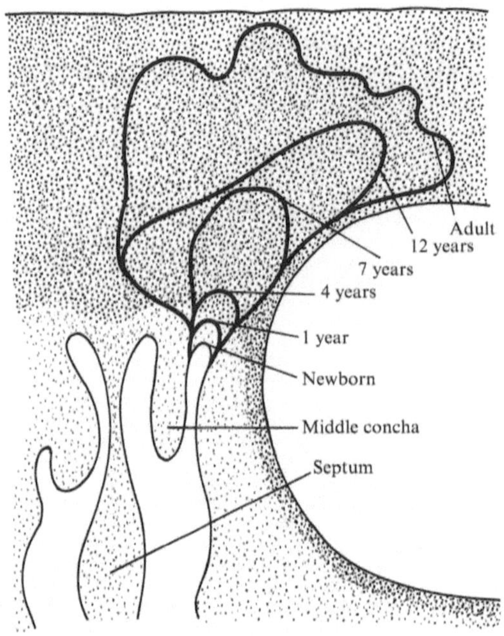

6.4 The Chronological Development of the Maxillary Sinus

The increase in the size of the maxillary sinus from infancy to adulthood is represented in the figure, together with the development of the alveolar recess following eruption of the third molar tooth. Beginning with the first premolar tooth, the dental roots can protrude into the maxillary sinus following development of the secondary (permanent) teeth. These teeth can be covered by an osseous plate of varying thickness. Bony dehiscences at the apices of the roots can also be found, such that the roots are covered only by the mucosa of the maxillary sinus. Consequently, surgery of the maxillary sinus may inadvertently result in injury to certain of the dental roots. This close relationship of the premolar and molar teeth to the maxillary sinus also explains complications of dental origin involving the sinus [1, 2].

Surgical Importance

Puncture, endoscopy and operations of the maxillary sinus; dental trauma and therapy.

References

1. Krmpotić-Nemanić J (1977)
2. Peter K (1925)

Plate modified from Peter [2]

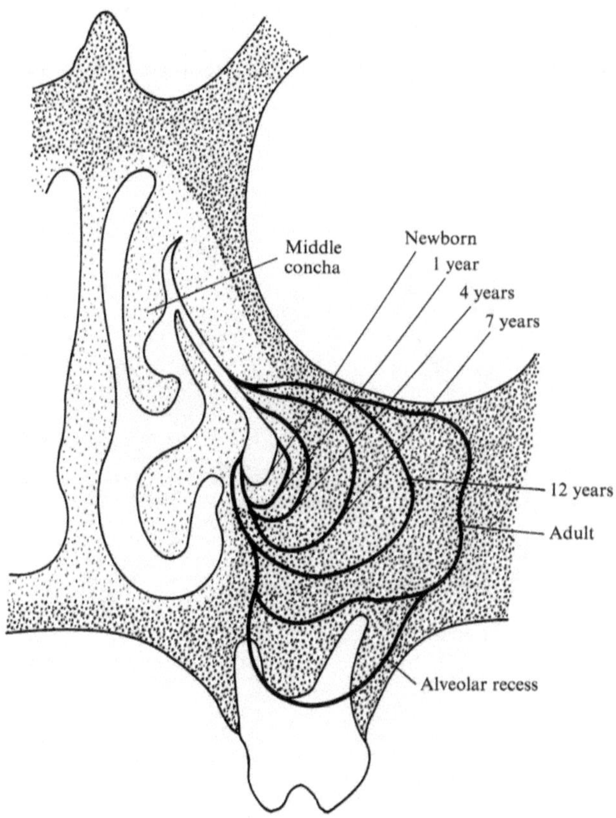

6.5 The Sphenoid Sinus in the Newborn Child

In the neonate, the body of the sphenoid bone has an anterior and a posterior ossification nucleus, which are separated by the craniopharyngeal canal. However, this bone does not have a sinus at birth. Later, its pneumatization proceeds from the ethmoid bone to Bertin's ossicle, which is situated between the sphenoid and ethmoid bones and is subsequently assimilated by the ethmoid bone [1].

In the newborn, the nasal region of the pharynx (the nasopharynx) is a relatively small structure, and has a 3–4 mm height and 15–18 mm length. The pharyngeal bursa is usually well-developed, in front of which lies the pharyngeal tonsils, or adenoids. The shape of the orifice of the Eustachian tube is a 4 mm long slit and is located on the lateral wall of the nasopharynx. Both the torus tubarius and the pharyngeal recess behind it (i.e., Rosenmüller's fossa) are well-developed. The palate has a high position, with the tubal orifice situated at the level of the hard palate. An invagination can be present on the posterior pharyngeal wall (overlying the arch of the atlas) and is *Luschka*'s pharyngeal bursa. The location of the bursa is marked by a depression in the adult. Tornwaldt [2] first described pathology of this bursa, with its closure resulting in the formation of a cyst (called *Tornwaldt*'s *cyst*).

Surgical Importance

Endoscopy and surgery of the nasopharynx, adenoidectomy.

References

1. Heiderich F (1934)
2. Tornwaldt GL (1885)

Plate adapted from Heiderich [1]

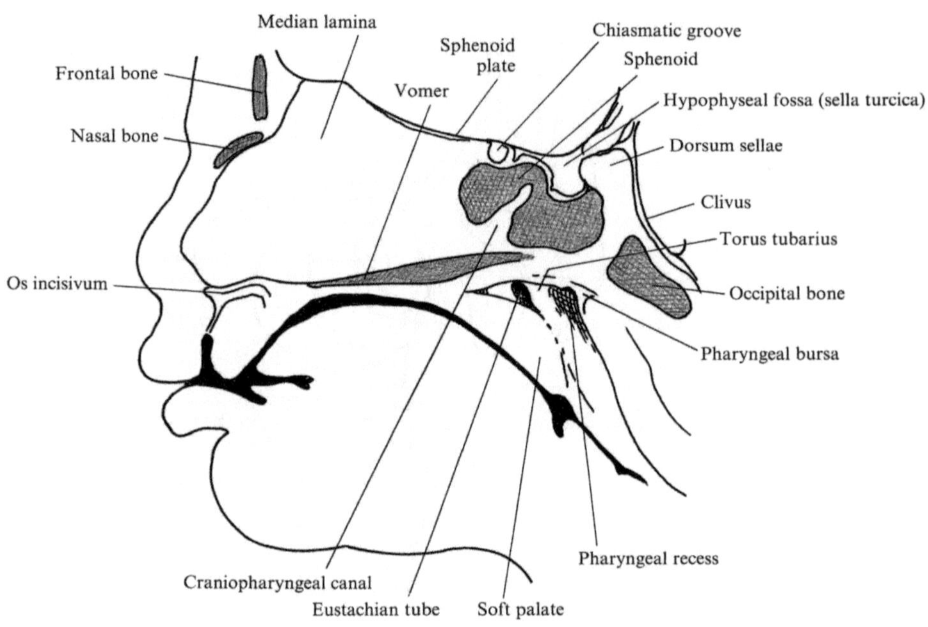

As with the other paranasal sinuses, the final dimensions of the sphenoid sinus vary from one individual to another. The increase in the size of the sinus is represented chronologically in the plate from infancy to full growth, with the numbers indicating the extension of the sinus in the corresponding year of life [1, 2].

The "paleosinus" develops anterior to what becomes the sphenoid sinus in Bertin's bones (*Bertin's concha*).

The degree of pneumatization of the sphenoid sinus is important in the transsphenoidal surgical approach to the hypophysis.

Surgical Importance

Endoscopy of the nasal cavity, surgery of the sphenoid sinus and skull base, endonasal pansinus surgery, transsphenoidal hypophysectomy, traumatology.

References

1. Krmpotić-Nemanić J (1977)
2. Peter K (1925)

Plate modified from Peter [2]

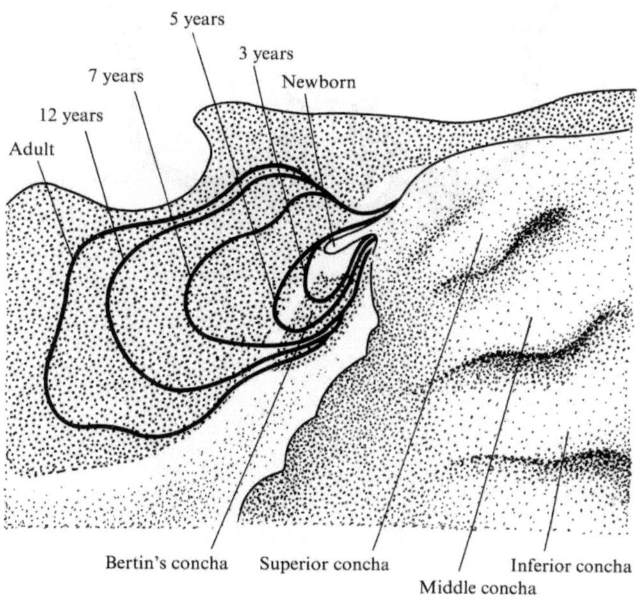

6.7 (Plates 1–2) Pneumatization of the Paranasal Sinuses

6.7 (Plate 1) Pneumatization of the Neonatal Facial Skeleton

The Plate depicts a frontal section through the nasal cavity and the paranasal sinuses in the infant (see Sect. 6.7, Plate 2). The ethmoid cells have already developed, and the anlage of the maxillary sinus appears as a small recess. The maxillary sinus reaches its full size only after the secondary teeth have completed their eruptions [1].

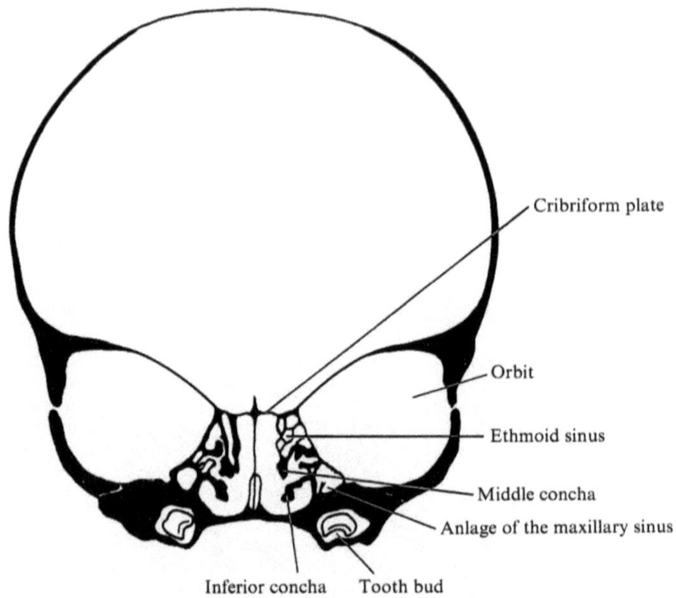

After complete pneumatization of the facial skeleton, the cribriform plate and the roof of the ethmoid sinus are situated at a lower level than are their locations in the newborn. On occasion, the cribriform plate can be situated at a much lower level than is the roof of the ethmoid labyrinth (see Sect. 6.18) [1].

Surgical Importance

Endonasal and extranasal surgery of the paranasal sinuses, surgery of the skull base and orbits.

Reference

1. Heiderich F (1934)

Plates adapted from Heiderich [1]

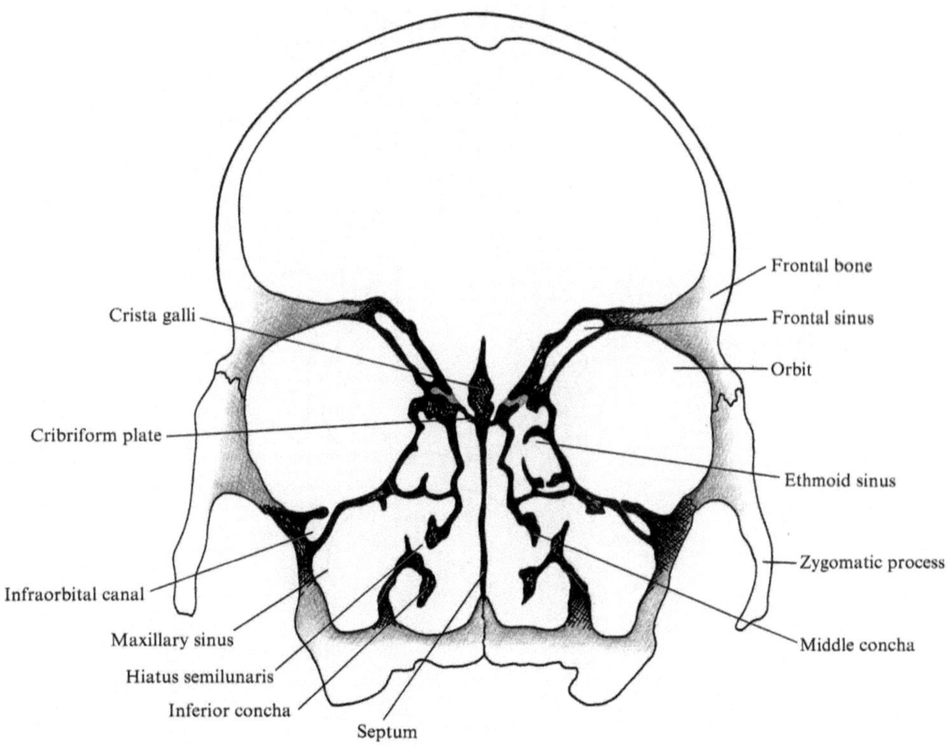

6.8 Projections of the Nasal Cavity and the Paranasal Sinuses in the Intact Facial Skeleton

As illustrated in the Plate, the projections seen are important for determining the best surgical approach to the nose and each of the paranasal sinuses.

Surgical Importance

Surgery of the paranasal sinuses and the base of the skull.

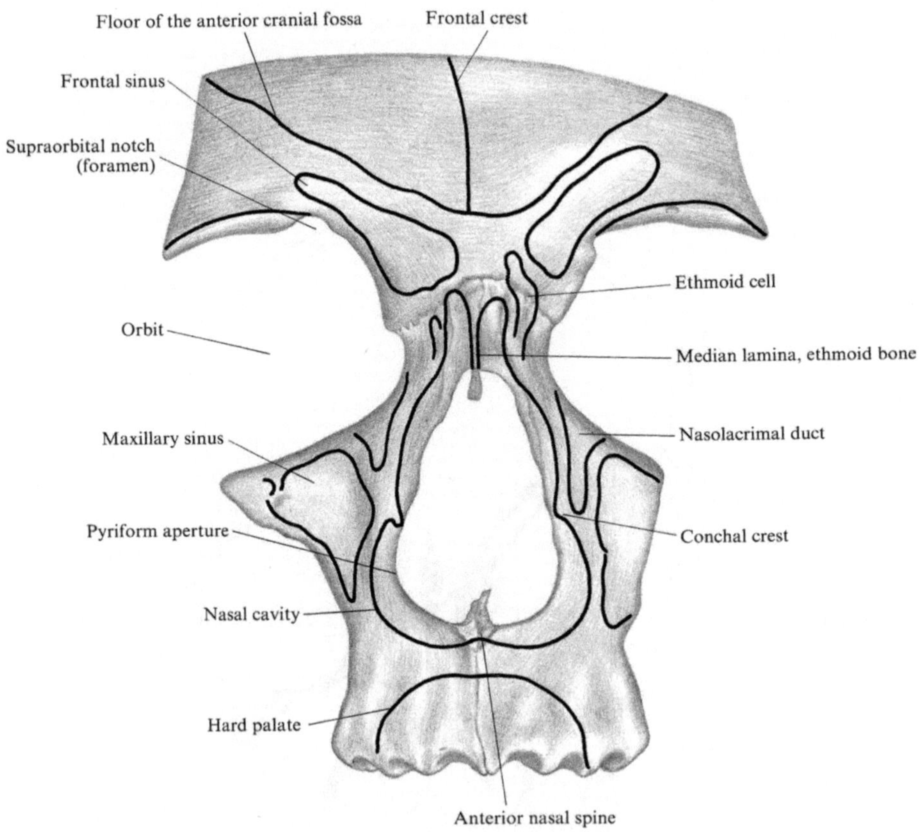

6.9 Anterior Projections of the Maxillary and Sphenoid Sinuses from Behind the Facial Bones

The various projections of the choanae, the maxillae, and the sphenoid sinuses are shown in their relationships to the facial bones, also illustrating various surgical approaches possible.

Surgical Importance

Surgery of the paranasal sinuses and skull base.

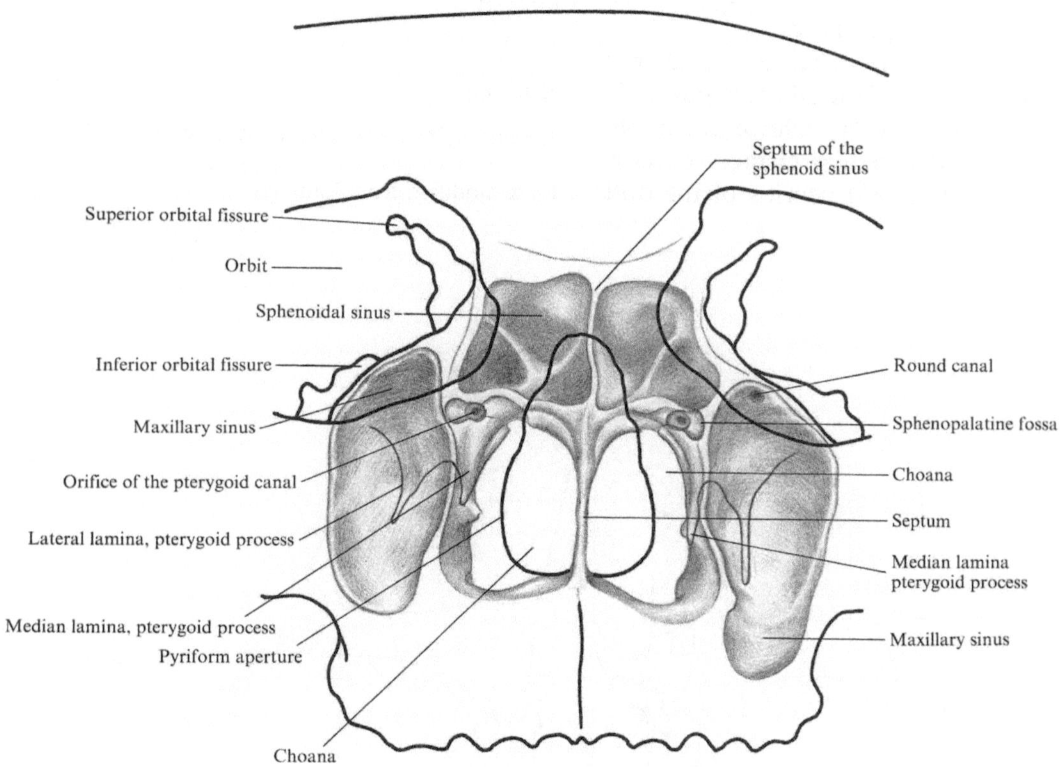

6.10 Various Projections of the Paranasal Sinuses onto the Skull Base (from the Anterior and Middle Cranial Fossae)

Variations in the sizes of the frontal, ethmoid and sphenoid sinuses are illustrated in relationship to the anterior and middle cranial fossae. Only rarely are the sinuses completely aplastic. If the frontal sinus extends into the lesser wings of the sphenoid bone and into the anterior clinoid process, the optic nerve can be surrounded by a pneumatic space and is fixed to the wall of the sphenoid by only a thin bony covering (see Sect. 6.11). In the presence of extreme pneumatization, extensions from *both* the ethmoid labyrinth and the sphenoid sinus may also surround the optic nerve.

If the optic nerve and artery are separated from the pneumatized spaces of the adjacent sinuses by only a thin osseous lamella, dehiscences in this layer may allow inflammatory changes resulting from sinusitis to produce a neuritis of the optic nerve. In such cases, there is also the possibility of injuring the optic nerve during endonasal or external surgery of the sinuses, and may occur in the presence of various neoplastic conditions (see Sects. 9.3–9.5).

Surgical Importance

Endonasal and external surgery of the paranasal sinuses, surgery of the skull base and of the hypopharynx.

Reference

1. Lang J (In: Lanz T v and Wachsmuth W 1979)

Plate modified from Lang [1]

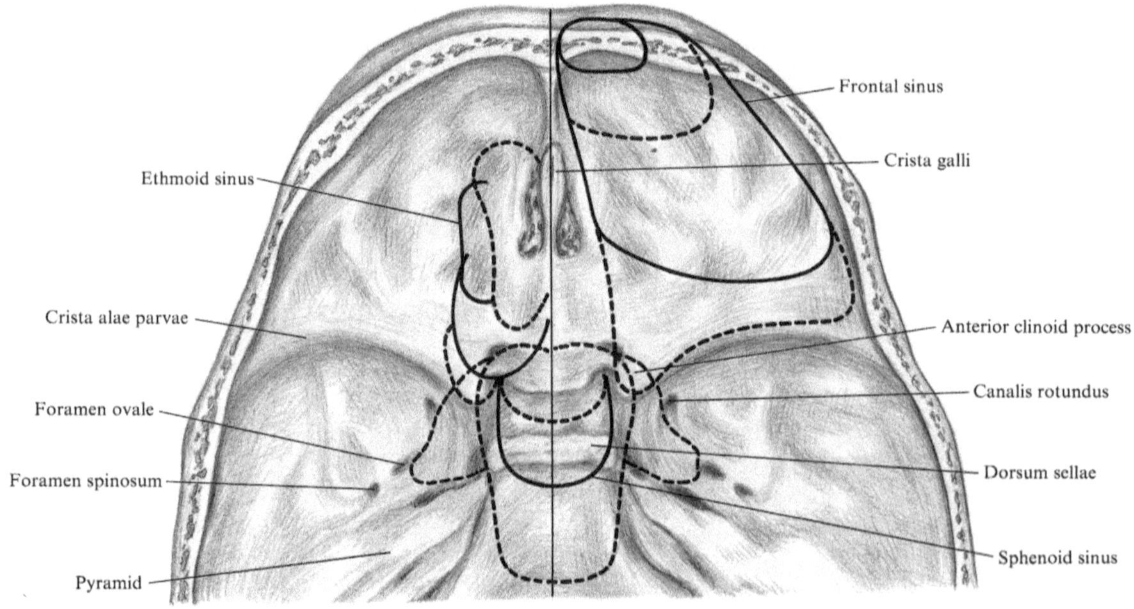

6.11 (Plates 1–2) Extensive Pneumatization of the Bone Adjacent to the Optic Nerve by an Ethmoid Air Cell

Plate 1 is the left side of a boney specimen as viewed from above and from the right. Plate 2 is a magnified view of the optic canal, as shown in the preceding Plate. In cases of extensive pneumatization of the posterior ethmoid bone, the optic nerve may be surrounded by only a thin osseous lamella that occasionally shows minute dehiscences. Pathologic processes can result in injury to the optic nerve in such cases [1, 2].

Surgical Importance

Endonasal and extranasal surgery of the paranasal sinuses, surgery of the skull base and the hypophysis

References

1. Denecke HJ (1953)
2. Krmpotić-Nemanić J (1977)

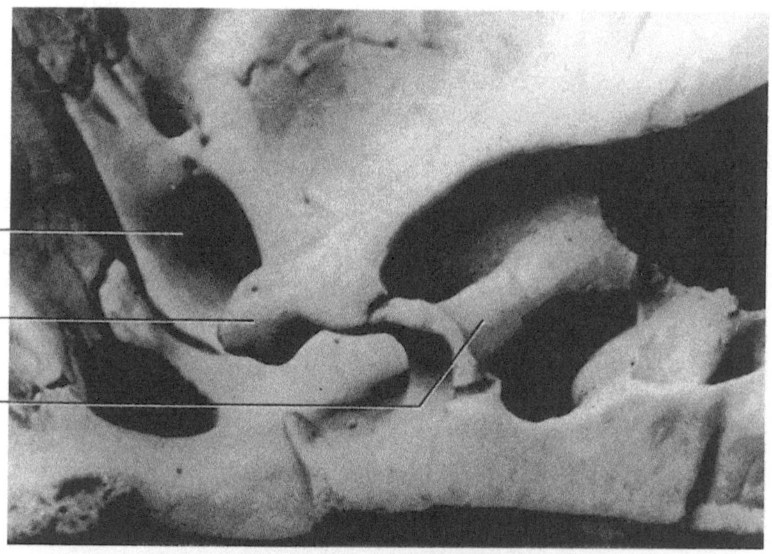

1

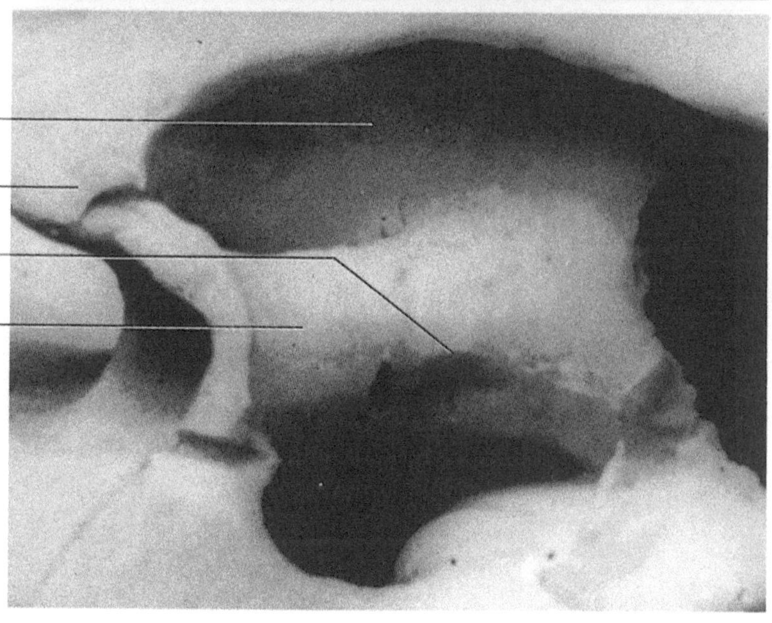

2

6.12 (Plates 1–3) The Relationship of the Sphenoid Sinus to the Sella Turcica and Clivus in the Presence of Varying Degrees of Sinus Pneumatization

The small sphenoid sinus (or "conchal" type) is depicted in Plate 1. In most cases, pneumatization of the sphenoid bone extends as far as the anterior wall of the sella turcica (Plate 2), allowing the hypophysis to be readily identified in the transsphenoidal approach to the gland. This contrasts significantly with the small sinus. When pneumatization is excessive (Plate 3), the adjacent vital structures – especially the internal carotid artery and the optic nerve – are covered either partly or completely by a thin layer of bone and may protrude into the sinus, making them prone to possible surgical injury (see Sect. 9.5).
A recess of the sphenoid sinus can extend into the clivus. Additionally, the most posterior part of the sinus can also be connected with the rest of the sinus by a narrow opening [1], and can render the diagnosis and surgical treatment of a diseased sinus difficult (see Sect. 6.13).

Surgical Importance

Transnasal, transseptal, and transethmoidal surgery of the sphenoid sinus; transsphenoidal hypophysectomy; traumatology

Reference

1. Jovanović S et al. (1954)

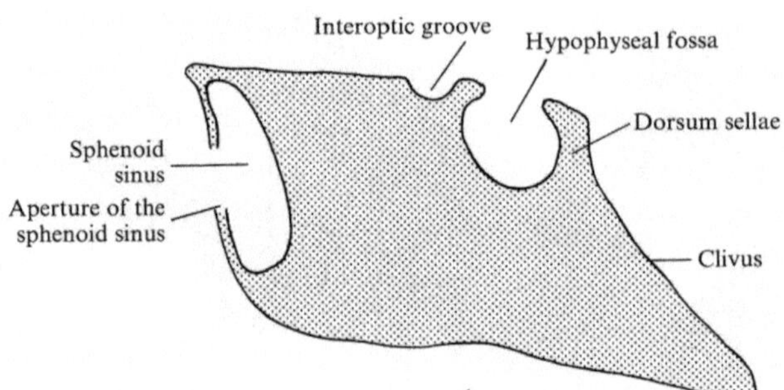

1

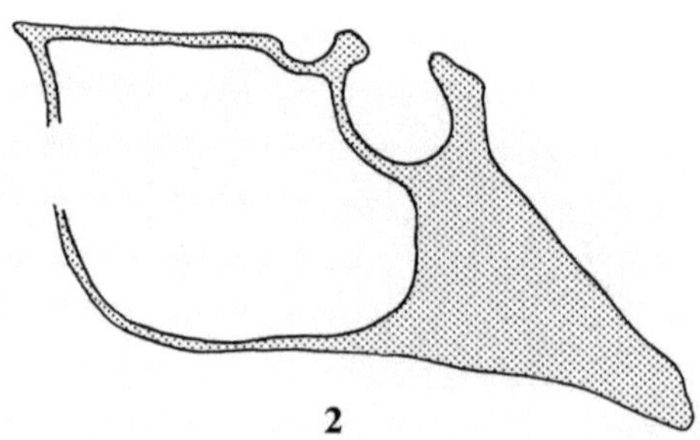

2

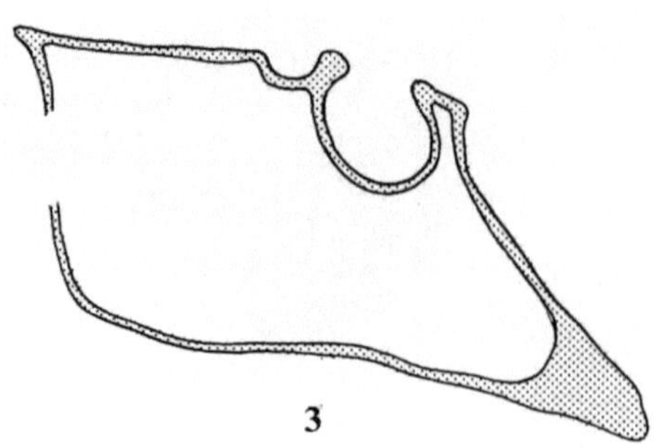

3

6.13 The Pneumatized Clivus and an Intact Intersphenoidal Septum

The orifices of the sphenoid sinus are 1.2–2.3 cm distant from the anterior wall of the sella turcica [2] (average distance, 1.5 cm). In many cases, the air space of the sphenoid sinus can encompass the entire clivus. However, the greater the degree of pneumatization, the thinner will be the bony plate separating the sinus from the posterior cranial fossa (inclusive of the brainstem and the artery [3]. According to Fujii et al. [2], the optic nerve, internal carotid artery and maxillary nerve may protrude into the lateral wall of the sphenoid sinus (Sect. 9.5). The bony plate separating these structures from the sinus can have a thickness of less than 0.5 mm. Not infrequently, these structures can be separated from the air spaces of the sinus only by the dura and the sinus mucosa. As a consequence, the vital structures present can be injured during surgery of this region.

Surgical Importance

Surgery of the sphenoid sinus and of the base skull, transsphenoidal approach to the clivus and hypophysis, aneurysms of the basilar artery and the internal carotid artery

References

1. Berendes et al. (1977)
2. Fujii L et al. (1979)
3. Krmpotić-Nemanić J (1977)

Plate from Krmpotić-Nemanić in Berendes, Link, Zöllner [1]

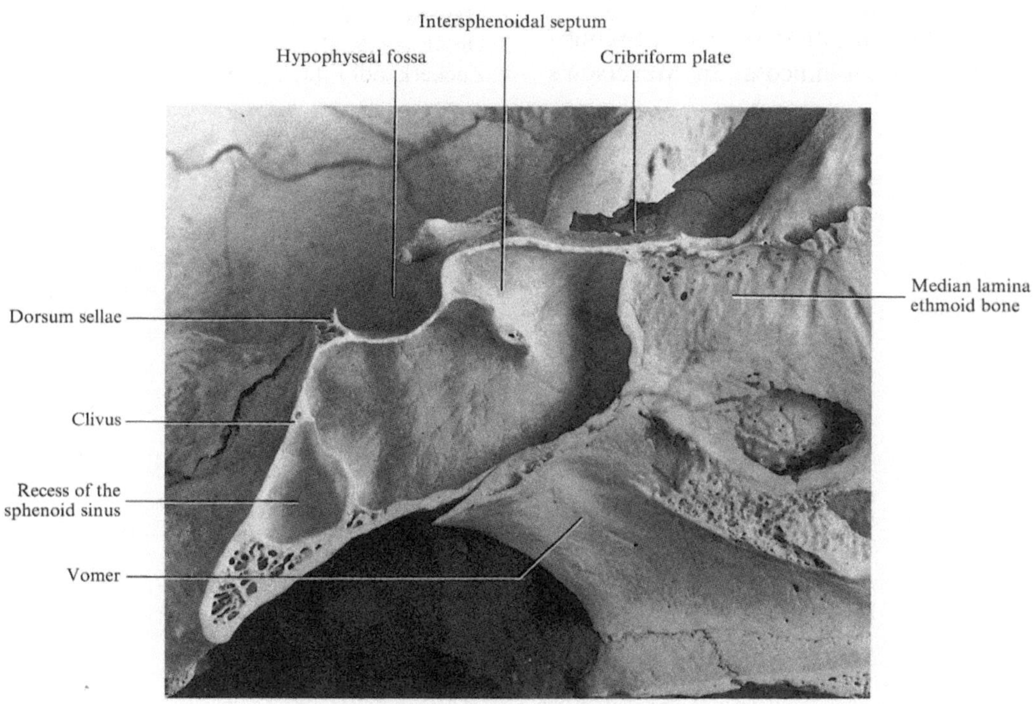

6.14 The Bony Relationships of the Hiatus Semilunaris of the Maxillary Sinus

A portion of the medial wall of the maxillary sinus has been removed. The hiatus of the maxillary sinus is covered above by the uncinate process of the ethmoid bone [1, 2]. The maxillary and the ethmoid processes of the inferior concha are located below. Posterior to the hiatus is the maxillary process of the palatine bone. The actual orifice of the maxillary sinus, hiatus semilunaris is created by the mucosa covering the adjacent bony elements and is thus a narrow lateral slit between the uncinate process and the ethmoid bulla. Lateral to the hiatus lies the ethmoid infundibulum, from which the clinically designated maxillary canal communicates above with the frontal sinus and below (medially) with the ostium of the maxillary sinus. The maxillary canal can be 3–6 mm wide and 6–8 mm long. According to Onodi [3], the duct between the maxillary sinus and the nasal cavity can vary from 3 to 19 mm in length and from 3 to 5 mm in width.

In addition to the maxillary canal and the hiatus semilunaris, 10 % of all cases show one or two accessory ostia (the so-called *fontanellae of Zuckerkandl* [4]). These latter openings are situated on the medial wall of the maxillary sinus in the reduplication of the adjacent mucosa. In Meyerson's cases [2], 30.7 % of cases showed one accessory ostium and 1/114 specimens had two accessory ostia. The diameter of these ostia have varied from 6.5 to 10.5 mm, and can be larger than the normal ostia. The locations of these ostia are generally situated behind and below the uncinate process. Paradoxically, the human maxillary ostium and hiatus semilunaris are situated poorly for allowing any spontaneous discharge of sinus secretions, in that they are located at the uppermost extension of the sinus.

A completely developed maxillary sinus can have frontal, alveolar, zygomatic, and palatine recesses (see Sect. 6.15.2).

Surgical Importance

Endoscopy of the nasal cavity and maxillary sinus, inferior meatal antral fenestration.

References

1. Draf W (1978)
2. Meyerson CM (1932)
3. Onodi A (1922)
4. Zuckerkandl E (1882)

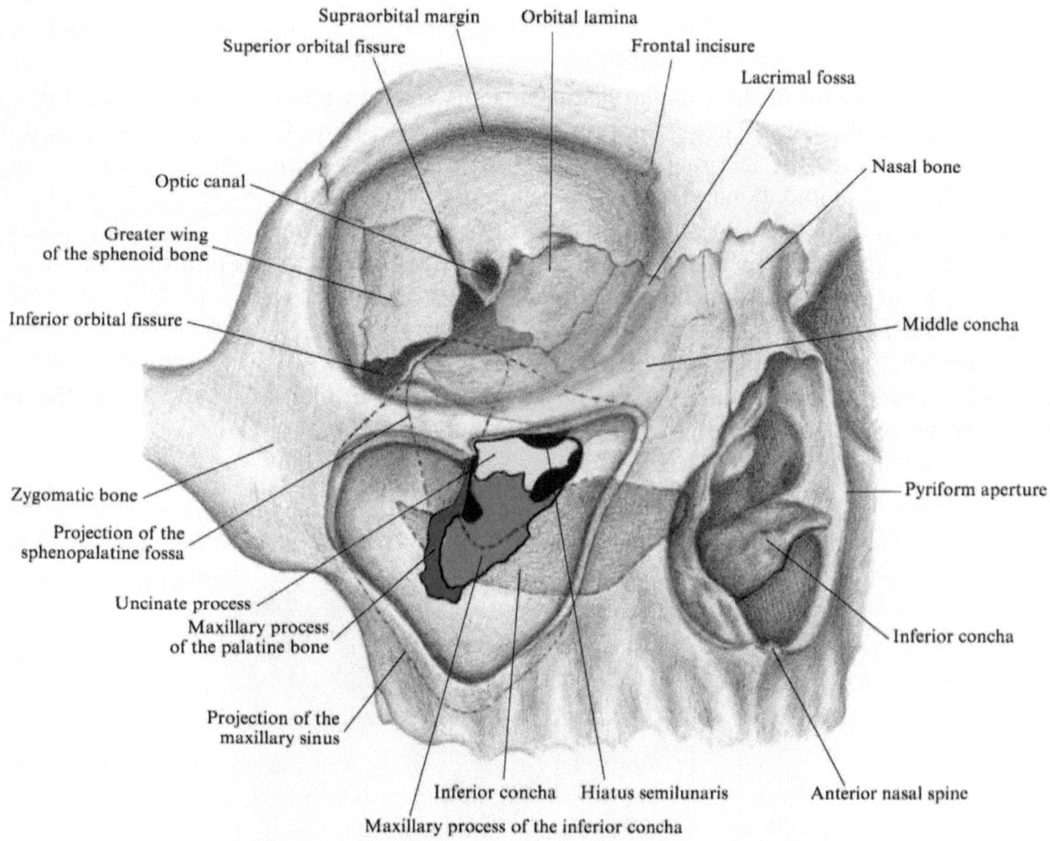

Supraorbital margin
Orbital lamina
Superior orbital fissure
Frontal incisure
Lacrimal fossa
Nasal bone
Optic canal
Greater wing
of the sphenoid bone
Inferior orbital fissure
Middle concha
Zygomatic bone
Pyriform aperture
Projection of the
sphenopalatine fossa
Uncinate process
Maxillary process
of the palatine bone
Inferior concha
Projection of the
maxillary sinus
Inferior concha
Hiatus semilunaris
Anterior nasal spine
Maxillary process of the inferior concha

6.15 (Plates 1–5) Internal Septa of the Maxillary Sinus and Their Possible Variations

6.15 (Plate 1) Partial Perpendicular Septation of the Maxillary Sinus (Medial View)

The normal anatomic hiatus of the maxillary sinus can assume different shapes. Its upper margin is in contact anteriorly with the lacrimal bone and posteriorly with the ethmoid bone. The *punctum convergii* is located in the latter part of the convergence of these two parts of the margin and is illustrated in the specimen (taken from the collection of D. Perović of the Department of Anatomy, University of Zagreb). From this punctum, a partial – more or less perpendicular – septum can develop into the maxillary sinus, somewhat dividing the sinus into an anterior and a posterior compartment. A septum of this type may cause difficulties in orientation during surgical exploration of the sinus. Additionally, Water's (antero-posterior) X-ray views of the maxillary sinus may falsely suggest that this formation is a pathologic condition.

Surgical Importance

Preoperative diagnosis, surgery of the maxillary sinus, antral puncture.

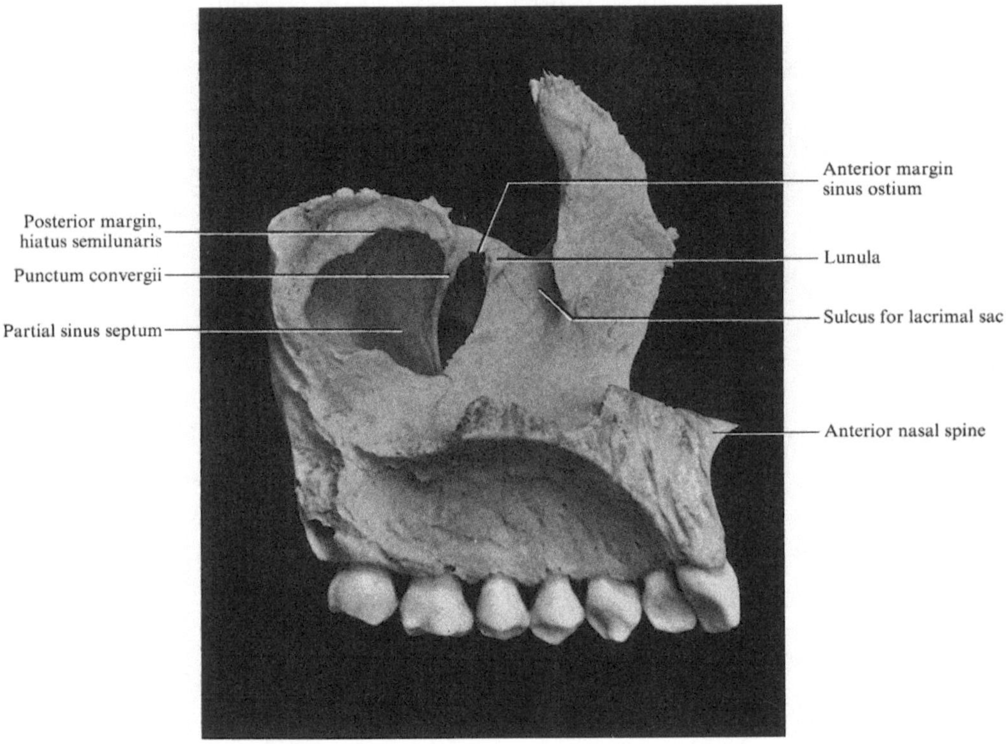

6.15 (Plate 2) The Basal Perpendicular Maxillary Septa

Underwood's septa are osseous ridges delimiting the root compartments of the premolar and molar teeth [1], and can also lead to false interpretations during endoscopy of the maxillary sinus. The roots of the premolar and molar teeth can occasionally protrude into the maxillary sinus, such that pathologic conditions of the teeth may also involve the sinus (see Sect. 6.4). During dental extractions, the maxillary sinus can be entered (to cause sinusitis), or the dental root may break free and slip into the sinus.

Surgical Importance

Endoscopy and surgery of the maxillary sinus, (e.g., Caldwell-Luc procedures), (upper) dental disease and therapy.

Reference

1. Underwood AS (1910)

Plate from Collection of D. Perović

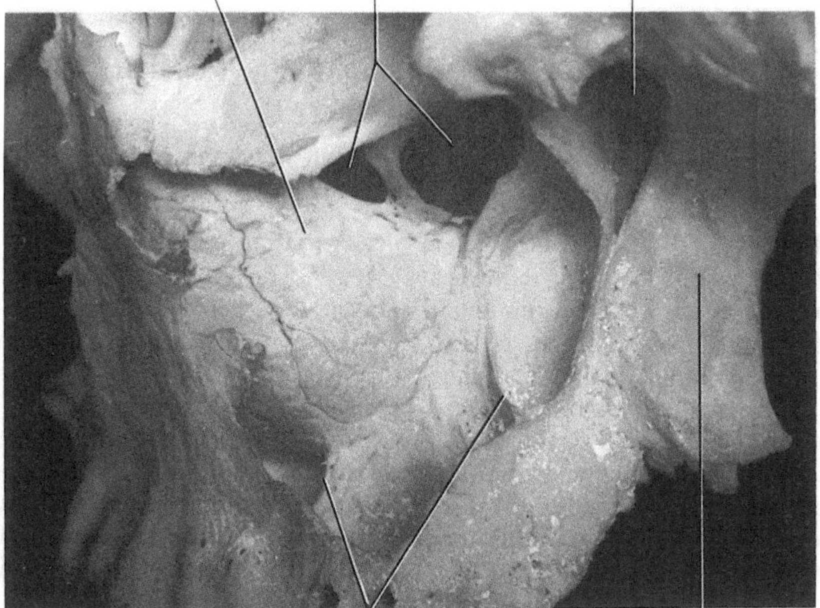

6.15 (Plates 3–4) Partial Horizontal Septa

The third type of a possible partial septa in the maxillary sinus is a horizontally-located maxillary process from the palatine bone and the inferior concha [1]. If these occur, neither process will follow the normal development of the sinus in its inferomedial direction [see 6.15 (Plate 3)]. While the floor of the sinus is gradually lowered as development progresses, both processes will maintain their primary horizontal positions [see Sect. 6.15 (Plate 4)]. Such horizontal septa can cause difficulties in diagnosis, as well as impede sinus drainage during planned antral punctures.

Surgical Importance

Antral endoscopy, puncture, and surgery.

Reference

1. Nemanić G (1968a,b)

6.15 (Plate 5) Complete Septation of the Maxillary Sinus (Left Lateral View)

The specimen depicted in Plate 5 has been taken from the Perović collection. A complete vertical septum in the maxillary sinus will create an anterior and a posterior sinus space [1, 2]. In general, the larger part is the anterior, while the smaller is the posterior (or the accessory maxillary sinus). Complete septation of the maxillary sinus manifests itself in two ways. In the Schlungbaum type [3], the two sinuses lie in an anteroposterior direction. The anterior sinus opens into the middle meatus, while the posterior sinus opens into the superior nasal meatus through a bony hiatus which is delimited by structures similar to those of the normal hiatus semilunaris (i.e., the maxillary process of the middle concha and the posterior maxillary process of the palatine bone). This anatomic association indicates that this accessory sinus develops in the cartilaginous nasal capsule at the same time as does the normal sinus. Either of the "two" maxillary sinuses can show isolated signs of pathology, and thus must be inspected, irrigated or treated separately from each other. This anatomical variation can be seen *only* on a lateral X-ray view in which the septum is tangential to the X-ray beam. If surgery of the maxillary sinus reveals the sinus cavity to be very shallow in an antero-posterior direction, the possibility of an accessory sinus space should be considered.

A second type of accessory sinus results from excessive pneumatization of the posterior part of the maxilla by a large ethmoid air cell [1]. In this case, a pathologic process involving the ethmoid bone will also involve this accessory sinus.

Surgical Importance

Endoscopy of the nasal cavity and maxillary sinus, endonasal and external surgery of the maxillary sinus and the ethmoid bone.

References

1. Perović D (1954)
2. Schlungbaum (1921)

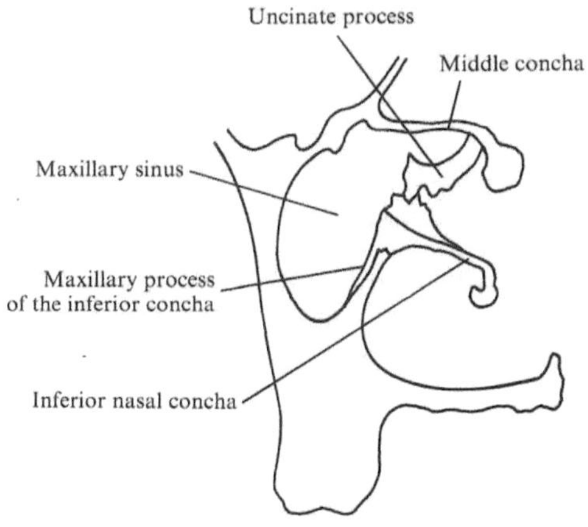
Uncinate process
Middle concha
Maxillary sinus
Maxillary process of the inferior concha
Inferior nasal concha

3

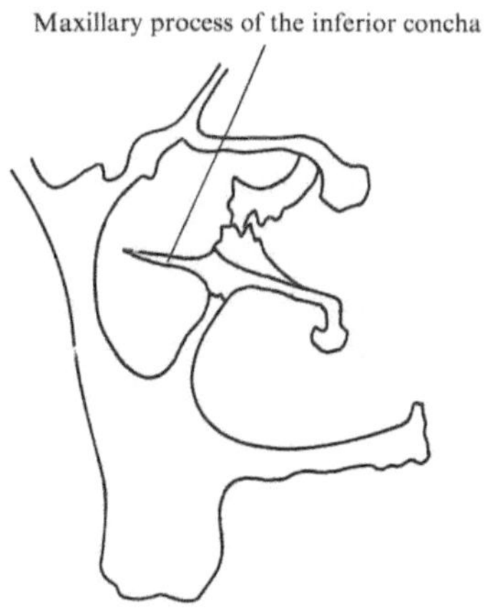
Maxillary process of the inferior concha

4

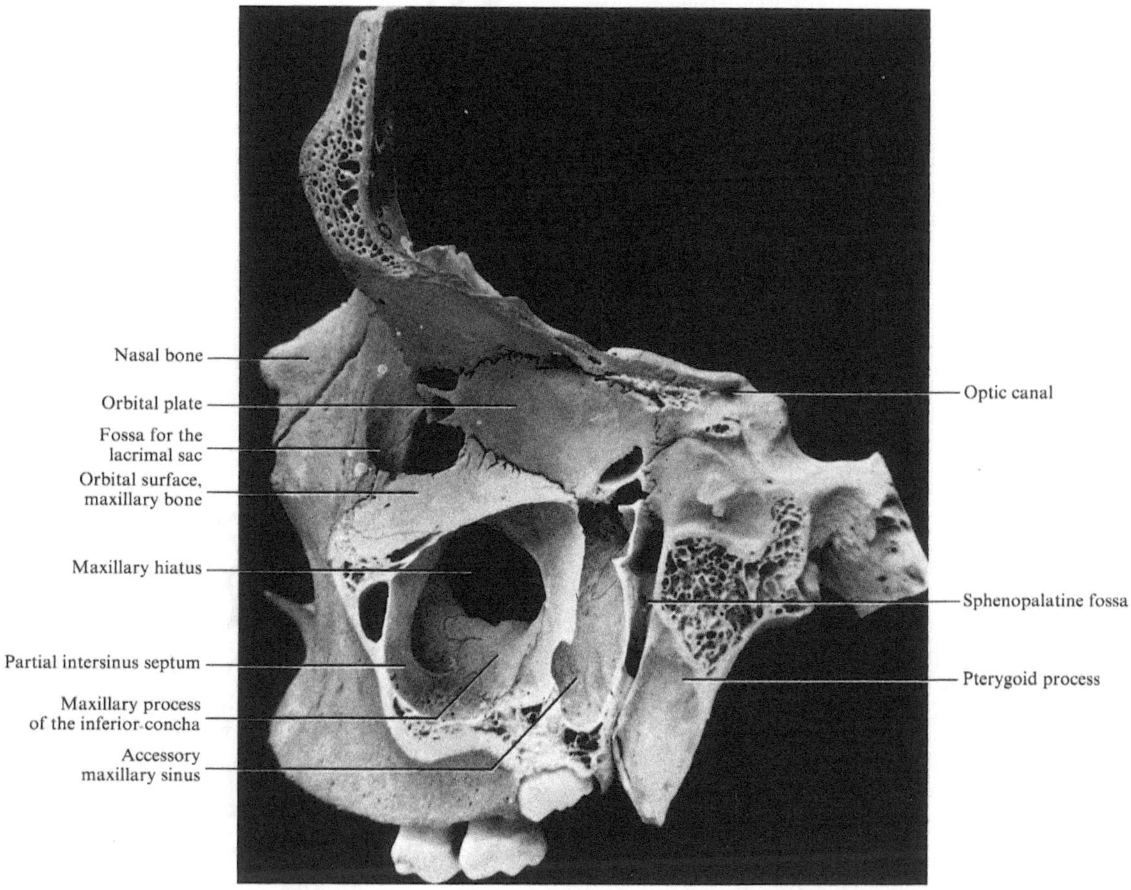
Nasal bone
Orbital plate
Fossa for the lacrimal sac
Orbital surface, maxillary bone
Maxillary hiatus
Partial intersinus septum
Maxillary process of the inferior concha
Accessory maxillary sinus
Optic canal
Sphenopalatine fossa
Pterygoid process

5

6.16 (Plates 1–3) Haller's Cells and the Transmaxillary Approach to the Ethmoid Bone

Haller's cells are large ethmoidal cells (Plate 6.17), and represent a point of access between the inferior part of the ethmoid bone and the posterosuperior part of the nasal surface of the maxilla behind and above the hiatus of the maxillary sinus (Plate 1) [2–5]. These cells can facilitate doing transmaxillary surgery of the ethmoid bone. However, if these cells are missing, the contact between the maxilla and the ethmoid is reduced to the bone joining the medial and lateral walls of the ethmoid labyrinth (Plate 2).
In extreme cases, the medial wall of the sinus can be displaced laterally, increasing the possibility of injury to the orbital contents during transmaxillary ethmoidectomy (Plate 3).
These relations can be seen on plain view or tomographic X-ray pictures of the sinuses.

Surgical Importance

Transmaxillary and endonasal ethmoidectomy, transmaxillary and transethmoidal decompressions of the orbit in endocrine ophthalmopathy.

References

1. Berendes, Link, Zöllner (1977)
2. Draf W (1982)
3. Krmpotić-Nemanić J (1977)
4. Perović D (1954)
5. Rudež V (1960)
6. Haller A v (1749)

Plates 1–3 by Krmpotić from Berendes, Link, Zöllner [1]

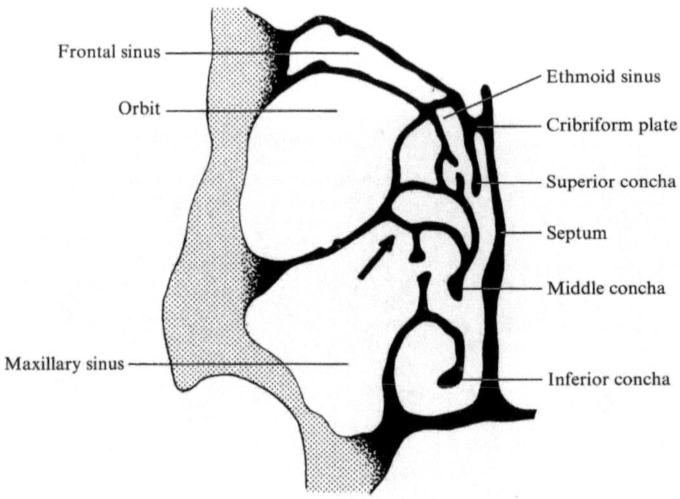

1

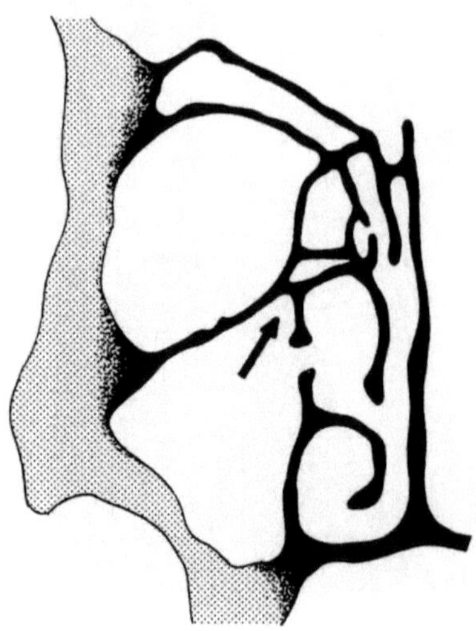

2

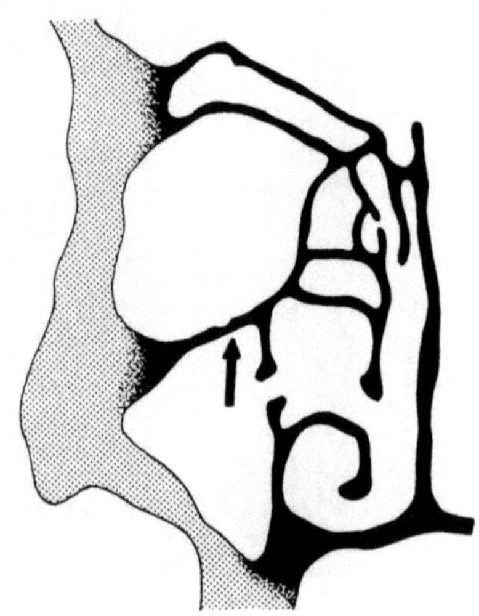

3

6.17 So-Called Haller's Cells in the Prepared Ethmoid Bone

The Plate depicts the floor of the ethmoid labyrinth and typical Haller's cells in a specimen from the Perović collection (see text in Sect. 6.16).

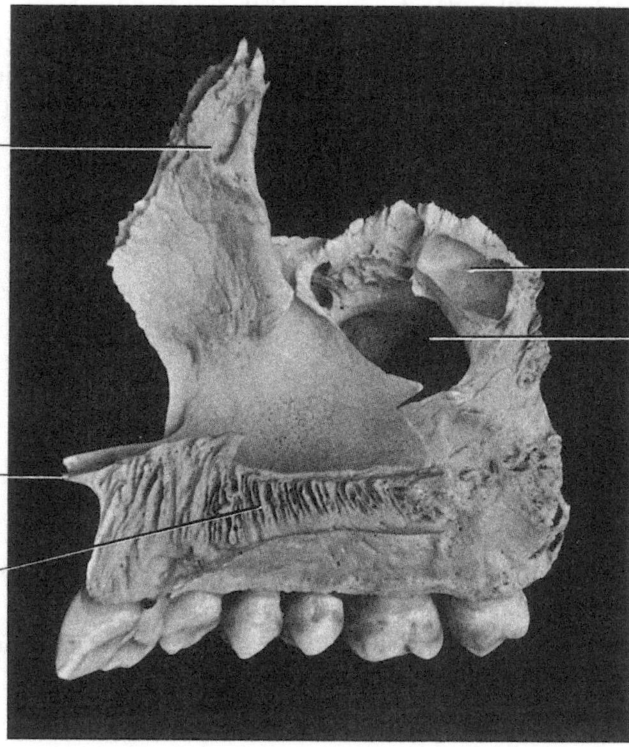

6.18 (Plates 1–2) Variations in the Relationship of the Lamina Cribrosa to the Roof of the Ethmoid Labyrinth

The cribriform plate always lies deeper than the roof of the ethmoid labyrinth and should be realized when doing surgery of the ethmoid bone. In order to avoid injuries to the cribriform plate, surgical instruments should *always* be kept lateral to the insertion of the middle concha. Normally, the anterior roof of the ethmoid labyrinth is formed primarily by the frontal bone (Plate 1). As previously described [2–3], the deeper the location of the cribriform plate, the larger will be the thin bony plate ascending from it toward the frontal bone (Plate 2).

When the cribriform plate is situated extremely deeply (ca. 20 % of cases), there is the possibility of injury to the endocranium even if the surgeon keeps his dissecting instruments lateral to the insertion of the middle concha. This is because the thin bony plate present is usually large in such cases, forming the medial part of the roof of the ethmoid labyrinth. Consequently, the use of an operating microscope and endoscope can be of great help in doing endonasal surgery of the ethmoid bone.

According to the statistics of Onishi [4] in Japan, lesions of the base of the skull were found in 0.15 % of 450000 cases undergoing ethmoidectomies, while acute meningitis occurred in 0.01 % of these cases. Systematic examinations of 34 anatomic specimens revealed that bony dehiscences were located principally in five sites: (a) the medial wall of the ethmoid bone, (b) adjacent to the bony orifice for the anterior ethmoid nerve, (c) the bony insertion of the medial concha, (d) the anterolateral part of the ethmoid roof, and (e) in the region of the foramen for the posterior ethmoid nerve.

If the median lamina or perpendicular plate of the ethmoid is fractured during septal surgery, the cribriform plate can be injured, resulting in possible hyposmia (with injury to the olfactory nerve) or a CSF rhinorrhea.

Surgical Importance

Mid-facial trauma; surgery of the ethmoid and nasal septum.

References

1. Berendes, Link, Zöllner (1977)
2. Keros P (1962)
3. Krmpotić-Nemanić J et al. (1972)
4. Onishi T (1981)

Plates 1–2 by Krmpotić from Berendes, Link, Zöllner [1]

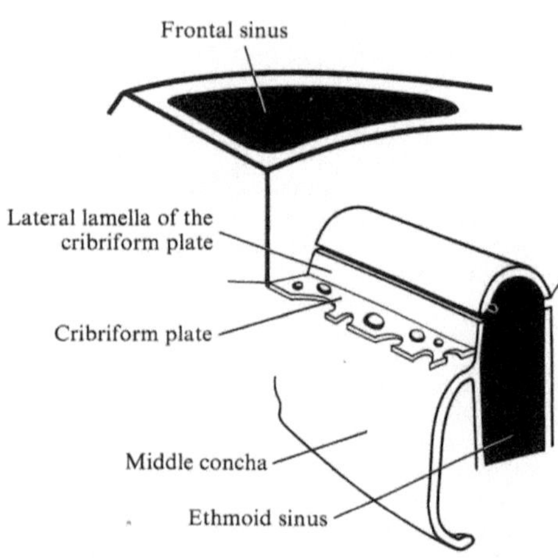

1

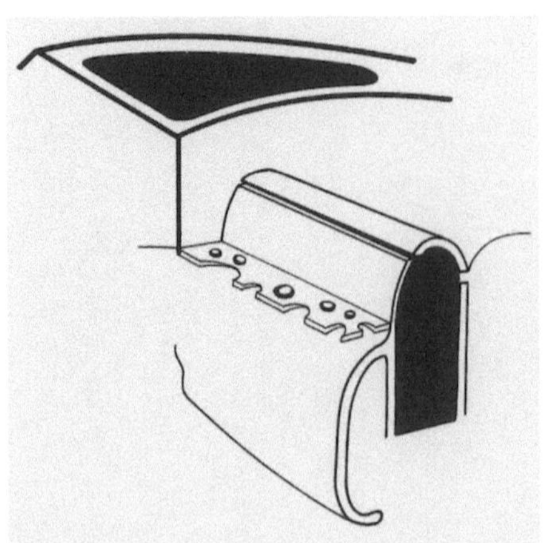

2

6.19 Topographic Relationships of the Frontal Sinus to the Ethmoid Labyrinth and the Middle Nasal Meatus (Medial View)

The frontal sinus may consist of one or more compartments, depending on whether or not the source of its pneumatization comprises one or more ethmoid cells [1, 3, 4]. If the frontal sinus is reduplicated anatomically, each sinus present will have a corresponding nasofrontal duct which can be identified endoscopically. According to Vogt and Schrade [7], there is a single communication between the frontal sinus and the nasal cavity in only one-third of cases.

The funnel-shaped orifice of the nasofrontal duct in the floor of the sinus has been designated the "*frontal infundibulum*" by Killian [2] (see Sect. 5.12). It should not be confused with the ethmoid infundibulum in the middle meatus between the uncinate process and the ethmoid bulla, since this latter structure is the caudal orifice of the nasofrontal duct [6]. Usually, the nasal part of the frontal bone is not pneumatized by one ethmoid cell (*Boyer's cell*) but by several cells called *frontal bullae*, which result in a tortuous nasofrontal duct. These bullae have to be removed in operations of the sinus together with the medial sinus floor in order to improve drainage [3, 4, 5].

Surgical Importance

Sinus endoscopy, endonasal and external surgery of the frontoethmoid sinuses, revision skull base surgery.

References

1. Draf W (1978, 1983)
2. Killian G (1903b)
3. Krmpotić-Nemanić J (1977)
4. Nikolić V (1963)
5. Sunarić R (1963)
6. Terracol JP et al. (1965)
7. Vogt K and Schrade F (1979)

Plate adapted from Nikolić [4]

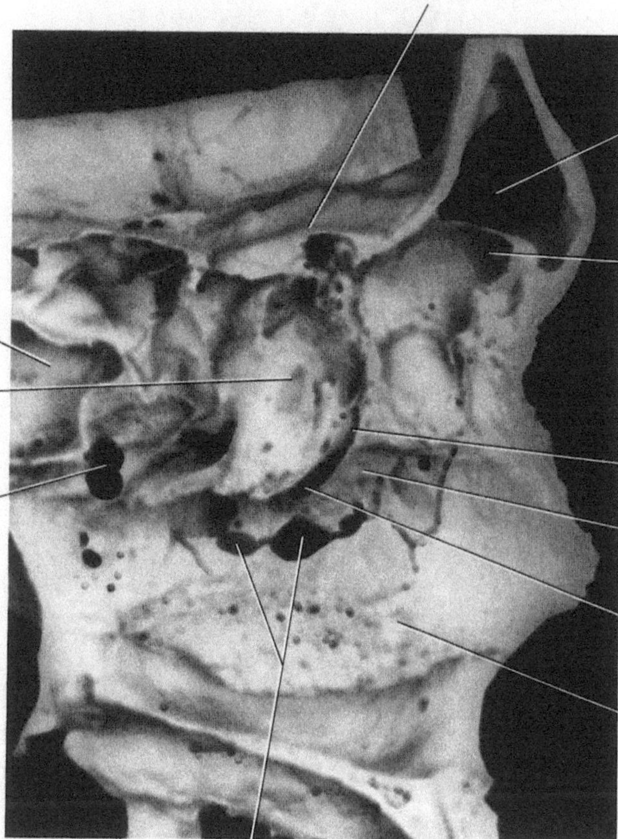

6.20 (Plates 1–3) The Surgically "Dangerous" Frontal Bone

When the interfrontal septum (*septum sinuum frontalium*) is asymmetrical, the olfactory crest can protrude as a bony ridge into the sinus. If this crest is removed during surgery, the olfactory groove and hence the endocranium are exposed, as illustrated in Plate 1.
A transverse section with normal relationships of the interfrontal septum, the floor of the sinus, and the crista galli are illustrated in Plate 2.
In the surgically "dangerous" frontal bone (Plate 3), the crista galli shows varying degrees of pneumatization. The interfrontal septum is situated laterally so that the olfactory groove protrudes as a ridge into the frontal sinus.

Surgical Importance

Traumatology, frontal sinusotomy, frontal sinus obliteration.

Referenes

1. Boenninghaus G (1923)
2. Denecke HJ (1953)

Plates 1–3 adapted from Boenninghaus [1] and Denecke [2]

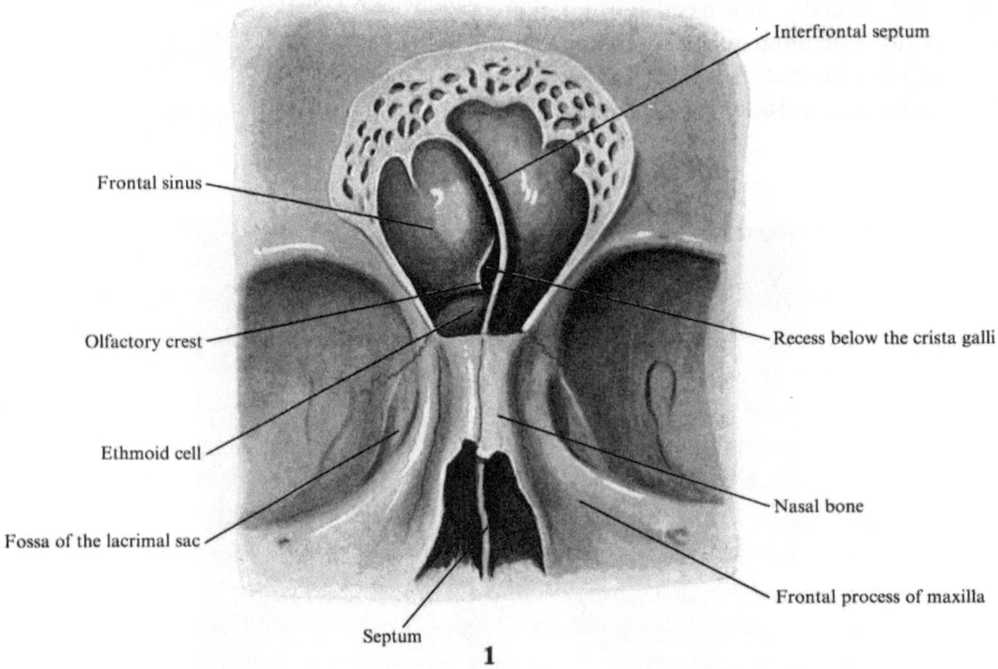

1

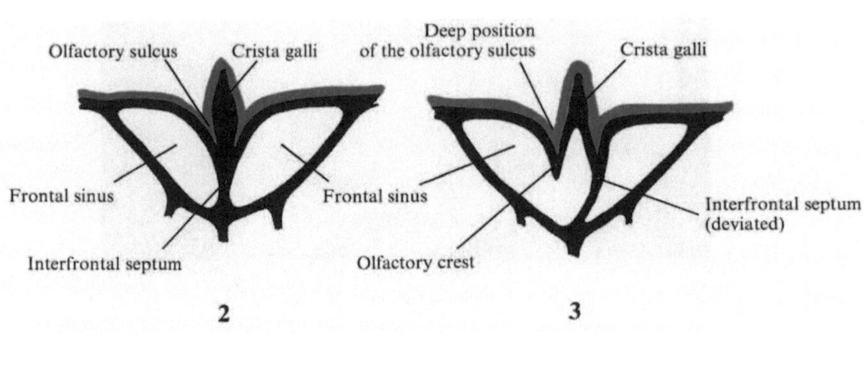

2 3

6.21 The Sphenopalatine Fossa (*Fossa Pterygopalatina*) and the Infratemporal Fossa (Lateral View)

After removal of the zygomatic arch, the ramus of the mandible, and part of the corresponding muscles of mastication, the principal vessels and nerves of the sphenopalatine and infratemporal fossae are exposed.

The maxillary artery is one of the terminal branches of the external carotid artery, and runs lateral or medial to the lateral pterygoid muscle across the infratemporal fossa [1]. At this latter site it gives off a number of branches, as discussed elsewhere [2, 5]. In the sphenopalatine fossa between the maxillary tuberosity and the pterygoid process, the maxillary artery gives off its terminal branches (see Sect. 6.23).

The location of the sphenopalatine foramen can vary considerably in its anatomic relationship to the middle concha [3, 4] (not represented, see Sect. 5.14).

Surgical Importance

Resections of the vidian nerve and the sphenopalatine ganglion, lateral approaches to the anterior and medial skull base, surgery of the retromaxillary space [5] and sphenopalatine fossa.

References

1. Denecke HJ (1953)
2. Lautenschläger A (1934)
3. Nikolić V (1967)
4. Stamm WK (1981)
5. Zehm S (1966, 1970)

Plate adapted from Lautenschläger [2]

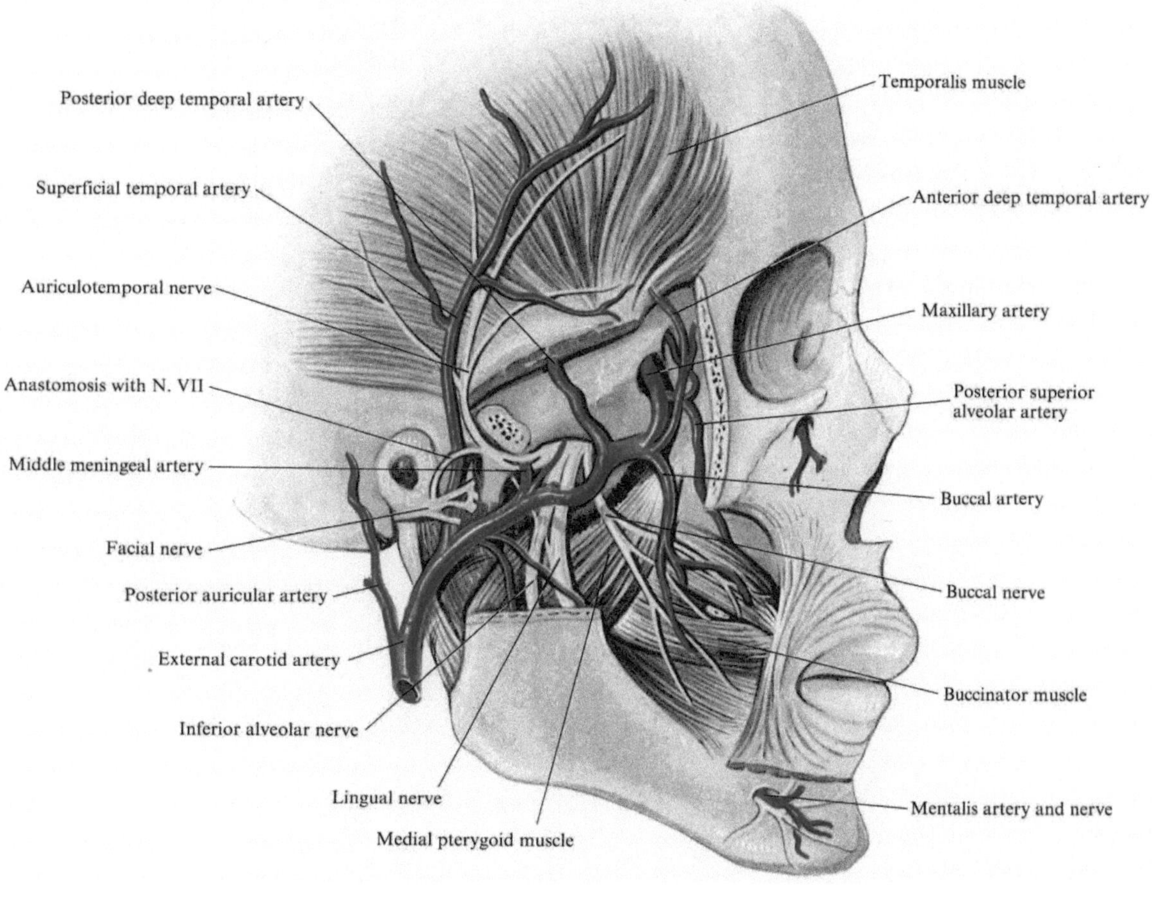

6.22 (Plates 1–4) Topography of the Sphenopalatine Fossa

6.22 (Plate 1) Projection of the Sphenopalatine Fossa Deep to the Zygomatic Bone and the Ascending Ramus of the Mandible (Lateral View)

The sphenopalatine fossa is a narrow space that is situated between the posterior wall of the maxillary sinus and the pterygoid process. Dorsal and lateral to it lies the infratemporal fossa. The latter fossa is bounded cranially by the infratemporal crest, medially by the lateral pterygoid process and laterally by the zygomatic arch. The pterygoid fossa is situated between the two laminae of the pterygoid processes (see Sect. 6.22.2). The sphenopalatine, infratemporal, and pterygoid fossae are also referred to as the retromaxillary space [1]. The posterior limit of this space comprises (from medial to lateral) the foramen spinosum, the styloid process, and the glenoid fossa [see Sect. 6.22 (Plate 2)].

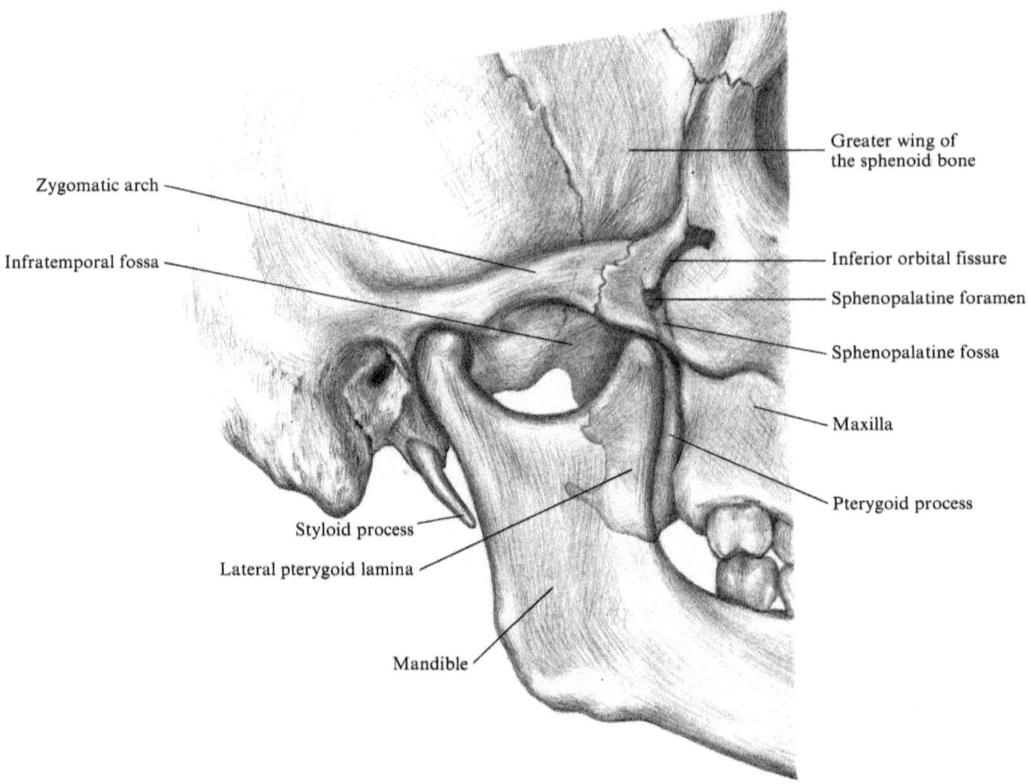

6.22 (Plate 2) The Sphenopalatine Fossa (*Fossa Pterygopalatina*) After Removal of the Mandible (Lateral View)

The medial boundary of the sphenopalatine fossa is the palatine bone. The fossa communicates with the orbit through the inferior orbital fissure and with the nasal cavity through the sphenopalatine foramen. The sphenopalatine ganglion and the terminal branches of the maxillary artery are situated in its superior part. Lateral to the sphenopalatine foramen in the sphenopalatine fossa are the orifices of the rotundus and pterygoid canals. The foramen rotundum (*canalis rotundus*) lies lateral and superior to the pterygoid canal (*canalis pterygoideus Vidii*).

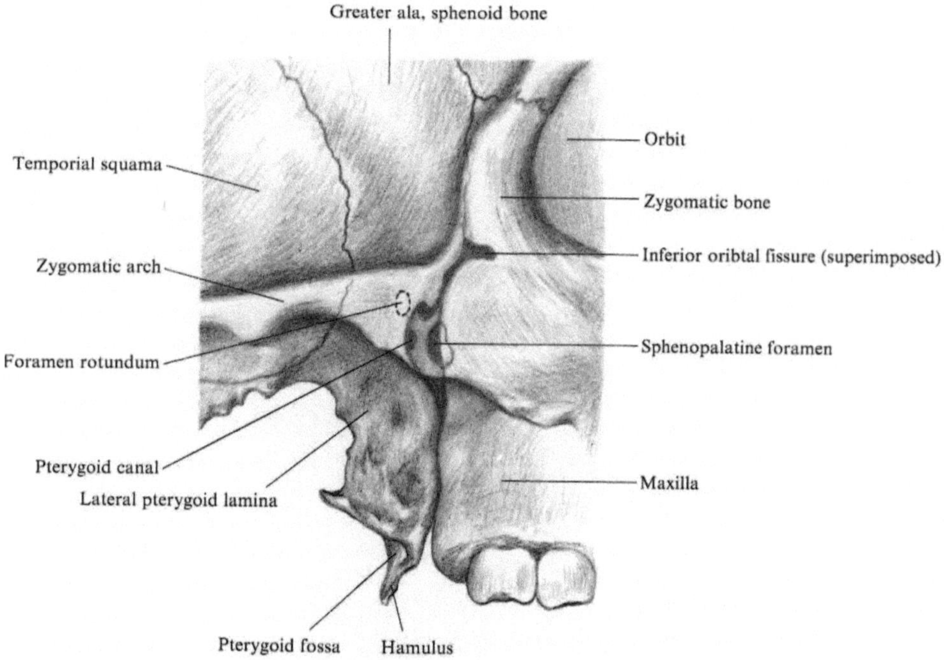

6.22 (Plate 3) The Sphenopalatine Fossa After Removal of the Zygomatic Bone, Greater Ala of the Sphenoid Bone, Zygomatic Arch, and Temporal Squama

The sphenopalatine foramen is bounded in front, below, and behind by the palatine bone (and the sphenopalatine incisure) and above by the sphenoid bone. The sphenopalatine artery and nasal nerves from the sphenopalatine ganglion pass through the foramen to the nasal cavity. The descending palatine artery and the greater palatine nerve traverse the greater palatine canal, while the lesser palatine nerves pass through the smaller palatine canals which extend from the sphenopalatine fossa to the hard palate. The pterygoid canal contains the nerve of the pterygoid canal and the pterygoid artery, and opens posteriorly into the sphenopalatine fossa. The nerve of the pterygoid canal consists of preganglionic parasympathetic fibers of the sphenopalatine ganglion and postganglionic sympathetic fibers which pass through the ganglion only (see Sect. 5.14, 6.21).

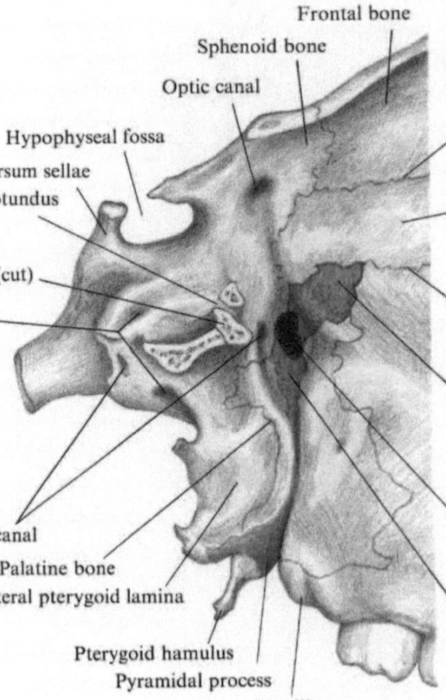

6.22 (Plate 4) Horizontal Section of the Sphenopalatine Fossa (Superior View)

The bony limits of the sphenopalatine fossa are clearly represented in Plate 4. As the result of excessive pneumatization, the sphenoid sinus can extend into the pterygoid process. The sphenopalatine fossa communicates laterally with the infratemporal fossa. This latter fossa is bounded anteriorly by the lateral wall of the maxillary sinus, laterally by the zygomatic arch and the coronoid process of the mandible (insertion point for the temporalis muscle), medially by the lateral pterygoid process, and superomedially by the inferior surface of the greater wing of the sphenoid bone. It contains the lateral pterygoid muscle, the maxillary artery, branches of the third division of the trigeminal nerve (including the inferior alveolar nerve and the lingual nerve), the middle meningeal artery, the auriculotemporal nerve (see Sect. 6.21), and the temporalis muscle. The infratemporal fossa is posteriorly partly delimited by the articular tubercle of the temporomandibular joint.

Surgical Importance

Surgery of the parotid gland, exposure of the retromaxillary space, transmaxillary approach to the sphenopalatine fossa, transantral ligature of the maxillary artery, Vidian neurectomy, lateral approaches to the orbits and to the anterior and medial skull base, traumatology.

Reference

1. Zehm (1966, 1970)

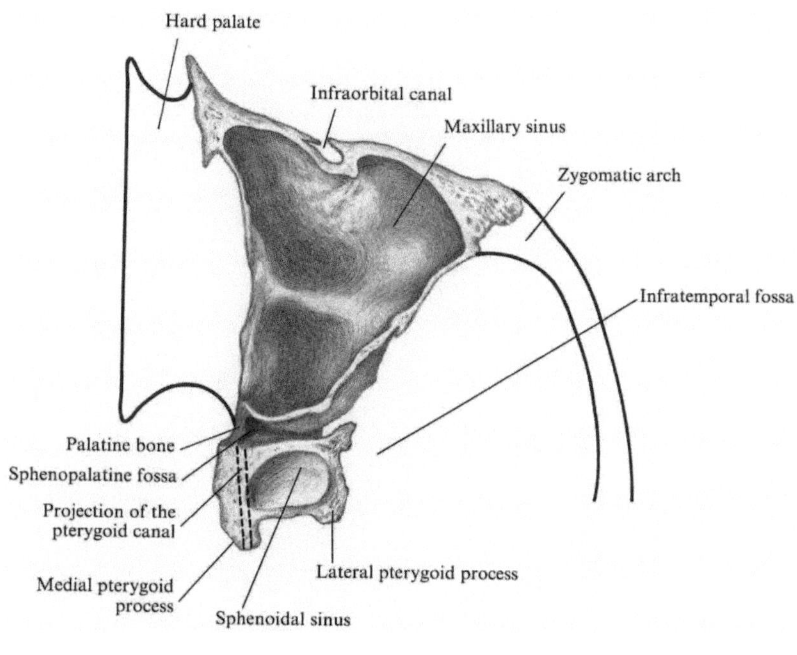

6.23 (Plates 1–6) Variations in Arteries and Nerves of the Infratemporal Fossa

As indicated in Plates 1–6, the location of the maxillary artery varies not only in its relationship to the muscles of mastication but also to the branches of the mandibular nerve [2].

In 66 % of cases, the maxillary artery lies superficially, i.e. lateral to the lateral pterygoid muscle (Plate 1) and lateral or medial to the buccal nerve (Plate 2) [3]. Less commonly, the artery lies medial to the lateral pterygoid muscle (Plate 3). In such instances, it can lie medial to the lingual and inferior alveolar nerves (Plate 4) or pass between the two nerves (Plate 5), in which case the inferior alveolar nerve crosses the artery from its lateral side. In the event of the latter occurrence, the artery can also traverse the inferior alveolar nerve (Plate 6).

Since the two parts of the lateral pterygoid muscle have an antagonistic action, Juniper [1] has suggested that the superior part of the muscle should be designated as the "superior pterygoid muscle."

Surgical Importance

Transantral surgery, radical maxillectomy, lateral approaches to the infratemporal and sphenopalatine fossae, anterior and medial portions of the skull base (including lesions of the naso- and oropharynx); dorsocaudal approach to the medial skull base.

References

1. Juniper RP (1981)
2. Krиźan Z (1960)
3. Krmpotić-Nemanić J (1977)

Plates 1–6 modified from Kriźan [2]

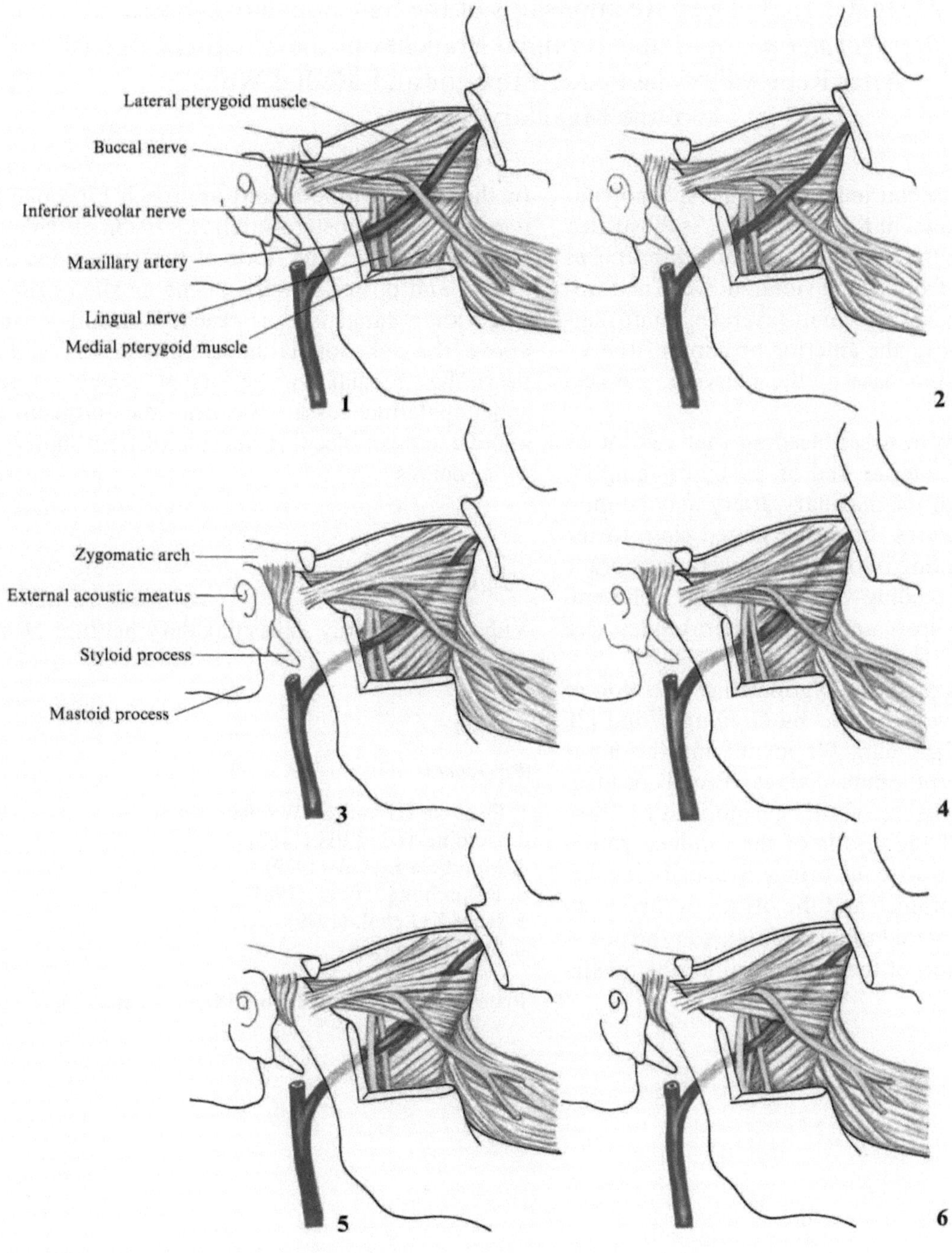

Lateral pterygoid muscle
Buccal nerve
Inferior alveolar nerve
Maxillary artery
Lingual nerve
Medial pterygoid muscle
1
2
Zygomatic arch
External acoustic meatus
Styloid process
Mastoid process
3
4
5
6

6.24 (Plates 1–2) Deep Relationships of the Sphenopalatine Fossa (*Fossa Pterygopalatina*) With the Terminal Branches of the Maxillary Artery After Removal of the Posterosuperior and Medial Walls of the Maxillary Sinus

The surgical vascular anatomy of the sphenopalatine fossa is shown in Plates 1 and 2. As illustrated in Plate 1, the maxillary artery and its branches are situated in front of the vidian nerve. The foramen rotundum lies 6–10 mm (average, 8 mm) lateral to and above the anterior opening of the vidian canal at the base of the pterygoid process [1, 2, 4].

In cases of severe nasal bleeding that cannot be controlled by anterior and/or posterior tamponades, ligature of the maxillary artery may be indicated [6]. Ligatures should be placed close to the site of the bleeding in order to prevent retrograde filling of the bleeding vessels from the collateral vessels of the ipsi- and contralateral sides (see Sect. 7.5).

In the transmaxillary-pterygopalatine resection of the vidian nerve described by Golding-Wood [2], a convenient landmark for identifying the nerve is the foramen rotundum. This can usually be identified after exposure of the sphenopalatine fossa and ligature of the vessels of the foramen rotundum. A vertical osseous bridge is usually located between this foramen and the orifice of the vidian canal. The pterygoid canal is situated in the posterior continuation of the medial wall of the maxillary sinus.

In the transantral-ethmoid approach [5], bone is removed at the postero-inferior portion of the ethmoid bone where the floor of the bone joins the medial and posterior walls of the maxillary sinus. The vidian canal is thus reached medial to and above the sphenopalatine ganglion. The vidian nerve lies medial to the major vessels of the sphenopalatine fossa, allowing the surgeon to avoid excessive bleeding and inadvertent injury to other nerves.

Surgical Importance

Vidian neurectomy, transmaxillary ligature of the maxillary artery.

References

1. Denecke HJ and Ey W (1980, 1984)
2. Golding-Wood PH (1962)
3. Mandiola E et al. (1979)
4. Rabischong P et al. (1981)
5. Rose KG et al. (1979)
6. Seiffert A (1928)

Plates 1 and 2 modified from Mandiola et al. [3]

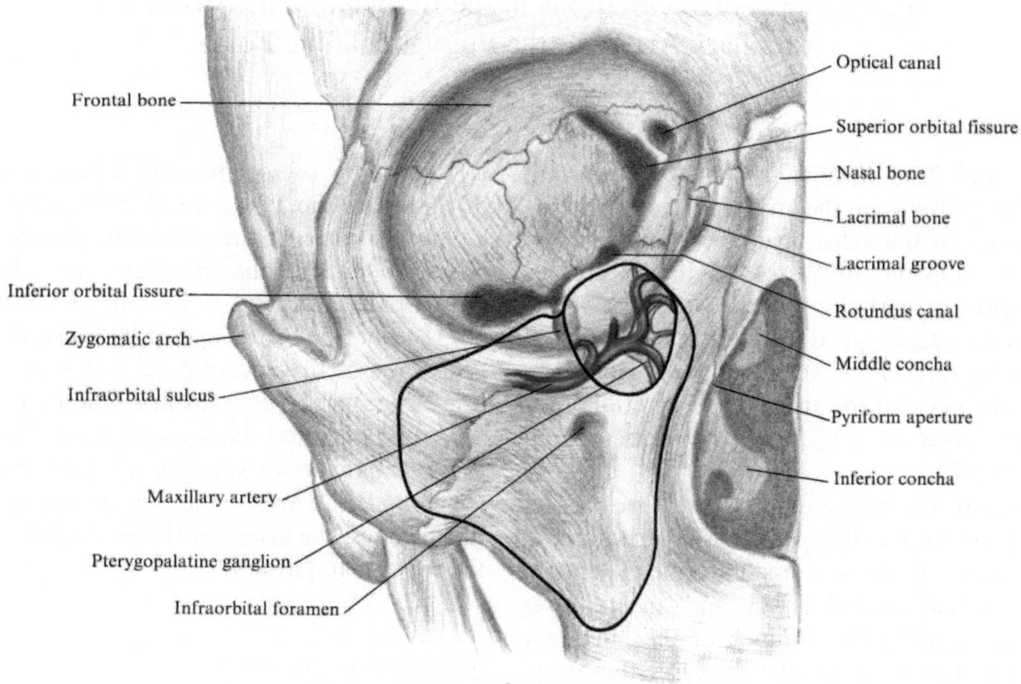
Frontal bone
Optical canal
Superior orbital fissure
Nasal bone
Lacrimal bone
Lacrimal groove
Inferior orbital fissure
Rotundus canal
Zygomatic arch
Middle concha
Infraorbital sulcus
Pyriform aperture
Inferior concha
Maxillary artery
Pterygopalatine ganglion
Infraorbital foramen

1

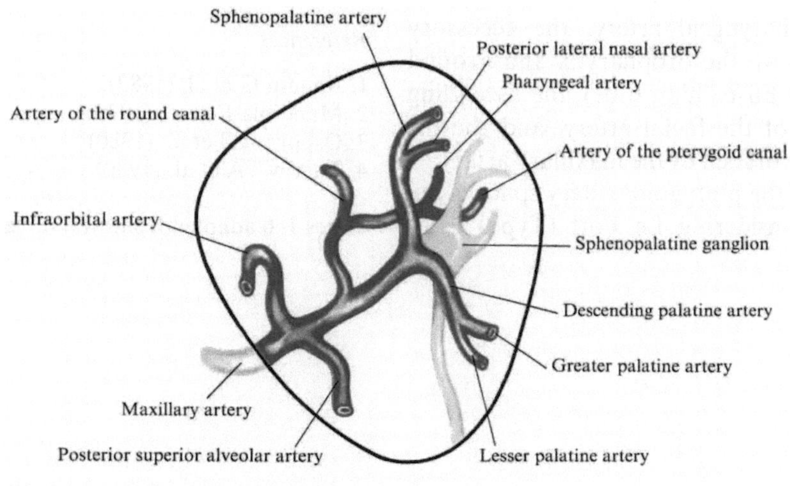
Sphenopalatine artery
Posterior lateral nasal artery
Pharyngeal artery
Artery of the round canal
Artery of the pterygoid canal
Infraorbital artery
Sphenopalatine ganglion
Descending palatine artery
Greater palatine artery
Maxillary artery
Posterior superior alveolar artery
Lesser palatine artery

2

6.25 (Plates 1–6) Variations in the Branching Patterns of the Maxillary Artery in the Pterygopalatine Fossa

Plates 1 and 2 in Sect. 6.24 depict the normal branching pattern of the maxillary artery in its relationship to the sphenopalatine ganglion. The schematic drawings of Plates 1–6 in the present series represent variations in the branches of the artery in the pterygopalatine fossa that are important for ligature of the vessel in this region. The average diameter of the artery at its point of entry into the fossa is 2.63 mm, and varies from 2.5–3 mm [4].

The artery of the pterygoid canal is in most cases a branch of the maxillary artery [3]. However, in 30 % of cases, it can be a small inconstant branch of the internal carotid artery and anastomoses within the canal with a branch of the maxillary artery. In such instances, the artery of the pterygoid canal passes through the foramen lacerum, giving off branches that supply the mucosa of the oropharynx. Regardless of its origin, the artery *always* gives branches to the sphenopalatine fossa and the oropharynx. Within the vascular plexus of the fossa, it anastomoses with branches from the ascending pharyngeal artery, the accessory meningeal artery (in the oropharynx and around the orifice of the Eustachian tube), the ascending palatine branch of the facial artery, and the descending palatine branch of the maxillary artery.

The extension of the pterygomaxillary space is important when considering Le Fort (Types I, II, III) maxillary osteotomies [1]. The average depth of this space (i.e., the area of contact between the anterior surface of the pterygoid process and the posterior wall of the maxillary sinus below the sphenopalatine fossa) is 15 mm, and varies from 11–18 mm. Before the age of 20 years, there is no synostosis in this location so that surgical osteotomy is generally unnecessary. The following vessels can be injured during actual maxillary osteotomy: the posterior superior alveolar arteries, the sphenopalatine artery, the greater and lesser palatine arteries, the artery of the pterygoid canal, and the infraorbital artery.

Surgical Importance

Surgical control of bleeding from the nose and nasopharynx, surgery of the base of the skull or nasopharynx, craniofacial surgery, vidian neurectomy.

References

1. Boudin G et al. (1982)
2. Mandiola E et al. (1979)
3. Osborn AG et al. (1980)
4. Turvey TA et al. (1980)

Plates 1–6 adapted from Mandiola et al. [2]

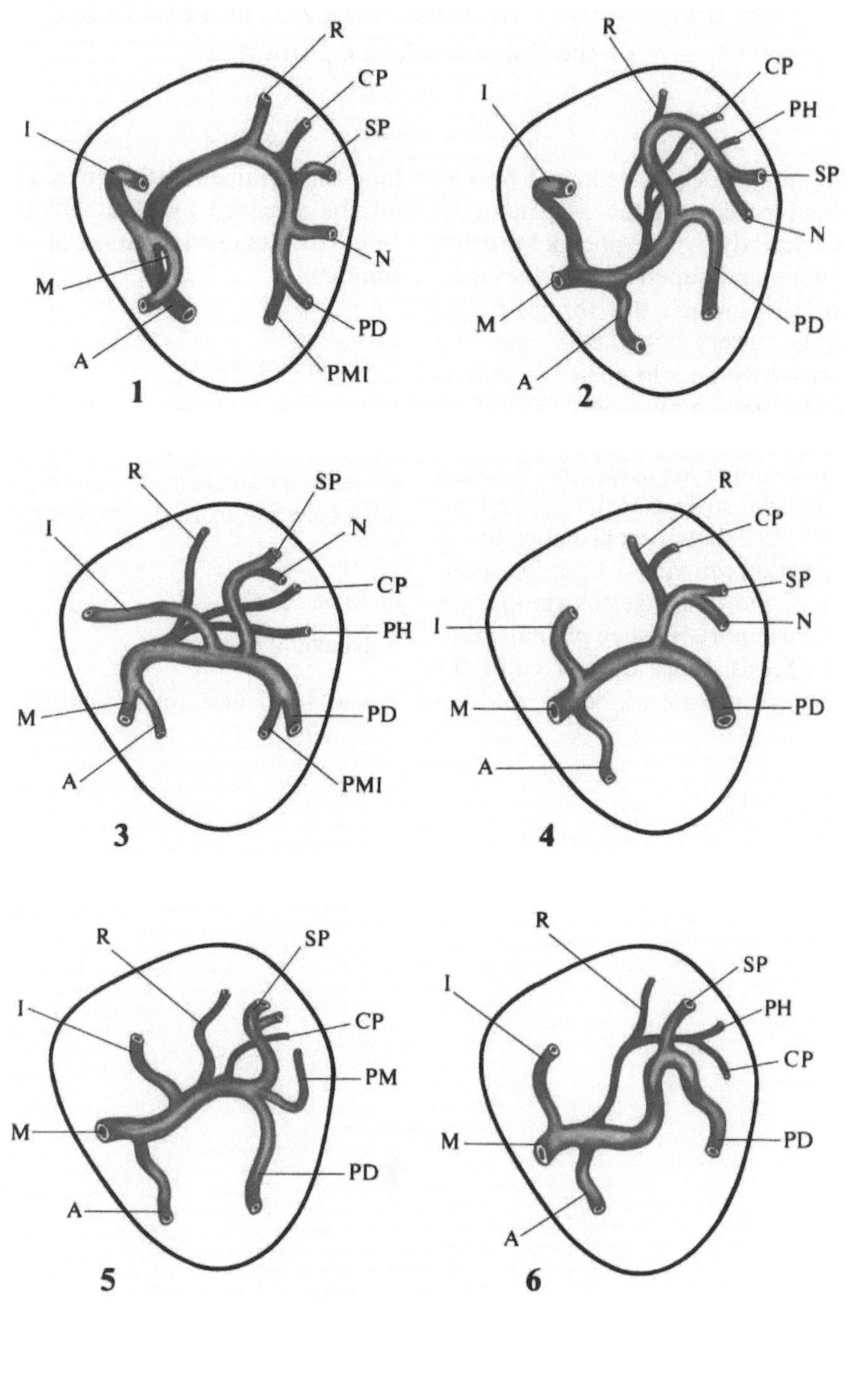

A	Posterior superior alveolar artery	*PH*	Pharyngeal artery
CP	Artery of the pterygoid canal	*PM*	Greater palatine artery
I	Infraorbital artery	*PMI*	Lesser palatine artery
M	Maxillary artery	*R*	Artery of the rotundus canal
N	Posterior lateral nasal artery	*SP*	Sphenopalatine artery
PD	Descending palatine artery		

6.26 (Plates 1–9) Variations in the Position and Shape of the Sphenopalatine Foramen

The sphenopalatine foramen is delimited anteriorly by the orbital process of the os palatinum inferiorly, and posteriorly by the sphenoid process of the palatine bone and superiorly by the body of the sphenoid [1]. Through this foramen pass the sphenopalatine artery (Sect. 5.8) and the nerves to the nasal cavity (i.e., the posterior-superior lateral and medial nasal nerves, posterior-inferior lateral nasal nerves and the nasopalatine nerve) (Sect. 5.9). In the majority of cases, the foramen lies behind the posterior surface of the middle concha (Sect. 6.24). As shown schematically in Plates 1–9, this position can vary, as can the shape and the number of the openings for the nerves. These variations are important when planning topical or local anesthesias. Since the nerves of the nasal cavity also pass in front of, below, and behind this foramen, satisfactory anesthesia can usually be achieved by anesthetizing a vertical zone *below* the assumed location of the sphenopalatine foramen.

Surgical Importance

Surgery of the septum and nasal cavity, therapeutic anesthesia for Sluder's neuralgia, surgery of the ethmoid and sphenoid sinuses, mid-facial trauma with bleeding from the sphenopalatine artery.

Reference

1. Nikolić V (1967)

Plates 1–9 adapted from Nikolić [1]

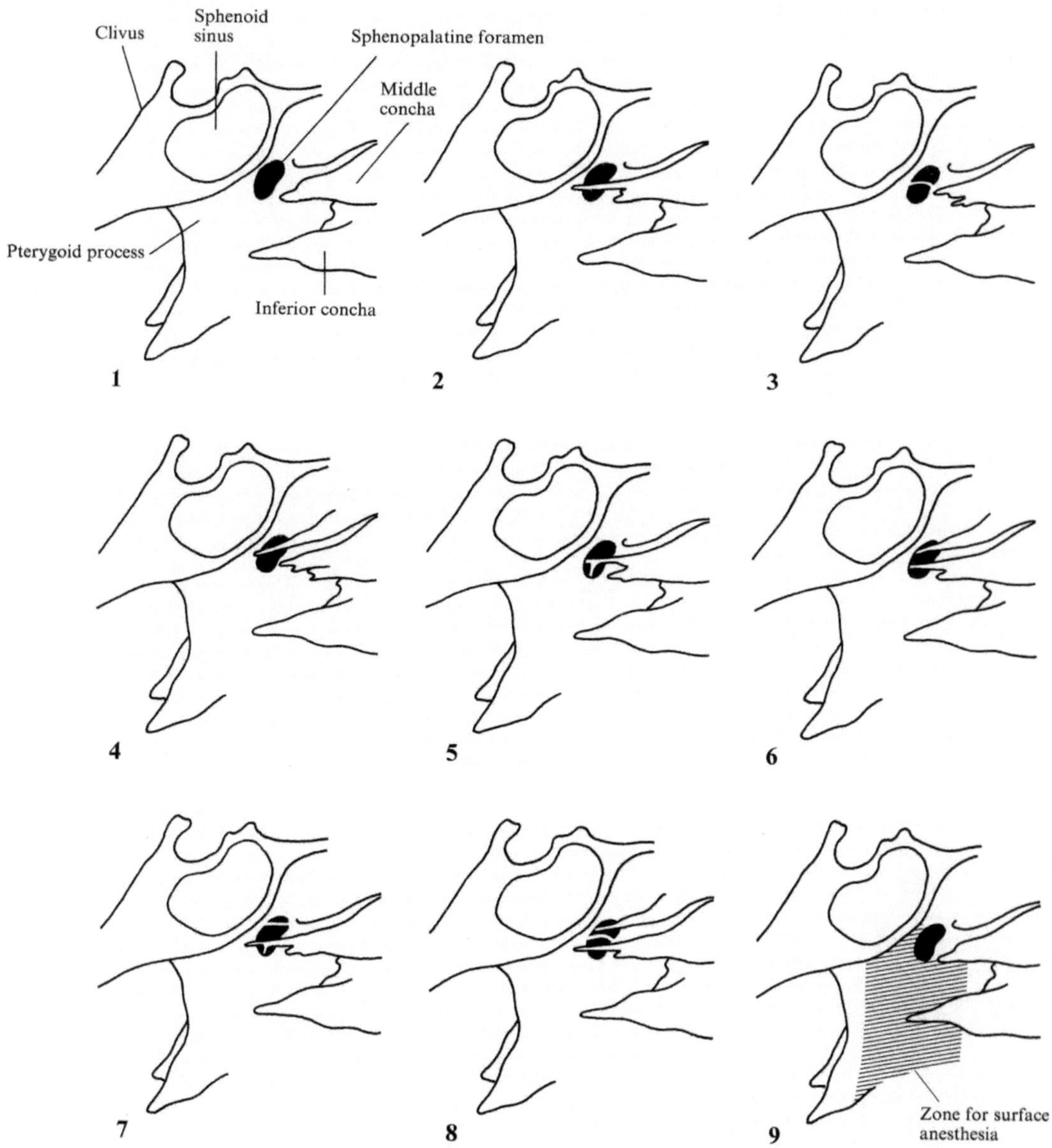
Clivus
Sphenoid sinus
Sphenopalatine foramen
Middle concha
Pterygoid process
Inferior concha
1
2
3
4
5
6
7
8
9
Zone for surface anesthesia

7 The Eye and its Orbital Adnexa

7.1 Sagittal Section of the Eye and the Orbital Adnexa (Lateral View)

The orbital contents are represented schematically in a parasagittal section showing their close relationships to the anterior cranial fossa superiorly and the maxillary sinus inferiorly. There is a direct communication between the orbit and the sphenopalatine fossa through the inferior orbital fissure.

Surgical Importance

Orbital complications in sinusitis, internal carotid-cavernous sinus fistula, transmaxillary approaches to the orbit and to the ethmoid bone, transfrontal approach to the orbit, treatment of blow-out fractures, decompression of the infraorbital nerve, transconjunctival approach to the orbital floor (in endocrine opthalmopathy), surgery of intraorbital tumors.

Reference

1. Pernkopf E and Ferner H (1963)

Plate adapted from Pernkopf

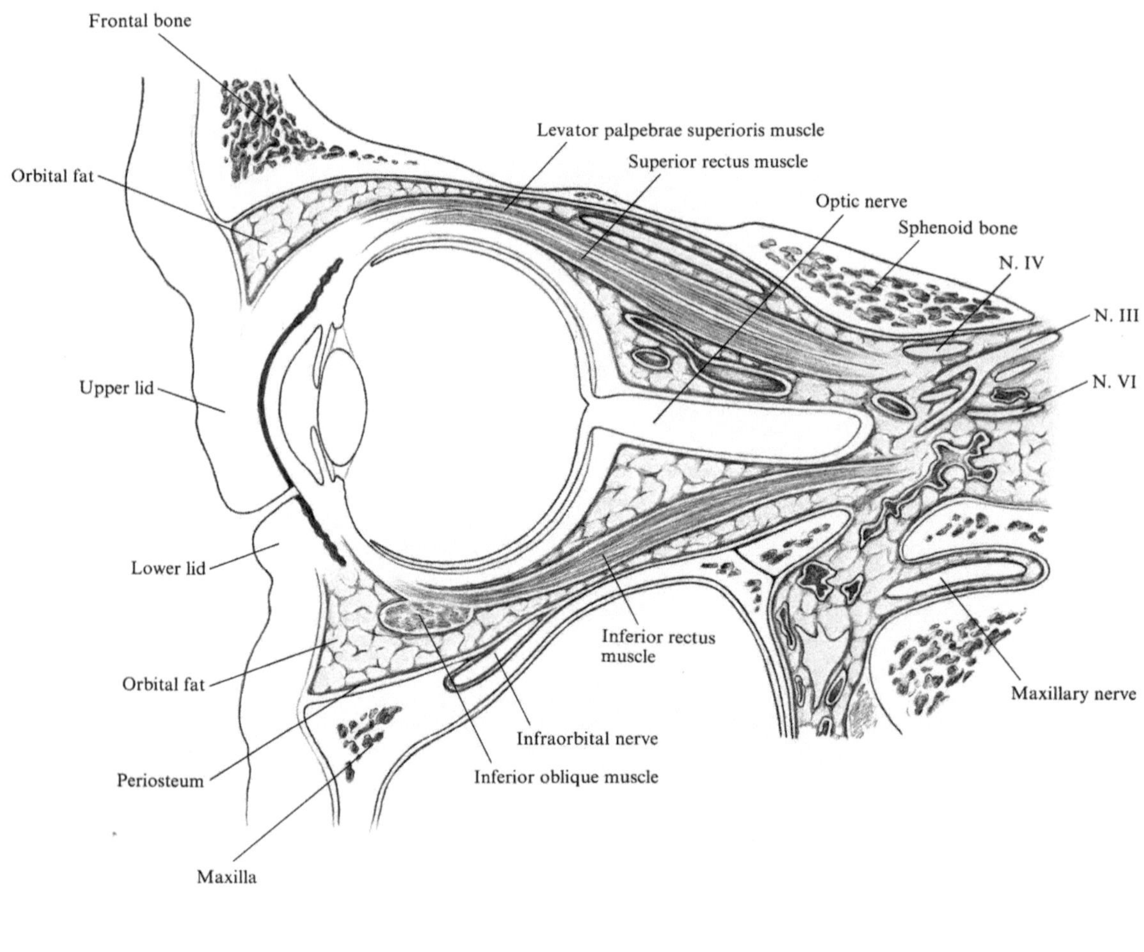

7.2 Topography of the Anterior Part of the Orbit and the Levator Palpebrae Muscle (Sagittal Section)

Plate 7.2 is a schematic representation of the most anterior part of the orbit and eyelids. The levator palpebrae muscle has two attachments, one to the tarsus and the other to the upper lid muscles and skin. These attachments are important in surgery to correct ptosis. The anatomy of the anterior-inferior orbit is also important for the transconjunctival approach to the orbital floor.

Surgical Importance

Ptosis surgery, surgical repair of orbital fractures, tumor surgery of the orbits (see Sect. 7.1).

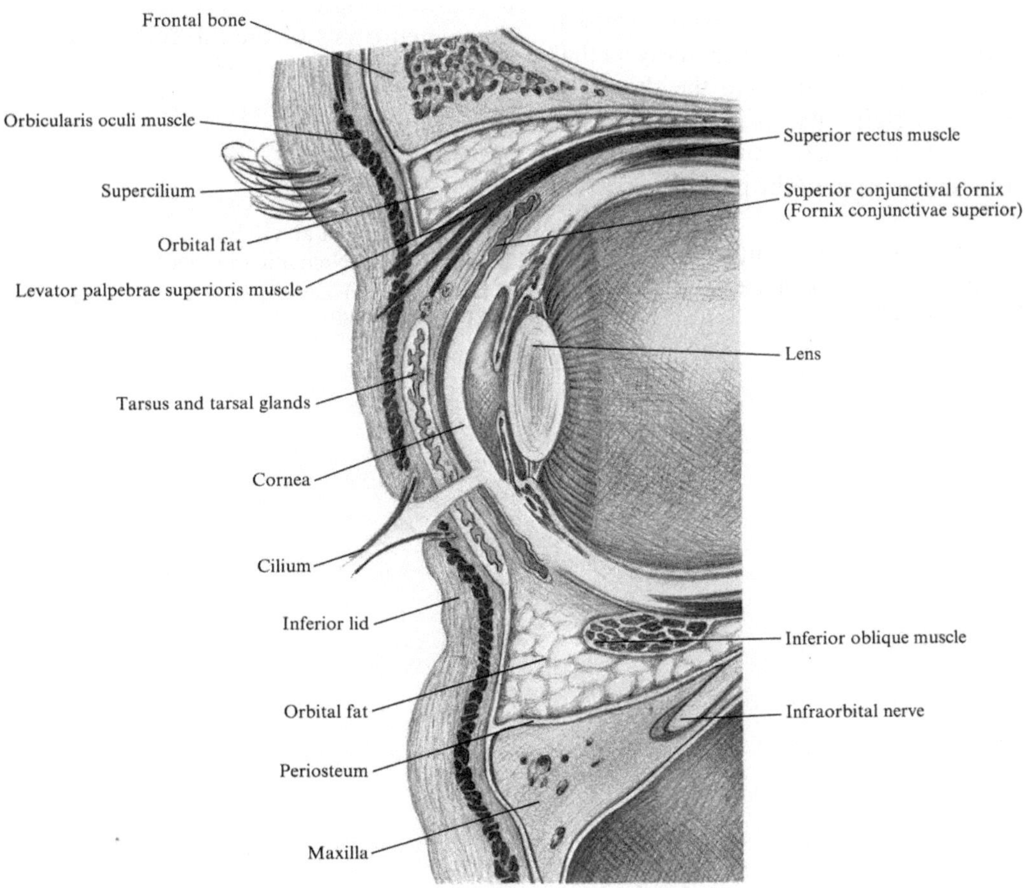

7.3 Vascular Relationships of the Orbital Arteries to the Maxilla and the Nasal Cavity

The facial artery lies in a bony groove along the frontal process of the maxilla. It anastomoses with the dorsal artery of the nose (*a. dorsalis nasi*) from the supratrochlear artery (the medial frontal branch of the ophthalmic artery) and with the infraorbital artery, which supplies the upper lip, the nose, and the lower eyelid.

The infraorbital nerve branches below the infraorbital artery, giving off inferior palpebral, external and internal nasal, and superior labial branches.

The ophthalmic artery enters the orbit through the optic canal and gives off the lacrimal artery, the supraorbital (lateral frontal) artery, the supratrochlear (medial frontal) artery, and the anterior and posterior ethmoid arteries, as well as small branches to the eye (see Sect. 7.5).

The infraorbital foramen is variable in shape and position, and may even have two or three external openings [1, 3]. Its position is independent of the shape of the facial skeleton. It opens in a medial direction with its opening inclined caudally so that it is not readily identified through the skin. The easiest way to determine its position is to bisect a line placed from the lateral canthus of the eye to the inferolateral margin of the piriform aperture [2].

The infraorbital sulcus is completely enclosed, forming a canal in 4.83 % of cases on the right and 5.27 % on the left [1].

The maxillary sinus is supplied by the sphenopalatine artery as well as by the anterior and posterior superior alveolar arteries (*aa. alveolares superiores anteriores et posteriores*).

The drainage pattern of the orbital lymphatics follows that of the orbital blood vessels. These lymphatics cross-anastomose with lymphatics from the maxillary sinus and the Eustachian tube through the pterygopalatine plexus. This lymphatic chain from the sinus, the Eustachian tube, and the nasopharynx is important in the propagation of inflammatory processes, as well as in the diagnosis and treatment of malignant tumors [4] (see Sects. 1.23, 24).

Surgical Importance

Surgery of the orbit and the maxillary sinus and bone, rhinoplasty, surgical control of nasal hemorrhage, neurolysis of the infraorbital nerve, tumor surgery involving the orbit.

References

1. Godziewski S et al. (1979)
2. Keros P and Nemanić G (1966)
3. Lang J (1981)
4. Mann W (1980)

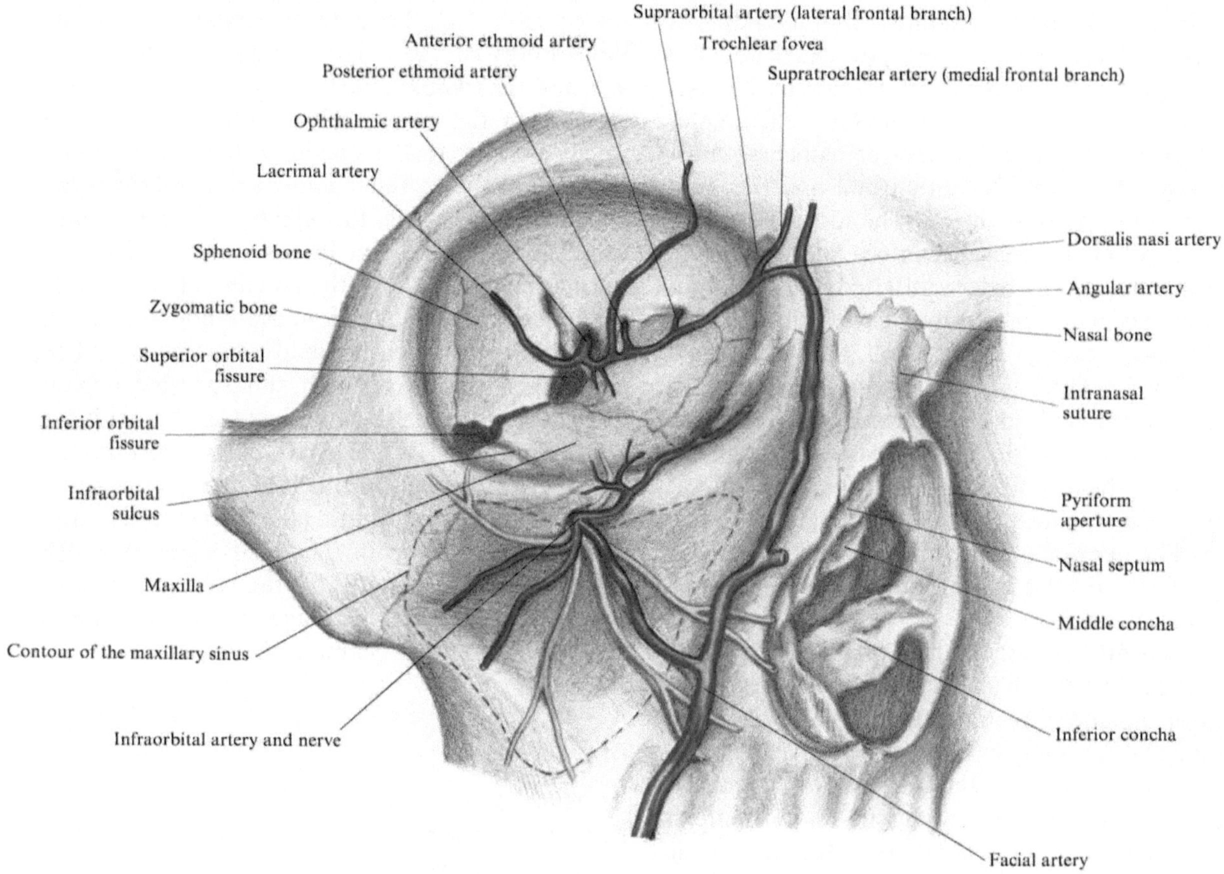
Supraorbital artery (lateral frontal branch)
Anterior ethmoid artery
Trochlear fovea
Posterior ethmoid artery
Supratrochlear artery (medial frontal branch)
Ophthalmic artery
Lacrimal artery
Sphenoid bone
Dorsalis nasi artery
Zygomatic bone
Angular artery
Nasal bone
Superior orbital fissure
Intranasal suture
Inferior orbital fissure
Infraorbital sulcus
Pyriform aperture
Nasal septum
Maxilla
Middle concha
Contour of the maxillary sinus
Infraorbital artery and nerve
Inferior concha
Facial artery

7.4 Topography of the Apex of the Orbit (Right Side, Anterior View)

The four recti muscles originate from a tendinous annulus which encircles the optic canal and the medial part of the superior orbital fissure. The lateral rectus muscle also originates from the greater wing of the sphenoid. The levator palpebrae muscle originates from the optic canal and the dura ensheathing the optic nerve. The anterior tendon of this muscle is separated into a superficial and a deep portion. The latter is inserted into the tarsus of the eye, while the former passes through the orbicularis oculi muscle to the skin (see Sect. 7.1, 2). The superior oblique muscle has its origin on the body of the sphenoid medial to the tendinous annulus. Its tendon passes through the trochlea, which is fixed to the medial wall of the orbit at the trochlear spine, thus changing its direction. The inferior oblique muscle has its origin on the inferior orbital margin lateral to the lacrimal fossa (see Sect. 7.2).

The recti muscles are innervated by the branches of the oculomotor nerve (*yellow*), with the exception of the lateral rectus muscle which is supplied by the abducens nerve (also *yellow*). The oculomotor nerve also supplies the inferior oblique muscle and the levator palpebrae muscle. Its radix brevis branch gives off parasympathetic fibers to the ciliary ganglion for innervation of the ciliary muscle and the pupillary sphincter. The branches of the oculomotor nerve pass through the pars nervosa of the tendinous annulus, while the optic nerve, the ophthalmic artery, and an accompanying vein pass through the pars optica of the annulus. Within the tendinous annulus are the nasociliary nerve (*green*, from V_1), the abducens nerve (*yellow*), the radix longa of the ciliary ganglion (*green*, from V_1), and the sympathetic branch to the ciliary ganglion from the carotid plexus (not shown). The trochlear nerve (*yellow*) which supplies the superior oblique muscle, the frontal nerve, and the lacrimal nerve (*green*) pass through the superior orbital fissure. This latter nerve brings secretory fibers to the lacrimal gland from the greater petrosal nerve, the sphenopalatine ganglion, the zygomatic nerve, and the zygomatico-lacrimal anastomosis. The nerve then passes through the superior orbital fissure (*not* through the annulus) together with the ophthalmic vein [1]. The length of the inferior orbital fissure in the newborn is 15.3 mm and its width is 5 mm [2, 3]. Its length later increases so that it becomes 29.3 mm in the adult while its width is 5.4 mm. The length of the superior orbital fissure also increases with age, and is 14.8 mm in the newborn and 19.5 mm in the adult.

Surgical Importance

Transmaxillary and transfrontal surgical approaches to the orbit, transmaxillary and transnasal surgery of the ethmoid sinuses.

References

1. Jo A and Trauzettel H (1974)
2. Lang J (1979)
3. Lang J and Schlehahn FA (1981)

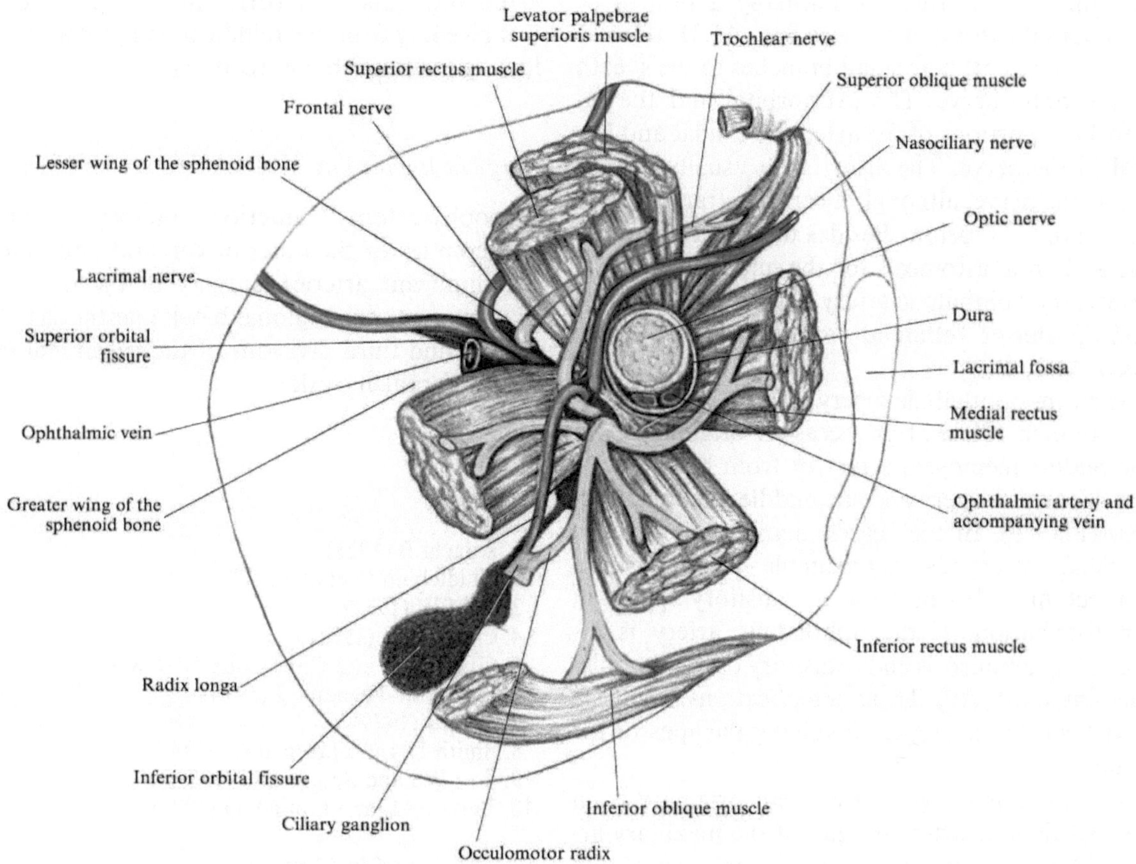
Levator palpebrae superioris muscle
Trochlear nerve
Superior rectus muscle
Superior oblique muscle
Frontal nerve
Nasociliary nerve
Lesser wing of the sphenoid bone
Optic nerve
Lacrimal nerve
Dura
Superior orbital fissure
Lacrimal fossa
Medial rectus muscle
Ophthalmic vein
Ophthalmic artery and accompanying vein
Greater wing of the sphenoid bone
Inferior rectus muscle
Radix longa
Inferior orbital fissure
Ciliary ganglion
Inferior oblique muscle
Occulomotor radix

7.5 (Plates 1–3) Variations of the Origin of the Ophthalmic Artery and its Branches (Right Orbit, Superior View)

The ophthalmic artery is normally a branch of the internal carotid artery (see Sect. 11.3). It occasionally gives off here small branches to the sheath of the optic nerve. The extraorbital and the intraorbital portions of the artery lie caudal and lateral to the nerve. The artery then usually crosses above the nerve, although it can also (rarely) cross the nerve from below. Besides the branches to the eye and an anastomosis for the middle meningeal artery, the ophthalmic artery gives off the anterior and posterior ethmoid arteries [2, 6–9] (see Sects. 11.3, 4).

The entire ophthalmic artery (Plate 1) or its lacrimal branch (Plate 2) on occasion originate from the middle meningeal artery or from both the internal carotid artery and the middle meningeal artery (Plate 3). In such cases, manipulation of the maxillary artery (as – for example – during Vidian neurectomy [3]) can cause a transitory spasm in its distribution. If the ophthalmic artery is affected, an amaurosis and transitory ophthalmoplegia may occur [10]. These complications have also been reported in surgery involving the apex of the orbit.

The middle meningeal artery can originate from the ophthalmic artery instead of the maxillary artery [4, 5]. In such cases, the foramen spinosum is missing. This is important in attempting to control bleeding from the middle meningeal artery following temporal bone fractures.

Surgical Importance

Hypophysectomy, anterior craniotomy, aneurismectomies of the anterior cerebral and anterior communicant arteries, surgery of the orbits, Vidian neurectomy, regional block anesthesias of the second and third divisions of the trigeminal nerve (at the foramen ovale).

References

1. Adachi B (1928)
2. Chidichino G et al. (1978)
3. Draf W (1982)
4. Guerrier Y (1981c)
5. Guerrier Y and Charissoux G (1981)
6. Krmpotić-Nemanić J (1978)
7. Lang J (1979)
8. Smith D and Larsen JL (1979)
9. Tan WS and Spigos DG (1979)
10. Vergnon L and Gotlieb O (1979)

Plates 1–3 modified from Adachi [1]

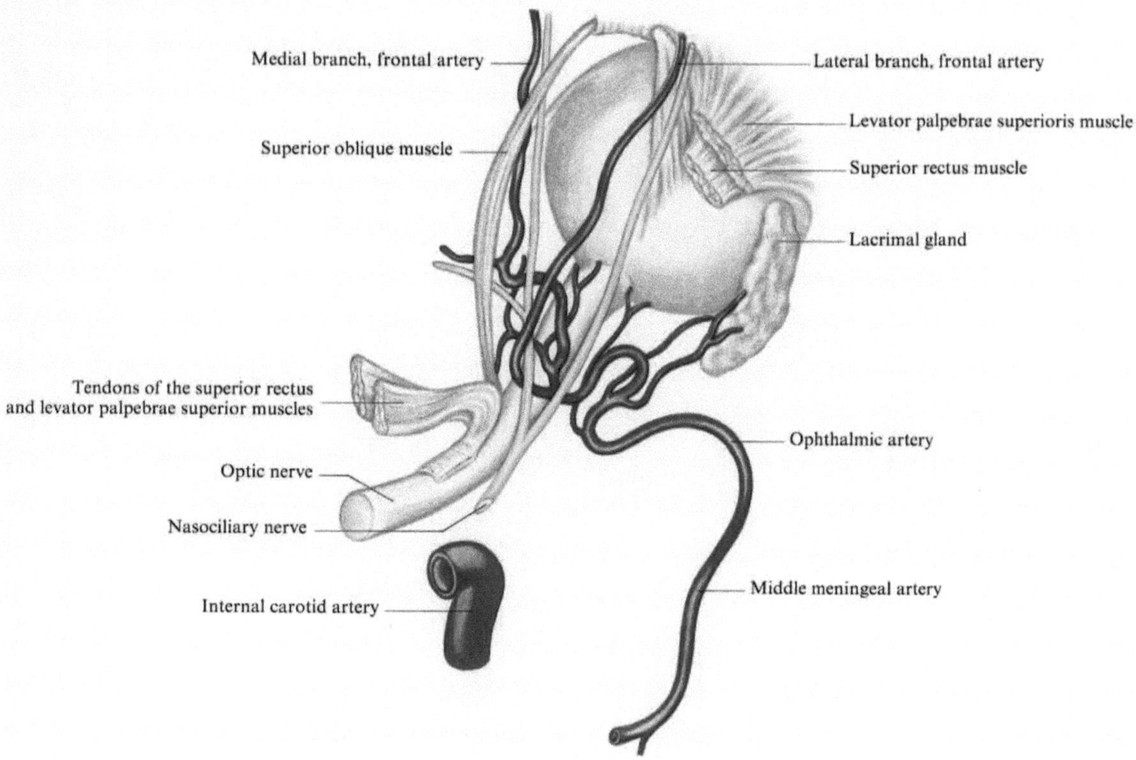

Medial branch, frontal artery
Lateral branch, frontal artery
Levator palpebrae superioris muscle
Superior oblique muscle
Superior rectus muscle
Lacrimal gland
Tendons of the superior rectus
and levator palpebrae superior muscles
Ophthalmic artery
Optic nerve
Nasociliary nerve
Middle meningeal artery
Internal carotid artery

1

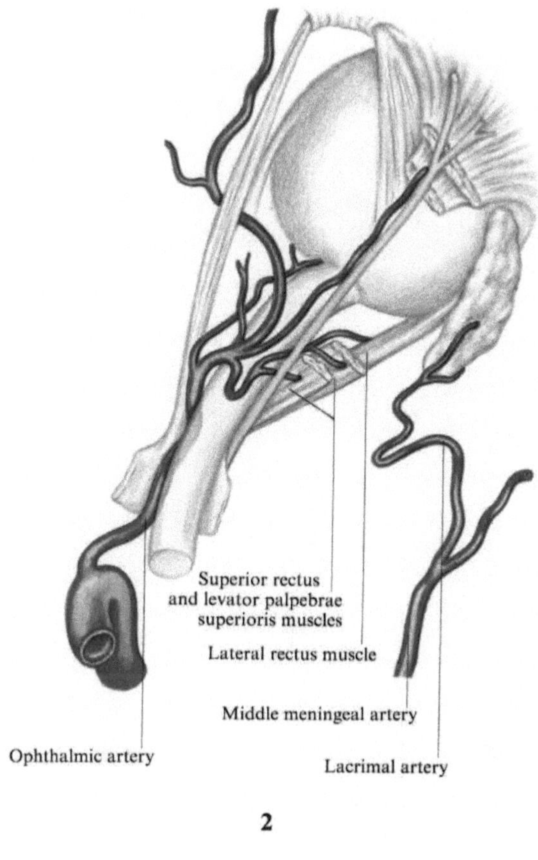

Superior rectus
and levator palpebrae
superioris muscles
Lateral rectus muscle
Middle meningeal artery
Ophthalmic artery
Lacrimal artery

2

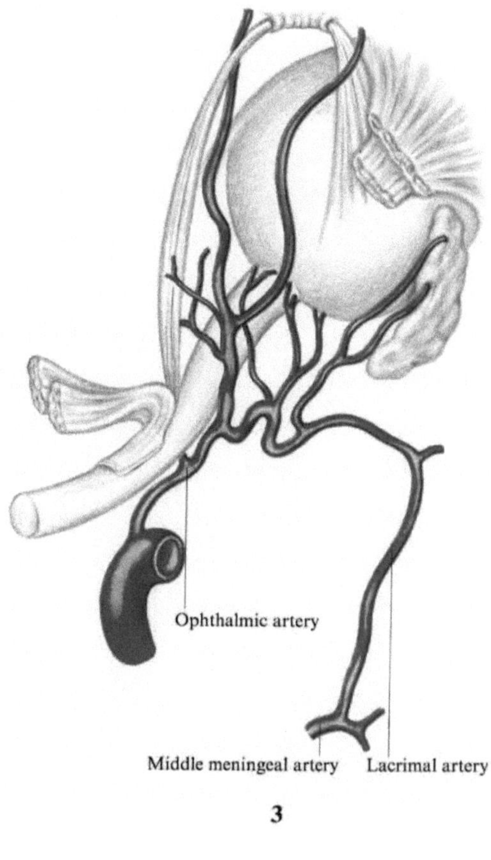

Ophthalmic artery
Middle meningeal artery
Lacrimal artery

3

8 The Ear

The *helix* of the auricle located superiorly begins with the crus helicis between the cymba and the cavum conchae, and continues supero-posteriorly as the posterior rim of the auricle to the cauda helicis above the lobule (see Sect. 8.2). The *spina helicis* extends anteriorly from the crus helicis over the anterior (supratragal) incisure. The *anthelix* is located between the cymba conchae and the helix, and extends superiorly as the superior and inferior anthelical crura. The *fossa triangularis* is located between these crura. The *scapha* is located between the helix and the anthelix, while the *cavum conchae* is located between the anthelix, the tragus, and the antitragus. The deep concavity between the anterior helical crus and the inferior anthelical crus is the *cymba conchae*. The *cavum conchae* lies behind the opening of the external auditory meatus.

The *supratragal incisure* is located between the tragus (and the supratragal tubercle) and the crus helicis. *Darwin's tubercle* is normally present in some ears, and is located on the posterior superior helical rim. The *posterior auricular sulcus* is formed between the anthelix and the antitragus.

The *auricular lobule* is composed of an inferior fold of soft tissue (beneath the antitragus and the intertragal incisure), and can vary considerably in its size and shape. It can also be split in some ears or even be missing.

Certain patients may have major malformations of the auricle, and demonstrate microtia or anotia. The auricular remnants in such individuals are represented by small appendices, the *appendices auriculae*.

Surgical Relevance

Otoplasty, reconstruction of the auricle.

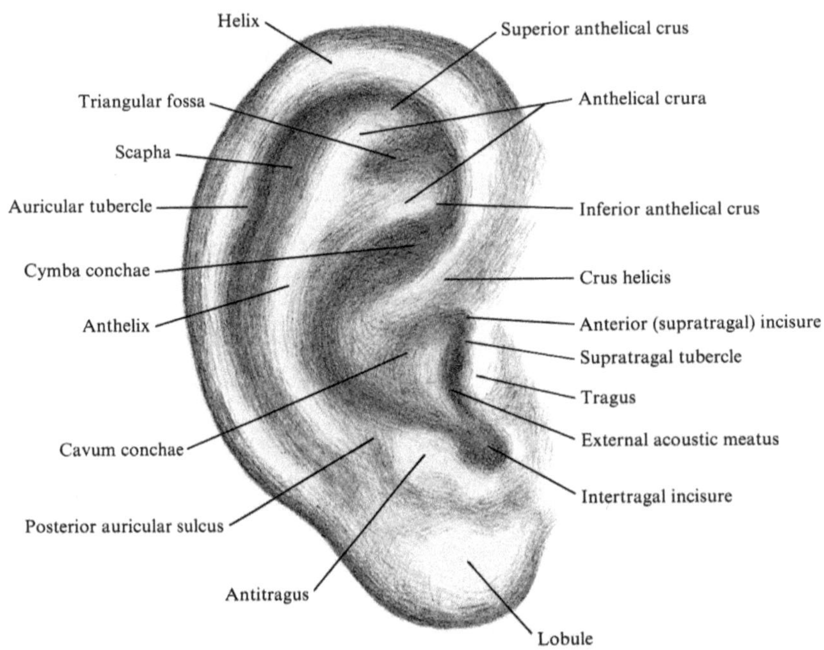

8.2 (Plates 1–2) The Auricular Cartilages (Lateral and Medial Views)

Many details can be observed on the cartilaginous infrastructure of the normal auricle. The muscles of the auricle are innervated by the facial nerve. According to Šercer [1], the anterior auricular ligament should be divided in otoplastic operations involving repositioning of the auricle. In doing microsurgery of the ear, an additional incision is made in the cartilage-free zone of the superior external acoustic meatus, which corresponds to the anterior (supratragal) incisure.

Surgical Importance

Otoplasty, (endaural) microsurgery of the ear, otopexy.

Reference

1. Šercer (1951)

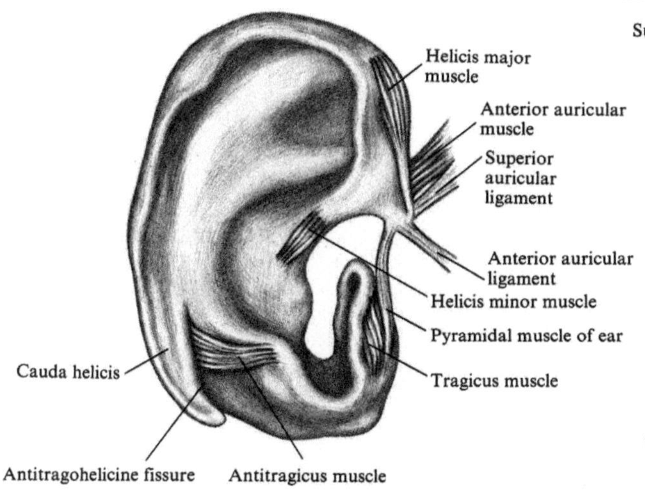

1

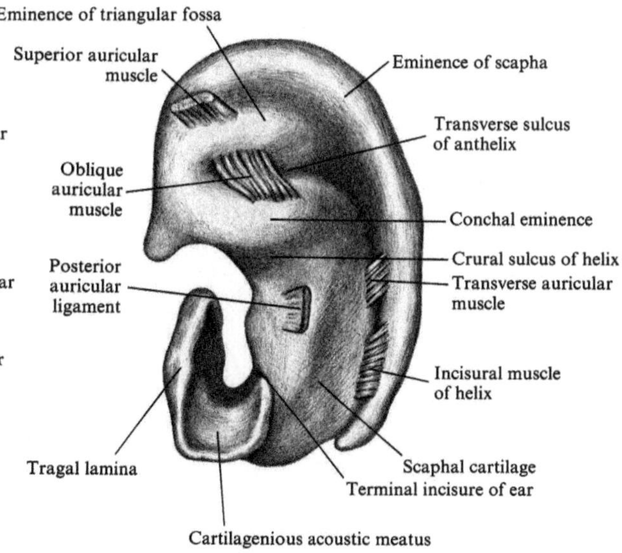

2

Plates 1 and 2 show the sensory innervation of the external ear. The delimitations between the innervated areas are not well-defined and overlap to various extents [1, 2]. Since the greater auricular nerve supplies the major part of the auricle, it should be preserved if possible in such cosmetic procedures as otopexy and rhytidectomy, and also in parotid surgery.

Surgical Importance

Hitzelberger's sign, stimulation of Arnold's nerve to produce cough following cleaning or trauma of the external acoustic meatus, referred otalgia (due to malocclusion, temporomandibular joint or cervical syndrome, laryngeal or pharyngeal carcinoma).

References

1. Hirsch C (1925)
2. Lang J (1979)
3. Lazorthes G et al. (1981)

Plates 1–2 modified from Lazorthes et al. [3]

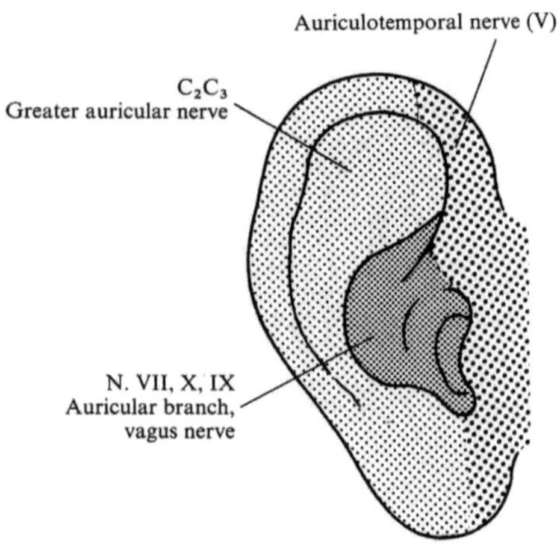

1

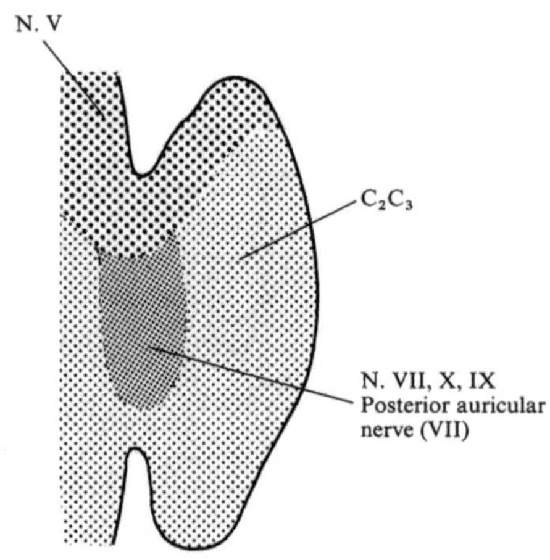

2

8.4 Right Temporal Bone and Petrous Pyramid (Lateral View)

The right temporal bone and petrous pyramid are depicted on the Plate. This shows the muscles inserting on the posterior and middle skull base, and the important nerves and vessels present.

Surgical Importance

Surgery of the lateral skull base, the ear, the parotid gland, the facial nerve, the mandible, the sphenopalatine and infratemporal fossae, the temporomandibular joint, and cervical fistulae.

Reference

1. Helms J (1978)

Plate from Helms [1]

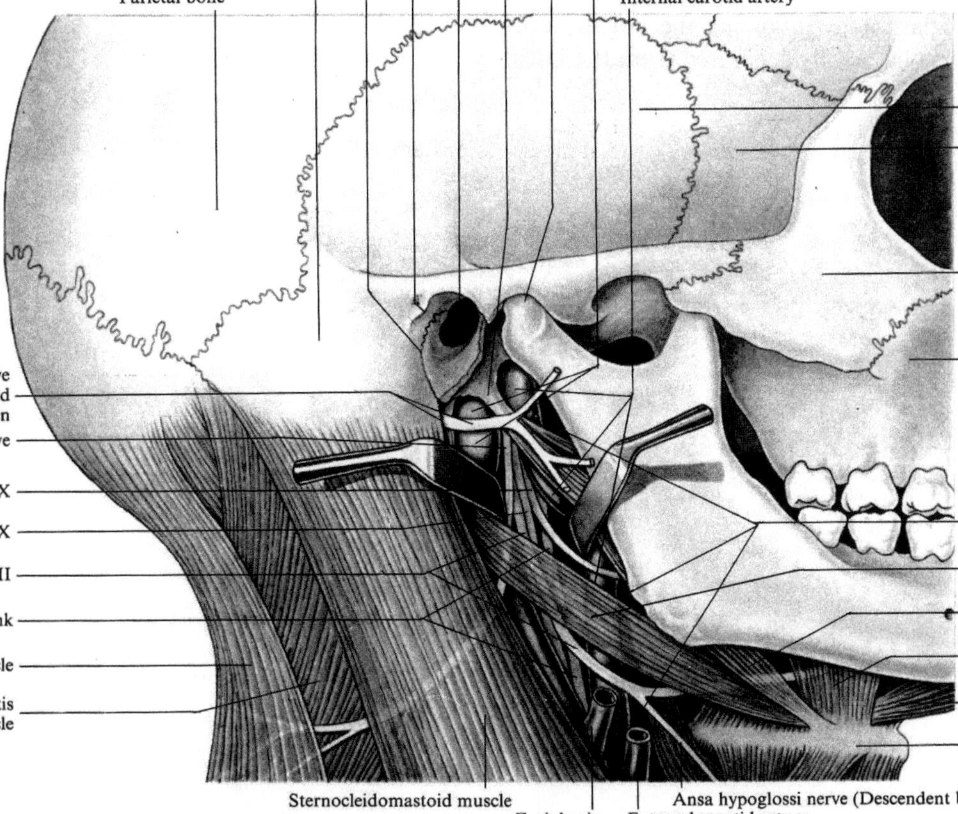

8.5 Pneumatization of the Middle Ear and the Adjacent Temporal Bone

The pneumatization of the middle ear varies considerably, from the sparse periantral pneumatic cells to large pneumatic spaces occupying the mastoid process, part of the temporal bone, the zygomatic arch, and the petrous apex. The actual degree of pneumatization can be seen on X-ray films. The so-called air cell chains (superior, inferior, anterior, and posterior) are very important in the infections and inflammations of the middle ear and their treatments. Both Stenger and Marx recognized the important role of these cell chains in determining the direction of disease spread, calling them *Heerstraßen* ("army routes"). The cells between the roof of the pyramid and the sigmoid sinus (sinodural angle; *Citelli* cells), the retrosinus cells (which can be overlooked in mastoidectomy), and the Pogány cells (situated on the facial canal above the stapedius muscle) are also important in the pathology and surgery of the middle ear. Unfortunately, pathology involving the Pogány cells cannot always be readily separated from the facial nerve.

In an extensively pneumatized mastoid, adjacent structures – such as the dura of the middle and posterior cranial fossa, the internal carotid artery, and sometimes also the facial nerve – are covered only by connective tissue instead of bone. In 35 skulls, Lang [1] found five bilateral and four unilateral defects of the tegmen which he believed could have been fatal if air were insufflated into the ear. These defects also facilitate the propagation of infection from the middle ear into surrounding structures.

Surgical Importance

Otologic surgery, antrotomy, mastoidectomy, facial nerve decompression, complications in acute and chronic otitis media, translabyrinthine approach to the petrous apex and internal acoustic meatus, surgery of the middle cranial fossa.

Reference

1. Lang (1983)

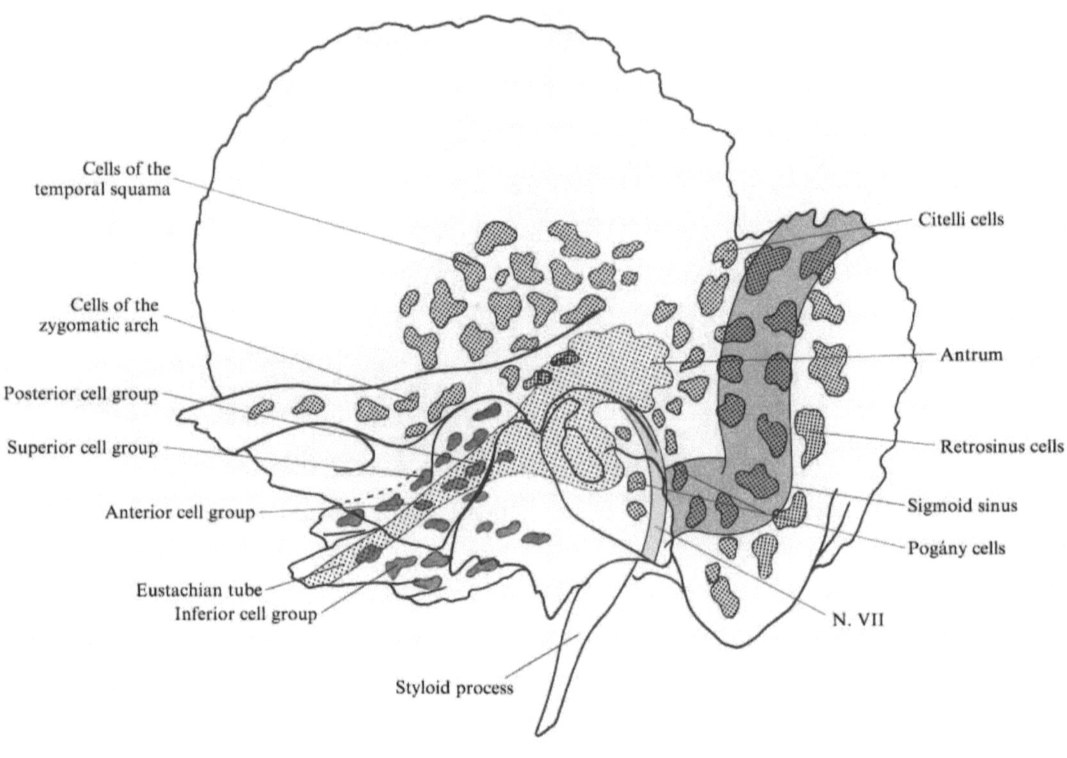

8.6 Spatial Relationships of the Anatomic Structures Within the Petrous Pyramid

The petrous pyramid lies in the base of the skull, and is inclined 45° medially and forward. The projections of the superior petrosal sinus and the internal carotid artery are located in the medial part of the pyramid and thus occur in front of the inner ear. The mastoid emissary vein varies in size and position, as does the jugular bulb. *Trautmann's triangle* is formed by the posterior free wall of the pyramid, and is located between the sigmoid sinus, the superior petrosal sinus, and the labyrinth. The endolymphatic sac lies in its inferior part.

Surgical Importance

Surgery of the petrous pyramid, the ear, the facial nerve, the infratemporal fossa, the jugular bulb, and the internal carotid artery.

Reference

1. Helms J (1978)

Plate from Helms [1]

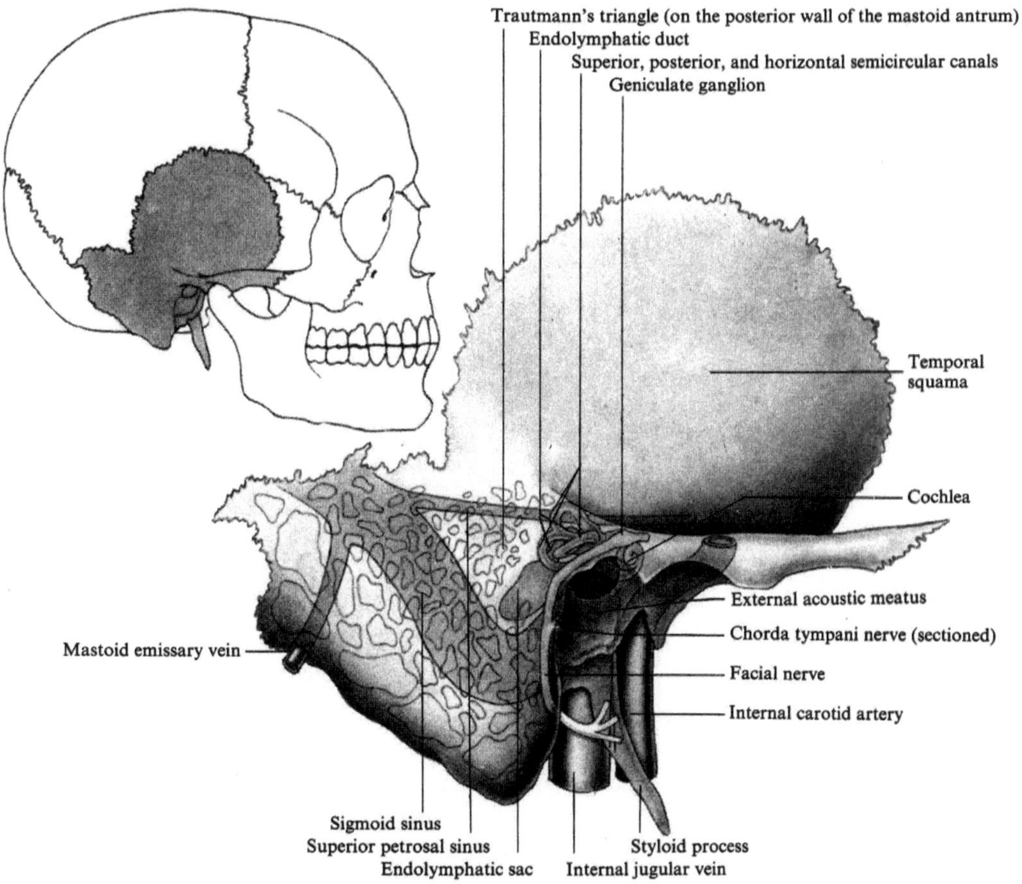

8.7 (Plates 1–2) The Temporal Bone

8.7 (Plate 1) Frontal Section of Körners Septum Through the Mastoid Antrum

Plate 1 is a frontal section through the antrum and the mastoid process. The temporal squama forms the anterior superficial part of the mastoid process. Its pneumatization originates from the so-called superior saccus. The petrous part forms the posterior and medial part of the mastoid process, which is pneumatized by the medial saccus. *Körner's septum* represents the bony partition at the site of union between the petrous and squamous portions of the temporal bone. This bony plate can be an obstacle to proper orientation when surgically approaching the antrum and the tympanic cavity [1].

Surgical Importance

Mastoidectomy, antrotomy, facial nerve surgery, tympanoplasty.

Reference

1. Proctor B and Nager ST (1982)

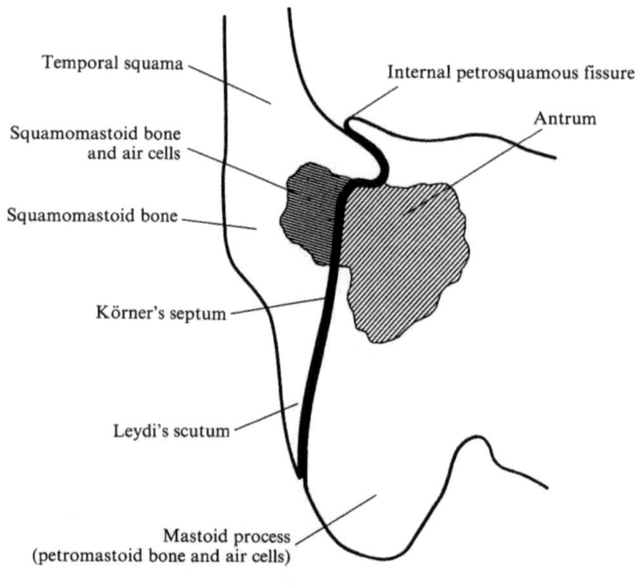

8.7 (Plate 2) Frontal Section of the Anterior Petrosquamous Suture (Anterior View)

Plate 2 is a frontal section at the level of the root of the zygomatic arch. The internal petrosquamous suture extends from the Glaserian fissure over the roof of the middle ear to the mastoid [1]. Körner's septum (the petrosquamous lamina) forms the border between the petrous and squamous parts of the mastoid process. The petrous part can cover the squamous part or result in its deflection into the middle ear. This latter occurrence may result in some fixations of the malleus. On frontal section, the petrous part displays two processes: the inferior one is the tegmental crest and penetrates between the temporal squama and pars tympanica to form the external petrosquamosal fissure or the (*Glaserian*) petrotympanic fissure with the adjacent bones. The chorda tympani nerve (in front) and the anterior tympanic artery and anterior ligament of the malleus (behind) pass through the petrotympanic fissure. The superior process forms the internal petrosquamosal suture with the temporal squama.

Surgical Importance

Local anesthesia of the ear (see also Sect. 8.7.1).

Reference

1. Proctor B and Nager ST (1982)

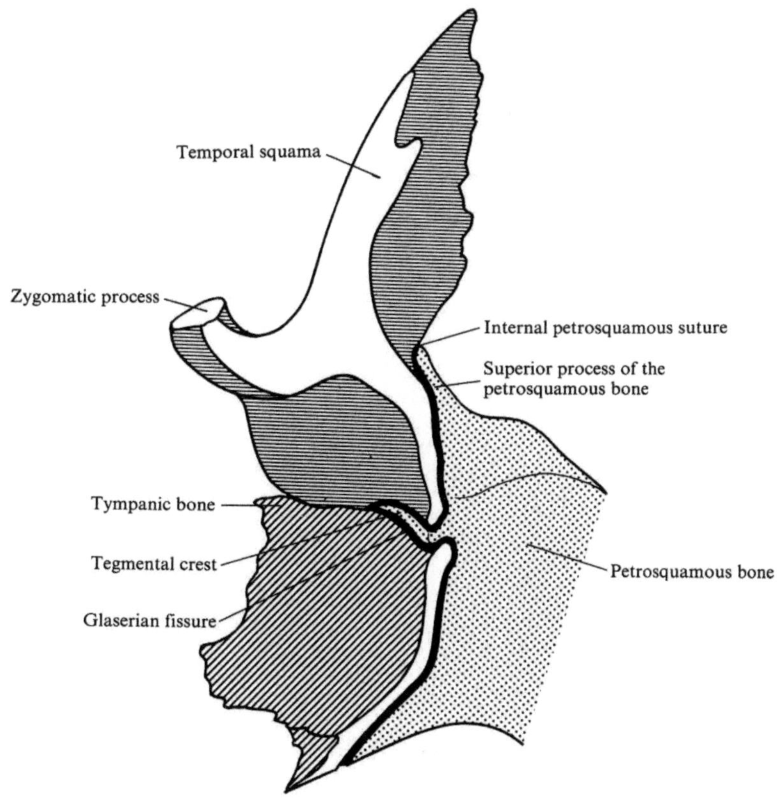

The distance of the sigmoid sinus from the posterior margin of the external acoustic meatus can vary considerably [1–3]. In extensive pneumatization of the pyramid, the sigmoid sinus is rarely an obstacle to temporal bone dissections (Plate 1). In contrast, an anteriorly placed sigmoid sinus (Plate 2) can hinder opening of the mastoid process for access to the antrum, tympanic cavity, and labyrinth, surgery of the internal auditory meatus and pontocerebellar angle, and the performance of transmastoid cisternoscopy [1–3].

Surgical Importance

Antrotomy, mastoidectomy, translabyrinthine surgery, thrombectomy from the sigmoid sinus or transverse sinus, cisternoscopy, operations on the internal auditory meatus and the cerebellopontine angle.

References

1. Denecke HJ and Guleke N (1963)
2. Nikolić V and Rudež V (1966)
3. Protor B and Nager ST (1982)

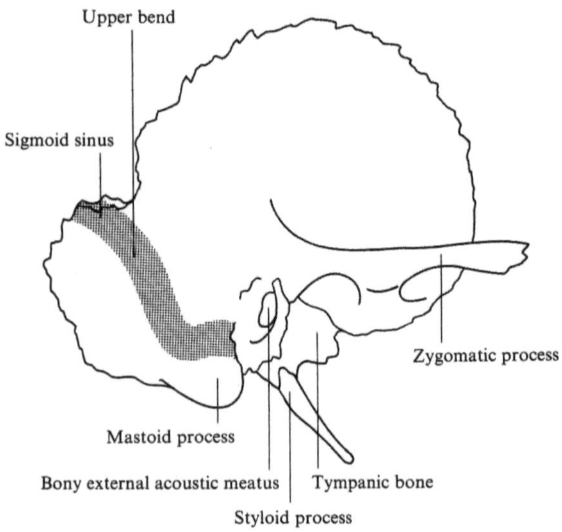

1

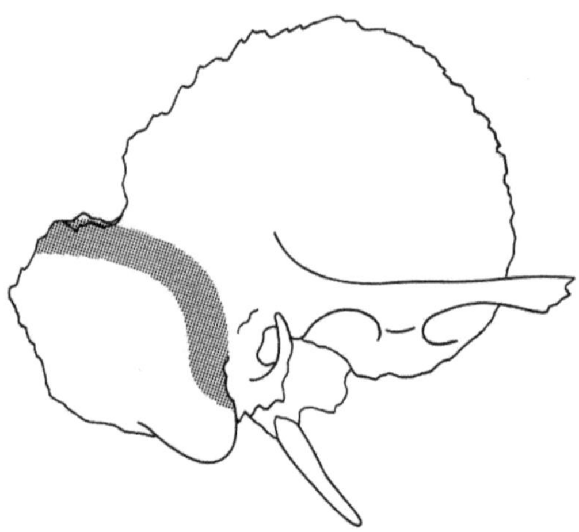

2

Plates 1–3 are posterosuperior views of the petrous pyramid and three variations in the location of the jugular bulb. Plate 1 depicts the normal relationship of the jugular bulb to the internal auditory meatus and the jugular foramen. In Plate 2, the jugular bulb is placed anterior to the labyrinth, and can be seen or even "glitter" through the bone of the middle cranial fossa (personal observation). In Plate 3, the jugular bulb is laterally displaced. In this latter case, the vessel can form the floor of the tympanic cavity as the result of a large bony dehiscence [2–4]. If the bulb impacts against the incudostapedial joint, conductive deafness may occur.
Anomalies in the size and position of the jugular bulb can be seen on mastoid X-rays, tomography, jugular phlebography, computed tomography, and densitometry, but may lead to a false diagnosis of tumor. Additionally, the internal carotid artery may also have an anomalous course to be situated in the floor of the middle ear, but this occurrence is not shown in the present Plates.

Surgical Importance

Surgery of the petrous pyramid (inclusive of glomus tumors), surgery of cerebellopontine angle tumors and tumors of the jugular foramen.

References

1. Helms J (1978)
2. Šercer A and Krmpotić J (1960)
3. Subotić R (1979)
4. West JM et al. (1974)

Plates 1–2 from Helms [1]

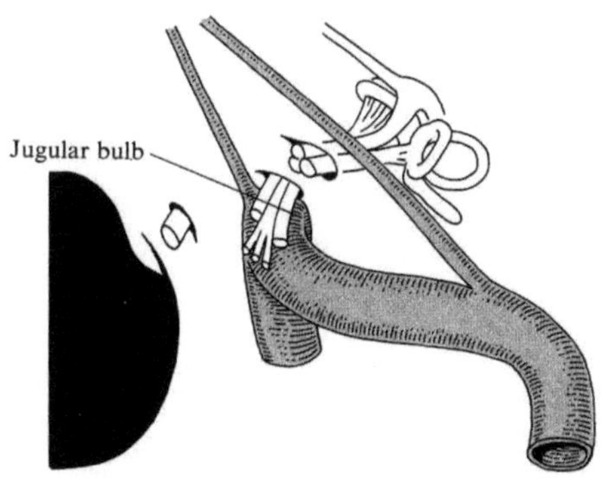

1

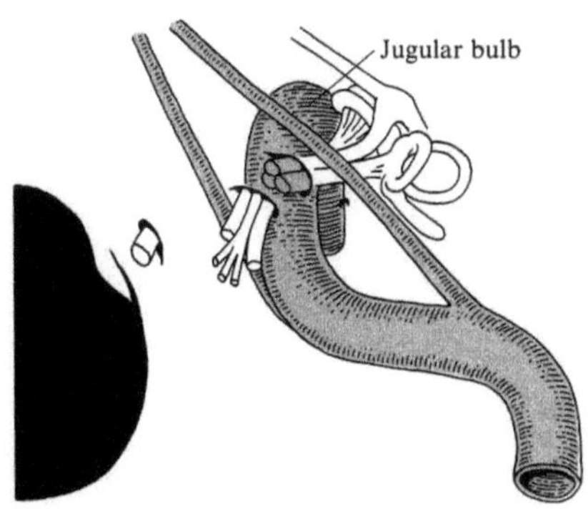

2

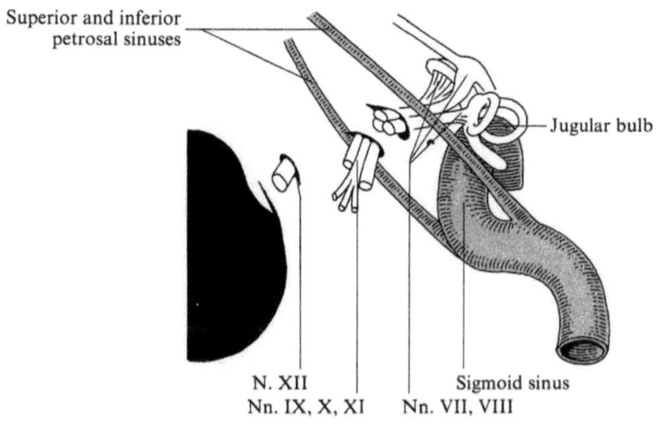

3

8.10 Schematic Illustration of the Blood Supply of the Middle Ear

A. carotis externa
  A. maxillaris
    A. tympanica anterior
    A. meningea media, A. tympanica superior
  A. pharyngea ascendens
    A. tympanica inferior
  A. auricularis posterior
    A. stylomastoidea
      A. tympanica posterior

A. carotis interna
  Rr. caroticotympanici

The Plate shows the derivation of the blood supply of the middle ear from the external and internal carotid arteries. The anterior tympanic artery, the middle meningeal artery and the superior tympanic artery are branches of the maxillary artery. The inferior tympanic artery is a branch of the ascending pharyngeal artery. The posterior tympanic artery is a branch of the stylomastoid artery, which is a branch of the posterior auricular artery. The only vessels derived from the internal carotid artery are the caroticotympanic branches. As shown by the Plate, the external carotid artery is the major source of the blood supply to the middle ear. In inflammations and tumors of the middle ear, the diameter of certain of the vessels can increase to impede any surgery performed [2].

Surgical Importance

Microsurgery of the ear, tumor surgery of the pyramid, facial nerve decompression.

References

1. Helms J (1978)
2. Lasjaunias P and Moret J (1978)

Plate from Helms [1]

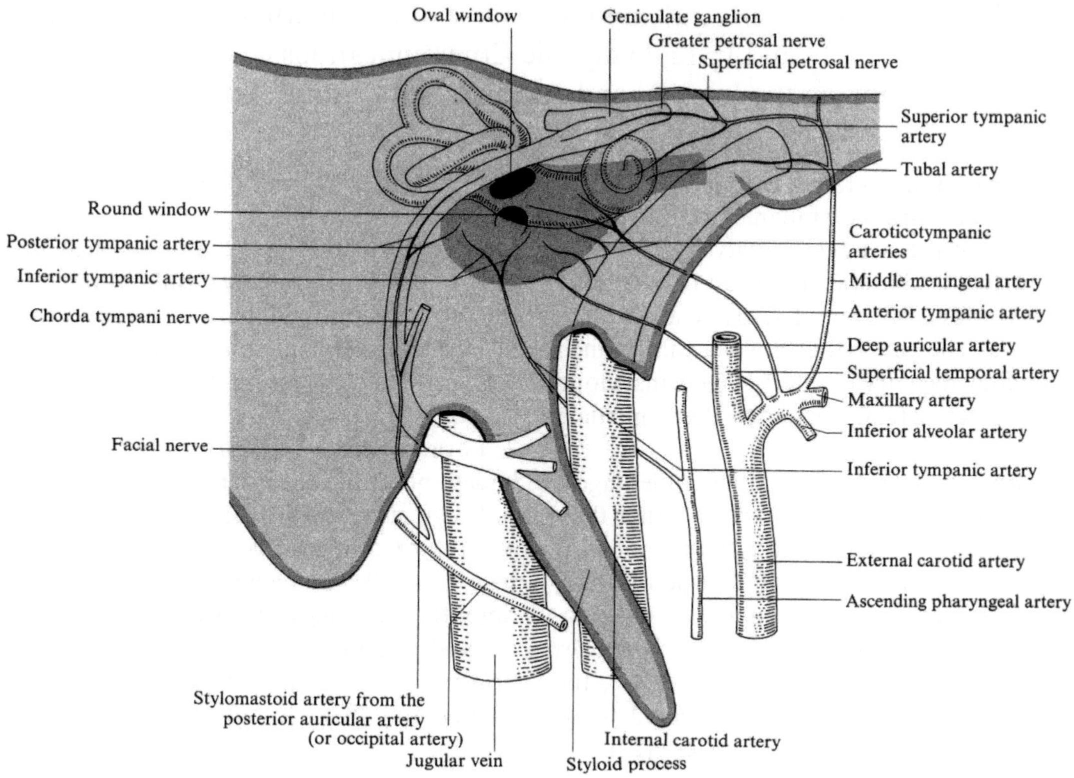

Oval window
Geniculate ganglion
Greater petrosal nerve
Superficial petrosal nerve
Superior tympanic artery
Tubal artery
Round window
Posterior tympanic artery
Inferior tympanic artery
Chorda tympani nerve
Caroticotympanic arteries
Middle meningeal artery
Anterior tympanic artery
Deep auricular artery
Superficial temporal artery
Maxillary artery
Inferior alveolar artery
Facial nerve
Inferior tympanic artery
External carotid artery
Ascending pharyngeal artery
Stylomastoid artery from the posterior auricular artery (or occipital artery)
Jugular vein
Internal carotid artery
Styloid process

8.11 (Plates 1–4) The Intratemporal Course of the Internal Carotid Artery and Intra-Tympanic Communications Between the Internal and and External Carotid Arteries

In Plate 1, the normal internal carotid artery gives off one or more caroticotympanic arteries to the anterior part of the tympanic cavity [1, 3, 5]. The stapedial artery is obliterated but can occasionally be seen as a strand of connective tissue at the arch of the stapes. The petrosal artery is a thin branch of the middle meningeal artery that runs along with the greater petrosal nerve to the geniculate ganglion.

Plate 2 shows the origin of the middle meningeal artery from the internal carotid artery rather than from the maxillary artery. Early in its course, it gives rise to the stapedial artery. The petrosal artery allows blood to flow retrogradely in the direction of the foramen spinosum and the branches of the middle meningeal artery. In this case, the caroticotympanic artery can be considered to be the point of origin of the stapedial artery [10].

Plate 3 shows the middle meningeal artery originating from the petrosal artery instead of the maxillary artery. The geniculate ganglion and the petrosal artery receive their blood supplies from an inferior tympanic artery, which is a branch of the occipital artery. In this illustration, the inferior tympanic artery is a large vessel which runs over the promontory in front of the stapes and not between its crura [11].

In Plate 4, the extra-cranial part of the internal carotid artery and the carotid foramen are missing. The course of the artery corresponds to an anastomosis between the inferior tympanic artery and the caroticotympanic artery [2, 4, 9]. From the origin of the latter vessel, the internal carotid artery takes its normal course up to the cavernous sinus. This anomaly can occur at the same time as persistence of the stapedial artery.

The vascular anomalies shown in Plates 2–4 are found occasionally during microsurgery of the ear [6, 7]. From a clinical standpoint, the following must be borne in mind:

1. Surgical sectioning of any atypical vessels in the tympanic cavity may result in impaired neurologic function (hemiparesis, cerebral infarction, facial paralysis), and should therefore be avoided – even when considering dividing arteries of small caliber.
2. Large vessels which protrude through any dehiscent bone into the tympanic cavity to cause possible impairment of hearing should be preserved and returned if possible to their original bony canals, the walls of which must then be reinforced.

Smooth pulsating protrusions indicate a vascular anomaly (involving the internal or external carotid artery or jugular bulb), while an irregular protrusion is more likely to represent a glomus tumor.

Surgical Importance

The "blue" tympanic membrane, surgery of the middle ear, glomus tumors, anteposition of the jugular bulb, pulsatile tinnitus.

References

1. Adachi B (1928)
2. Evans TH (1956)
3. Fischer A G T (1915)
4. Goodman RS and Cohen NL (1981)
5. Helms J (1978)
6. Killian FC et al. (1980)
7. La Scala G et al. (1978)
8. Lasjaunias P and Moret J (1978)
9. Poncet P and Miller P (1979)
10. Quisling RS (1980)
11. Storrs LA (1976)

Plates 1–4 adapted from Lasjaunias and Moret [8]

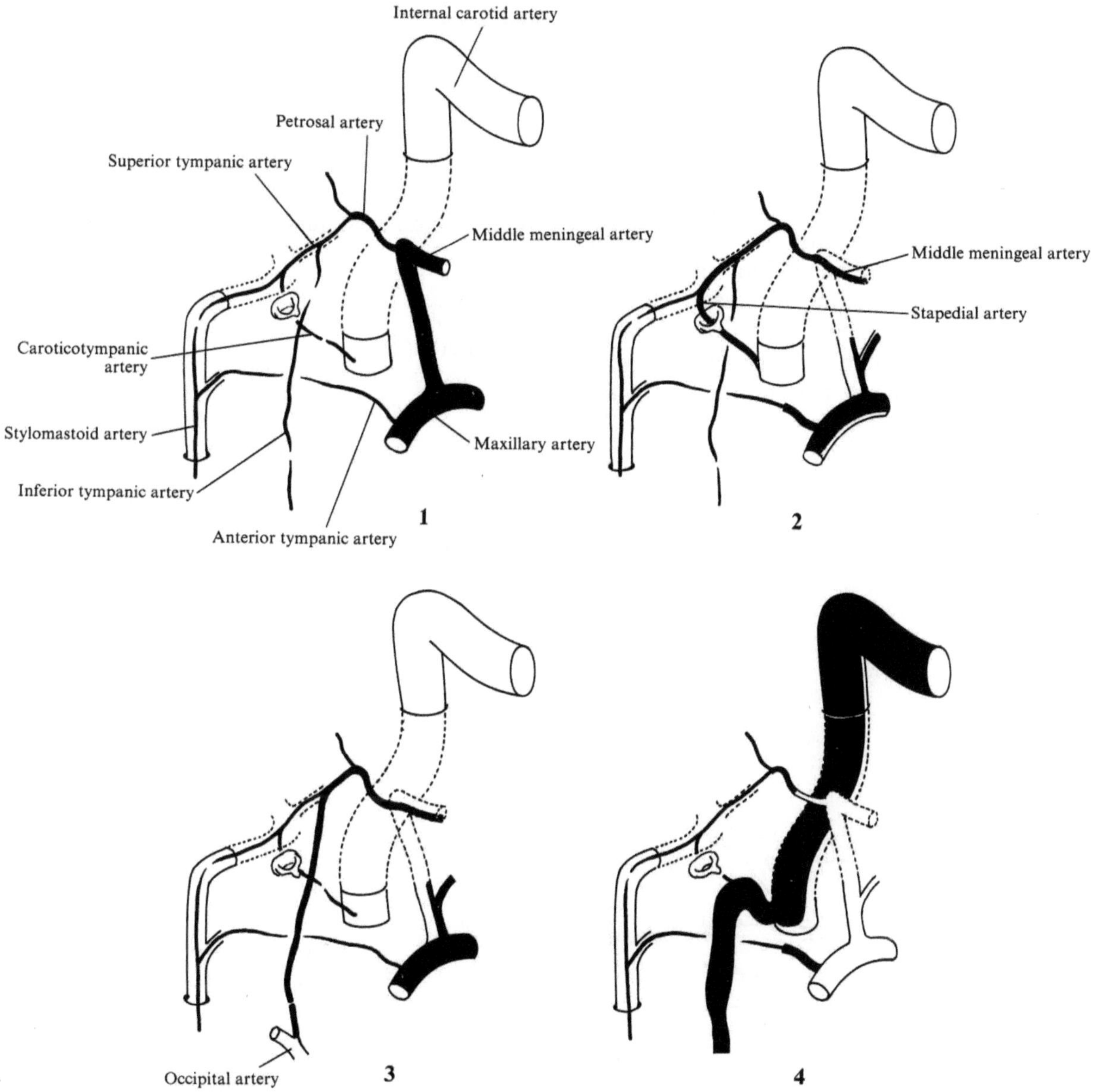
Internal carotid artery
Petrosal artery
Superior tympanic artery
Middle meningeal artery
Caroticotympanic artery
Stylomastoid artery
Inferior tympanic artery
Maxillary artery
Anterior tympanic artery
1
Middle meningeal artery
Stapedial artery
2
Occipital artery
3
4

In the 12–15 mm embryo, the stapedial artery is present as a branch of the internal carotid artery and runs between the stapedial crura to the petrotympanic fissure. The vessel then gives off supraorbital and infraorbital branches, with the mandibular branch as a common trunk [1–3].
In the 15–17 mm embryo, the external carotid artery forms an anastomosis with the mandibular branch of the stapedial artery. The infraorbital vessel lies medial to the mandibular nerve. Later, a lateral anastomosis forms to create a loop around the nerve. The medial part of the loop will obliterate while the lateral part persists. The anastomosis and the lateral part of the loop subsequently form the definitive maxillary artery. The middle meningeal artery is encircled by the roots of the auriculotemporal nerve and originates from the mandibular branch of the stapedial artery above the anastomosis and the common trunk.

Surgical Importance

Surgery for malformations of the facial and aural region, middle ear surgery.

References

1. Križan Z (1960)
2. Krmpotić-Nemanić J (1978)
3. Weidenbecher M (1981)

Plate adapted from Križan [1]

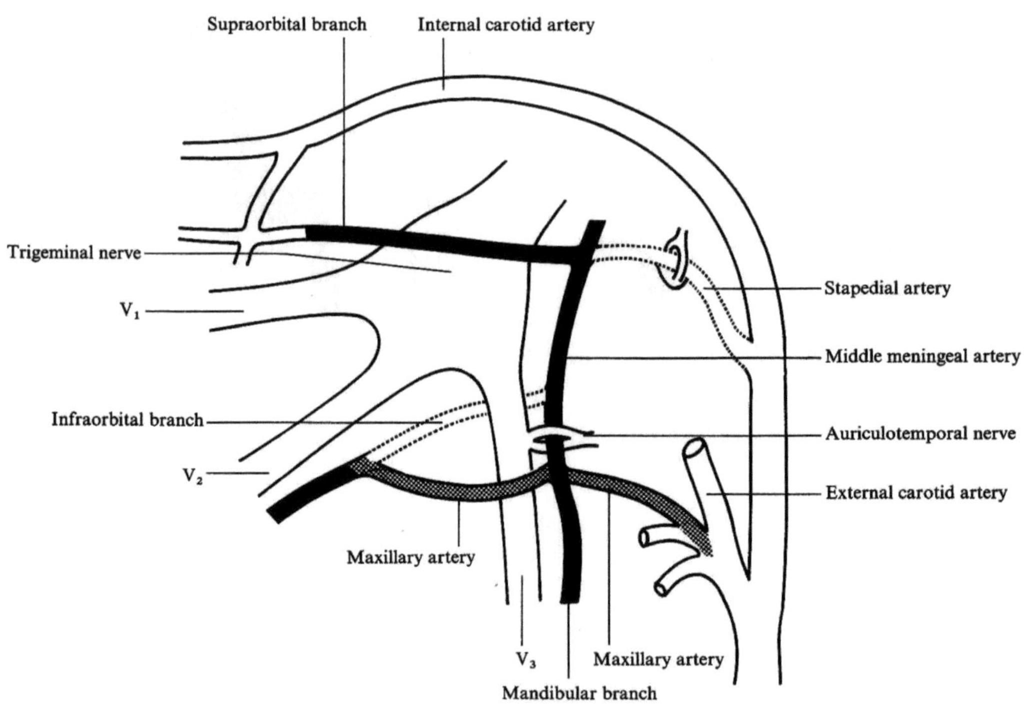

8.13 Topographic Relationships of the Facial Nerve to the Middle and Inner Ear Structures (Lateral View)

The Plate is a schematic representation of the facial nerve in relationship to the structures of the middle and inner ear. This illustration is useful only as an approximation of the nerve's course since the visual angles used during operations performed in this region must be constantly changed. When looking through the external auditory meatus, the incus and the head of the malleus project over the horizontal and superior semicircular canals. In the transmastoid view, these structures project in front of the canals close to the geniculate ganglion and the cochlea. During mastoid surgery – and particularly in surgery of the labyrinth and the internal auditory meatus, a three-dimensional orientation based on a two-dimensional situation is vital. However, orientation from a surgical atlas is insufficient for understanding surgical anatomy at the actual time of operation. If preservation of cochlear function is desired, orientation on sections through the pyramid is indispensable.

Topographic orientation of the ossicles to the inner ear structures also depends on the direction of the surgeon's visual axis (e.g., through the external acoustic meatus or through the mastoid). This should be kept in mind at surgery when looking for the horizontal semicircular canal and determining its relationship to the incus. The endolymphatic sac lies behind the inferior half of the posterior canal.

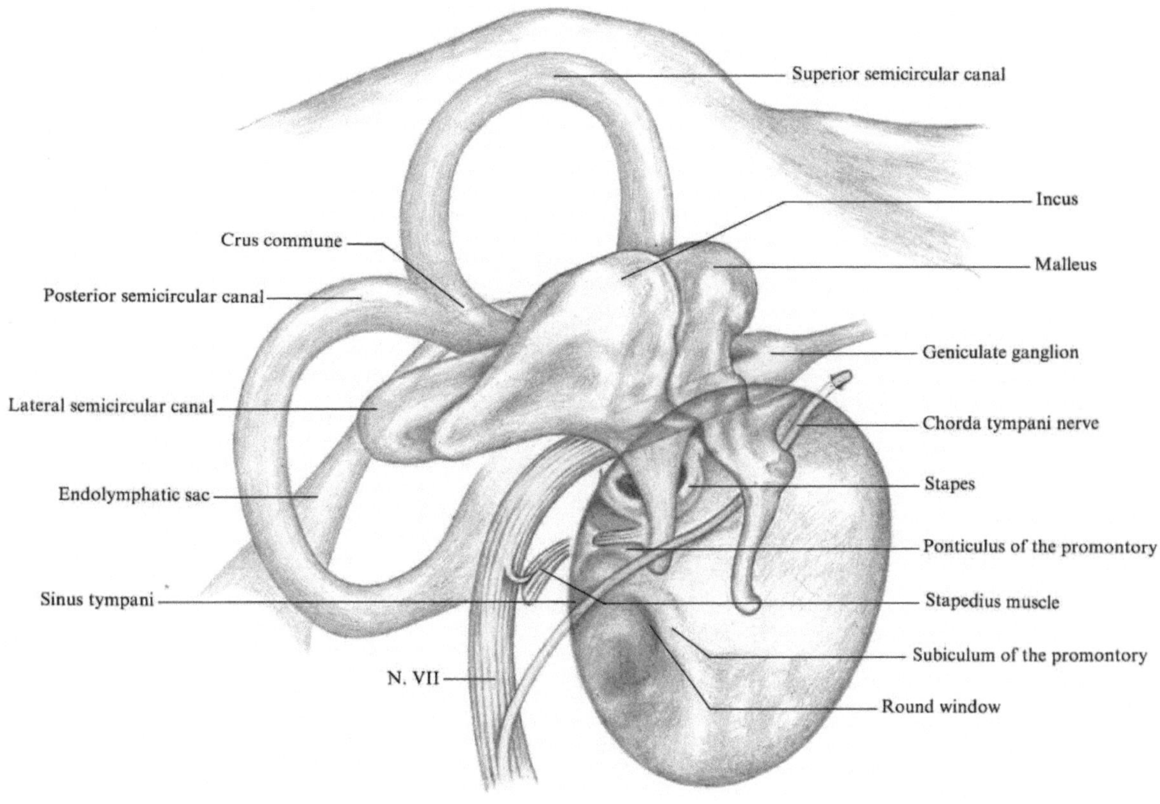

8.14 Topography of the Posterior Genu of the Facial Nerve and its Relationship to the Stapedius Muscle (See Sect. 8.13)

The recess in the posterior part of the tympanic cavity is of great importance in otologic surgery [1, 2]. The sinus tympani is normally situated behind the notch of the round window beneath the *ponticulus*. This latter structure is a bony ridge extending from the promontory to the pyramidal process. The sinus tympani can extend posteriorly below the facial nerve so that it can be approached surgically beneath the horizontal semicircular canal medial to the facial nerve. If a cholesteatoma extends into such an extremely large sinus, its excision can be technically difficult. The *subiculum promontorii* is a bony projection of the promontory over the notch of the round window, and on occasion must be removed to expose the round window membrane. The position of the stapedius muscle varies in relationship to the facial nerve, with the bulk of the muscle lying lateral or sometimes medial to the nerve (see Sect. 8.13).

Surgical Importance

Microsurgery of the middle and inner ear, the mastoid, the intratemporal portion of the facial nerve, and the internal auditory meatus; cholesteatoma surgery, tympanotomy to repair round window membrane rupture.

References

1. Gerhardt HJ and Otto HD (1981)
2. Marquet J (1981)

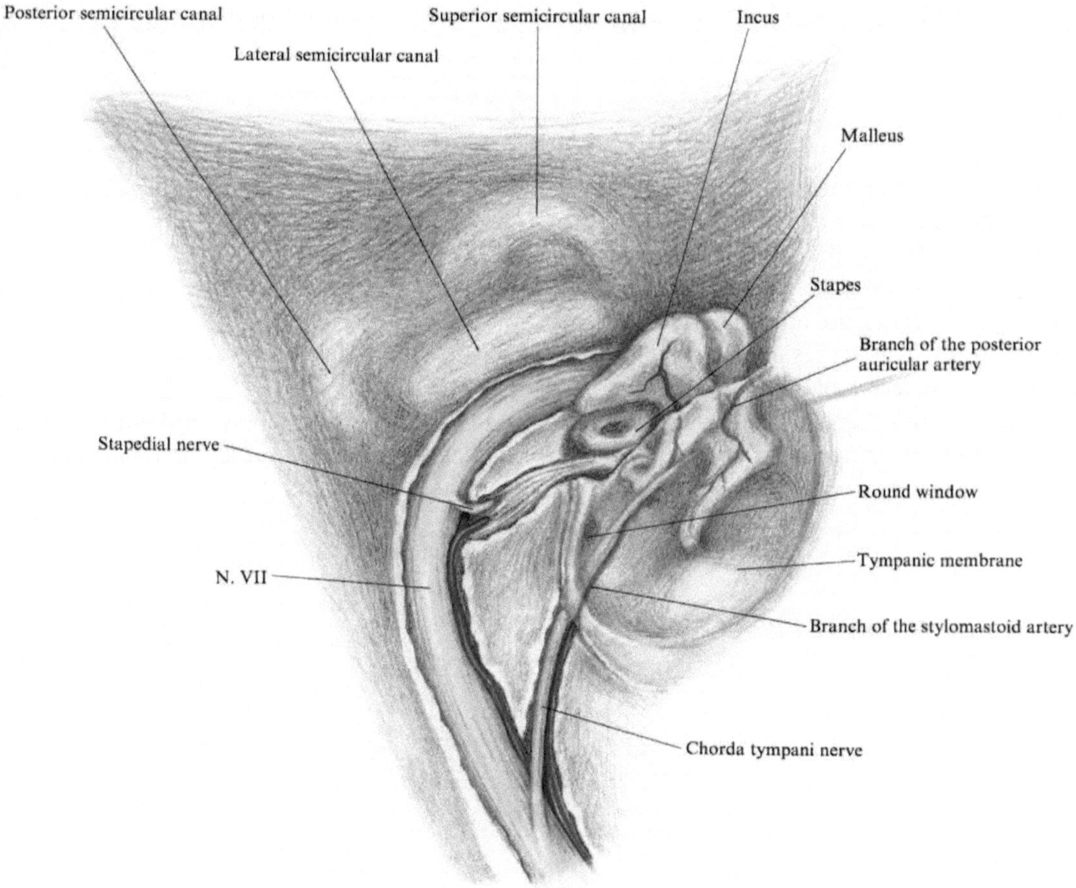

Posterior semicircular canal
Superior semicircular canal
Incus
Lateral semicircular canal
Malleus
Stapes
Branch of the posterior auricular artery
Stapedial nerve
Round window
Tympanic membrane
N. VII
Branch of the stylomastoid artery
Chorda tympani nerve

Plate 1 is a detailed illustration of a temporal bone specimen while Plate 2 is a schematic drawing of vital structures within the bone. The so-called "bony bridge" is a surgical artefact which appears when the lateral part of the bony external auditory canal is removed. This leaves only a ridge of bone containing the posterior part of the tympanic annulus. The posterior part of the tympanic membrane is fixed to the bridge. Additionally, the inferior part of the bridge lies in close proximity to the facial nerve.

Surgical Importance

Microsurgery of the ear: antrotomy, simple and radical or modified radical mastoidectomy.

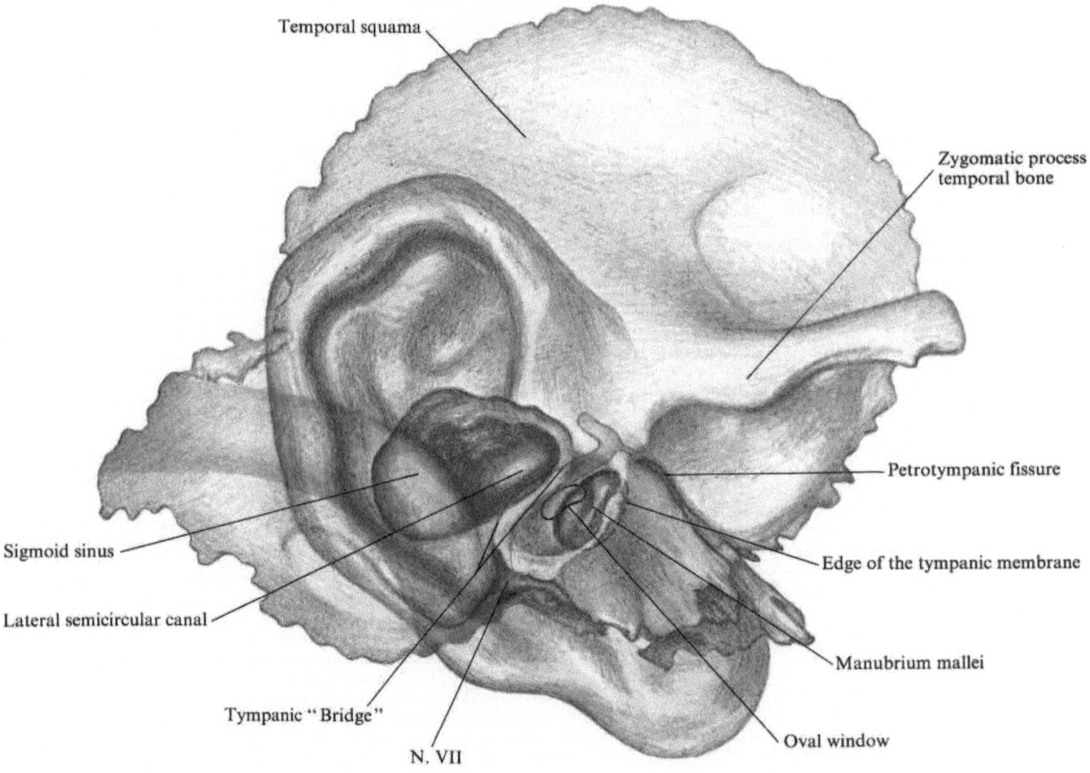

1

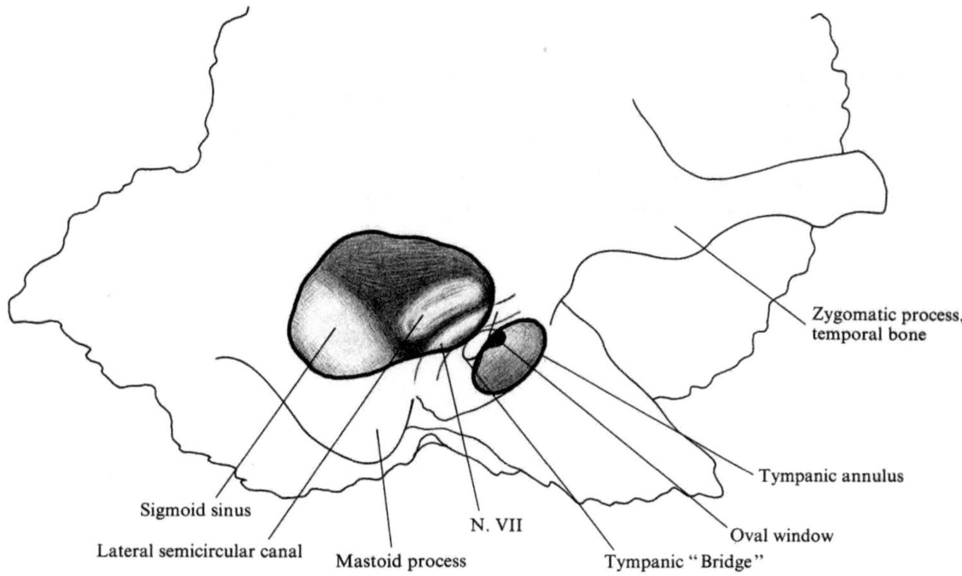
Zygomatic process,
temporal bone
Tympanic annulus
Sigmoid sinus
N. VII
Oval window
Lateral semicircular canal
Mastoid process
Tympanic "Bridge"

2

8.16 The Right Petrous Pyramid From Lateral and Below After Radical Mastoidectomy and Partial Labyrinthectomy

The semicircular canals and the facial canal have been opened and the posterior wall of the external auditory meatus removed. The transverse and sigmoid sinuses, the jugular bulb and the internal jugular vein, and portions of the carotid artery are schematically shown through the overlying bone.

Surgical Importance

Surgery of the sigmoid sinus and jugular bulb, surgery of the lateral intratemporal internal carotid artery, facial nerve decompression, labyrinthectomy.

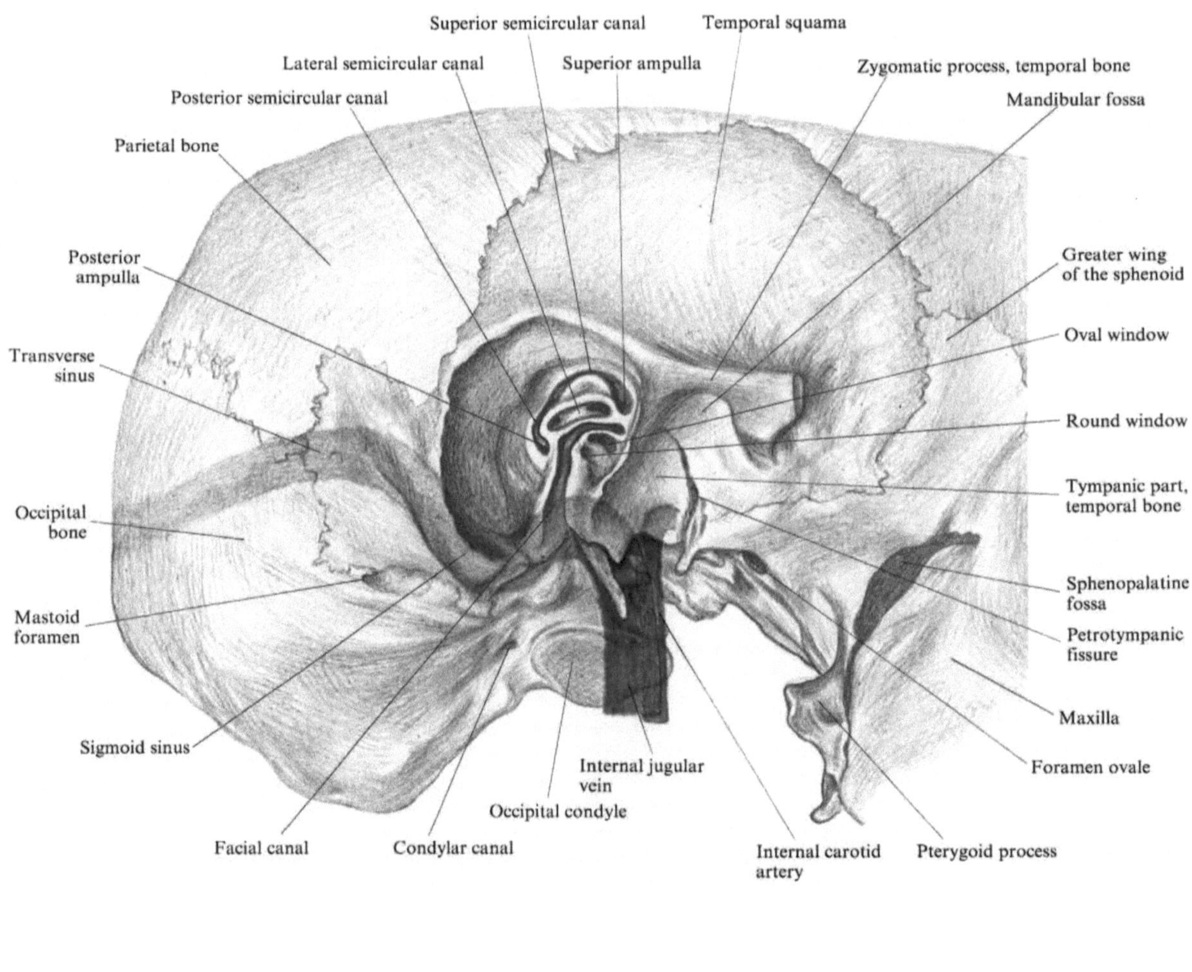

8.17 Lymphatic Drainage of the External Auditory Canal, Middle Ear, and Eustachian Tube

The lymph drainage from the external ear can be divided into three zones: the *anterior zone* (tragus and external meatus) drains to the preauricular and parotid nodes, the *inferior zone* (lobule, helix, cavum conchae, and inferior part of external auditory meatus) to the infraauricular nodes, and the *posterior zone* (upper part of meatus, anthelix, and auricle) to the retroauricular and deep cervical nodes. Most of the lateral portion of the tympanic cavity (and the tympanic membrane) are drained to the infraauricular nodes, while the medial part of the tympanic cavity and the Eustachian tube drain to the lateral retropharyngeal and deep cervical nodes (see Sect. 1.23). This large system for lymphatic drainage explains the poor prognosis for medially growing malignant tumors involving the ear. Lymph node metastases in such cases are not only difficult to diagnose, but are also sometimes impossible to remove completely [1].

Surgical Importance

Inflammatory and neoplastic processes of the external and middle ear and Eustachian tube (e.g., *otitis externa necroticans*), resections of the auricle or temporal bone.

Reference

1. Cachin Y et al. (1969)

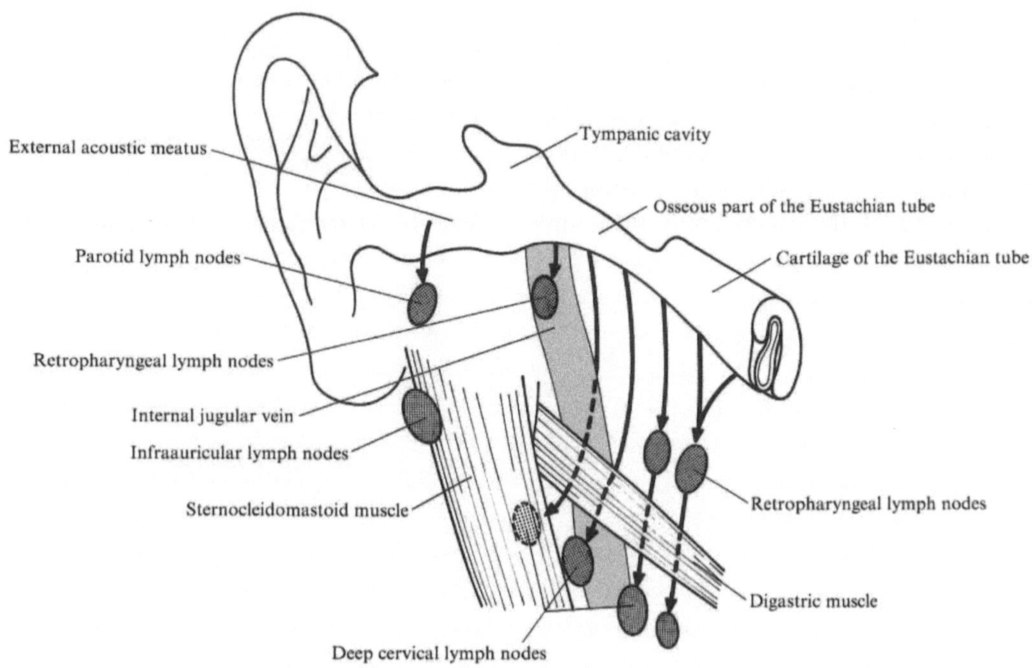

8.18 (Plates 1–15) Variations in the Intratemporal Course of the Facial Nerve

Plates 1–15 depict the various patterns seen clinically in the intratemporal course of the facial nerve. *Plate 1*: Normal course of the facial nerve; *Plate 2*: posterior displacement of the lateral genu of the facial nerve [3]; *Plate 3*: posterocaudal displacement of the lateral genu of the facial nerve [3]; *Plate 4*: displacement of the lateral genu of the nerve to the vertical part of the sigmoid sinus [3]; *Plate 5*: flattening of the lateral genu of the nerve with its displacement and narrowing of the notch of the oval window [3]. *Plate 6*: vertical course of the facial nerve from the geniculate ganglion across the promontory [3]; *Plate 7*: anteromedial displacement of the facial nerve in the notch of the oval window at the level of the stapes footplate (personal observation); *Plate 8*: loop formation of the facial nerve through the stapedial crura [7]; *Plate 9*: protrusion of the facial nerve through a dehiscent facial canal and dehiscent facial sheath in front of and behind the stapes (personal observation); *Plate 10*: reduplicated facial nerve passing around the stapes, with one part of the nerve running free through the tympanic cavity while the other (tympanic) part is situated in the normal facial canal [8]; *Plate 11*: divided course of the facial nerve in the mastoid process [5]; *Plate 12*: crossing of two divisions of the facial nerve in the mastoid bone and the stylomastoid foramen [3]; *Plate 13*: division of the facial nerve into two branches below the lateral genu [8]; *Plate 14*: division of the facial nerve into three branches in the mastoid [1]. *Plate 15*: displaced entry of the facial nerve lateral to the labyrinth. The geniculate ganglion is reached by following a line drawn parallel to the lateral semicircular canal. From this point, the course of the nerve is relatively normal [2].

Developmental anomalies in the course of the facial nerve have been reviewed by Gerhardt and Otto [4], who suggest that these occur as the result of a mutual shifting of the branchial arches in a partial hypoplasia.

Surgical Importance

Surgery of the middle and inner ear; intratemporal facial nerve surgery.

References

1. Botmann JWM and Jongkees LBW (1955)
2. Feldmann H (1975)
3. Fowler EP (1961)
4. Gerhardt HJ and Otto DH (1981)
5. Halbrock KH (1960)
6. Helms J (1978, 1980, 1981)
7. Johnson LG and Kingsley TC (1970)
8. Miehlke A (1973)
9. Proctor B and Nager GT (1982)
10. Samii M and Janetta OJ (1981)

Plates 1–15 from Helms [6] and Samii and Janetta [10]

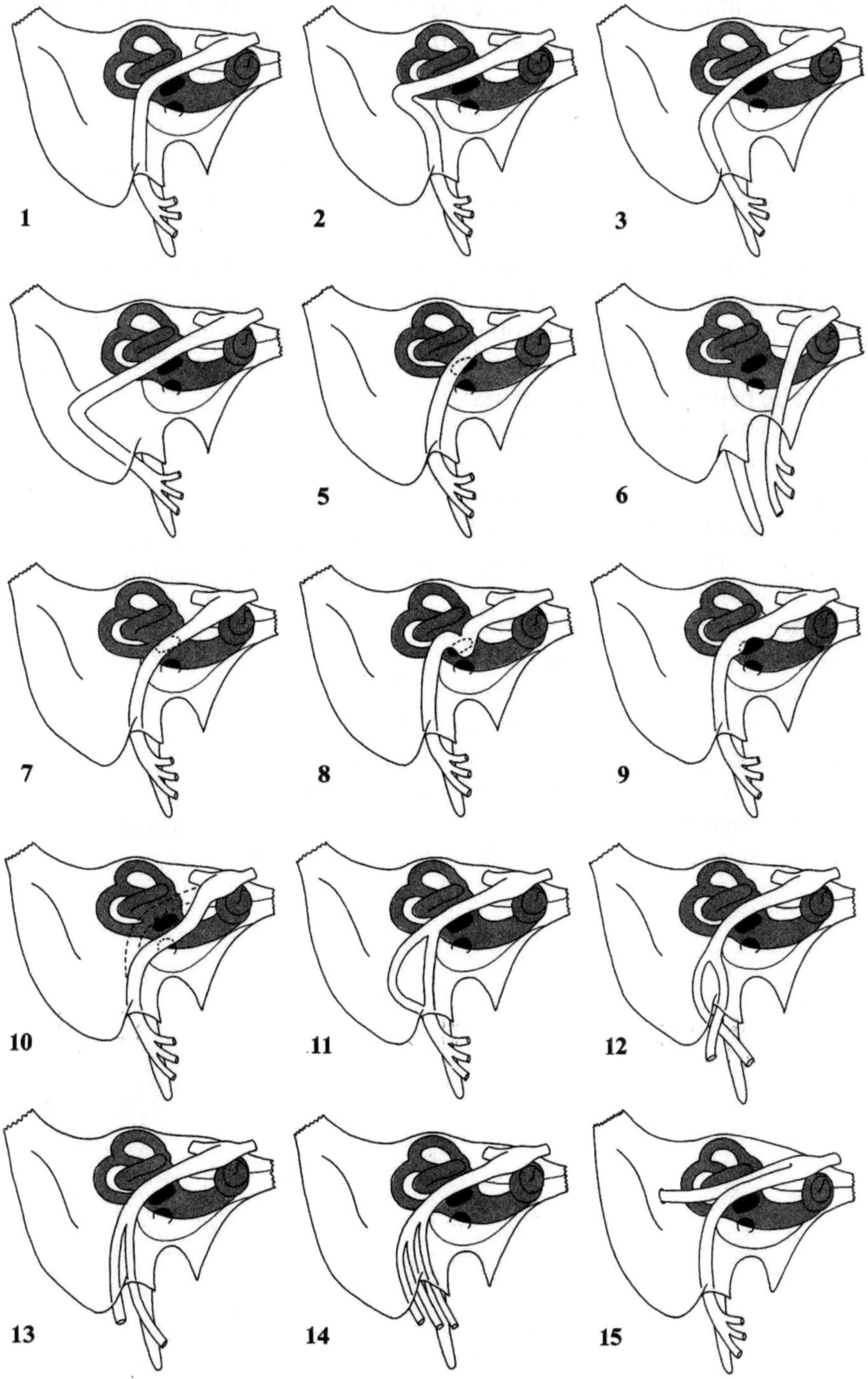
1
2
3
5
6
7
8
9
10
11
12
13
14
15

8.19 The Schematized Arterial Supply of the Intratemporal Facial Nerve (Lateral and Superior View)

The intratemporal facial nerve is supplied by an arterial arcade that is formed by an anastomosis between the petrosal artery and the stylomastoid artery. A small pregeniculate portion of the nerve is supplied only by the petrosal artery. The existence of any direct communications between the arteries and veins along the facial nerve has not yet been clarified [3]. The petrosal artery originates from the middle meningeal artery in the region of the foramen spinosum or 15–20 mm above it. In certain cases, this vessel may be reduplicated. On rare occasions, it originates from the internal carotid artery, giving off branches to the dura and the subarcuate fossa. The vessel may also anastomose with the superior tympanic artery. In the Fallopian canal, the artery can further branch into many small or two larger vessels. In the region of the stylomastoid foramen, the facial nerve is supplied by the stylomastoid artery, which ramifies in the sheath of the nerve into many small or two larger branches and then anastomoses with the petrosal artery in the tympanic cavity (see Sect. 8.30, 11.16) [2, 4–6]. When there is a persistence of the stapedial artery, the blood supply of the facial nerve may also be derived from this vessel.

The axonal tissue of the facial nerve has a relatively poor vascularity [1], even though the interfascicular vessels anastomose with the segmental vessels of the epineurium.

Surgical Importance

Intratemporal and middle cranial fossa surgery, facial nerve decompression.

References

1. Anson et al.
2. Lang J (1979)
3. Maffei G and Zinni C (1967)
4. Mercier P et al. (1982)
5. Minatogawa T et al. (1980)
6. Sciascia R et al. (1973)

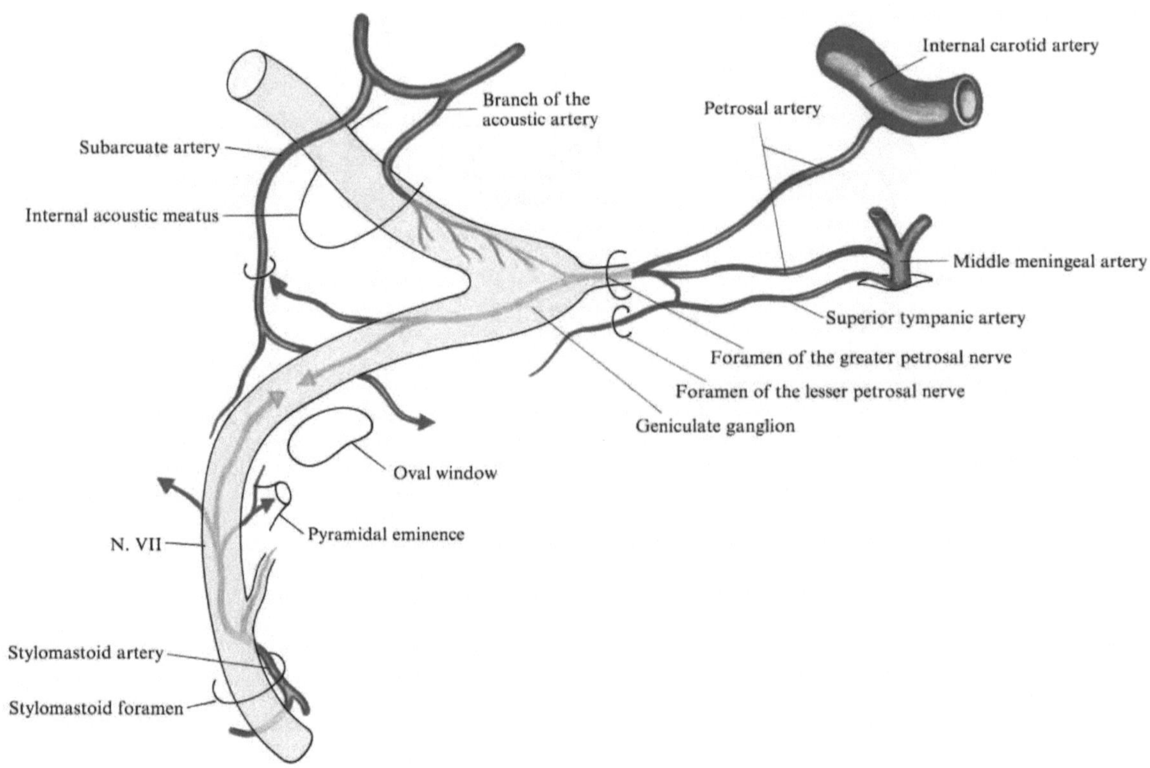

8.20 (Plate 1–4) Variations in the Course of the Tympanic Nerve

As shown in the Plates, the tympanic nerve originates from the inferior ganglion of the glossopharyngeal nerve and reaches the tympanic cavity through the tympanic canal. This canal is located within a bony septum between the internal jugular vein and internal carotid artery. The tympanic nerve may also (although rarely) be formed jointly from the glossopharyngeal and vagus nerves.
In 52% of cases, the tympanic nerve runs as a single trunk to the level of the round window (Plate 1). In 34% of cases, it gives one branch to the hypotympanum (Plate 2) and two branches in 10% of cases (Plate 3). In 2% of cases, there are two canals in the promontory that are located in front of the round window (Plate 4) [1]. In one specimen, the nerve and a branch from the sympathetic carotid plexus were found to run together to the tympanic cavity (not represented).

In the upper part of the promontory and on the roof of the tympanic cavity, the tympanic plexus is formed by the tympanic nerve, the caroticotympanic nerves, and a communicant branch from the facial nerve. This plexus is also known as the *anastomosis of Jacobson* and gives rise to the lesser petrosal nerve.

Surgical Importance

Surgery of the middle ear, interventions in disturbances of function of the parotid and submandibular glands.

Reference

1. Donaldson I (1980)

Plates adapted from Donaldson [1]

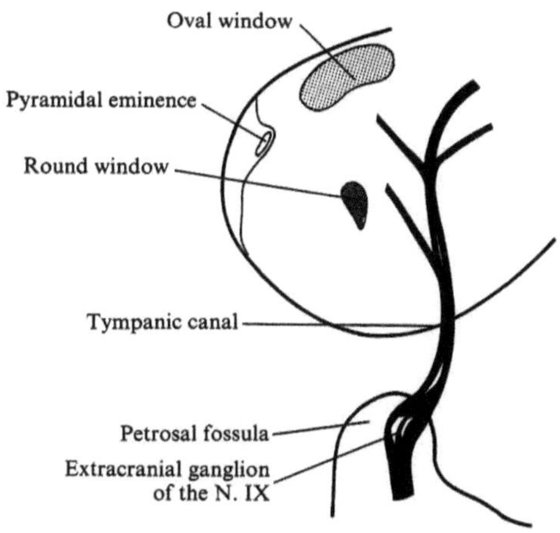

1

2

3

4

8.21 (Plates 1–2) Positional Relationships of the Stapes Foot Plate to the Structures of the Vestibule

The relationships of the structures of the membranous labyrinth to the oval window are represented schematically. The saccule is shown in the anterior half of the window while the utricle is located more posteriorly. The cochlea lies close to the inferior margin of the window. In manipulations the stapes footplate (as during stapedectomy), the utricle and saccule can be injured. Consequently, any stapes prosthesis is best placed posteroinferiorly in the oval window to avoid any vital structures in the vestibule. If the oval window must be enlarged during surgery (especially to remove a "floating" foot plate), the posteroinferior part of the bony rim should be removed to protect the underlying labyrinth.

The schematic drawing in Plate 2 is a view through the notch of the oval window and the vestibule in the region of stapedial crura. The distances of the foot plate from the saccule and utricle are indicated, and are approximately 1 mm and 1.25 mm respectively.

Surgical Importance

Stapedectomy, stapedotomy

Reference

1. Anson BJ and Donaldson JA (1973)

Plates 1 and 2 modified from Anson and Donaldson [1]

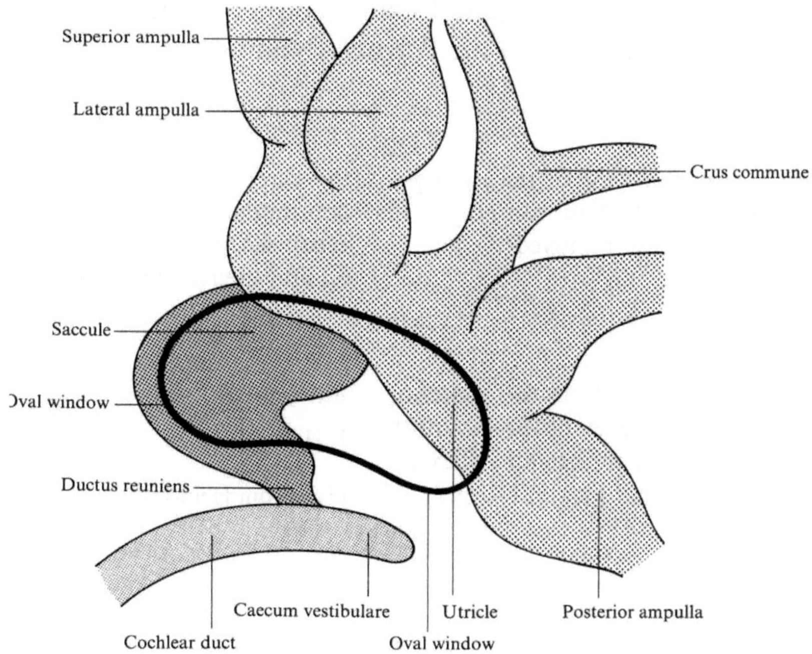
Superior ampulla
Lateral ampulla
Crus commune
Saccule
Oval window
Ductus reuniens
Caecum vestibulare
Utricle
Posterior ampulla
Cochlear duct
Oval window

1

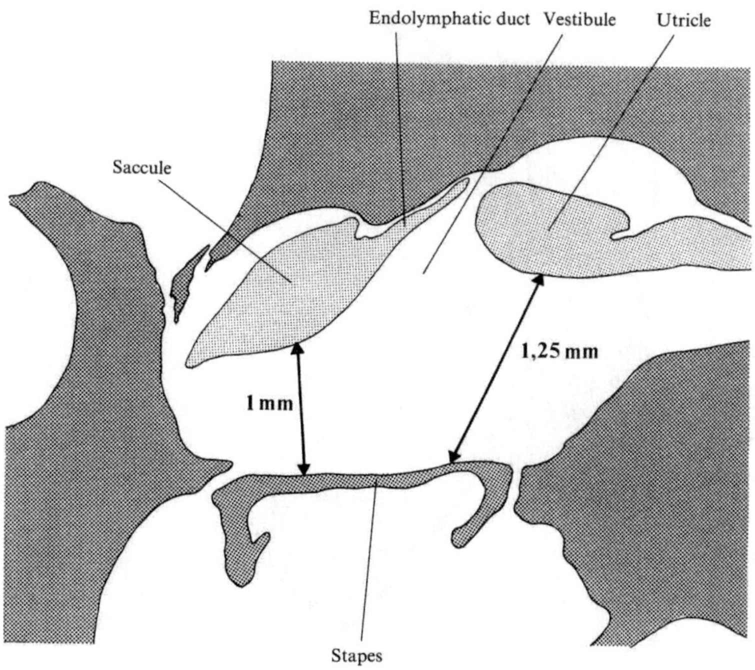
Endolymphatic duct
Vestibule
Utricle
Saccule
1,25 mm
1 mm
Stapes

2

8.22 Topographic Relationship of the Labyrinth to the Petrous Pyramid

The Plate shows a posterior view of the right petrous pyramid. Portions of the tentorium cerebelli and the falx cerebri have been removed.

Surgical Importance

Surgery of the petrous pyramid via the posterior and middle cranial fossa, surgery of the labyrinth and internal auditory meatus, transtemporal extradural exposure of the internal meatus, surgery of the cerebellopontine angle (resection of acoustic neurinomas, vestibular neurectomy (Menière's disease), saccotomy, surgery of the Gasserian ganglion and greater petrosal nerve.

Reference

1. Helms J (1978)

Plate from Helms [1]

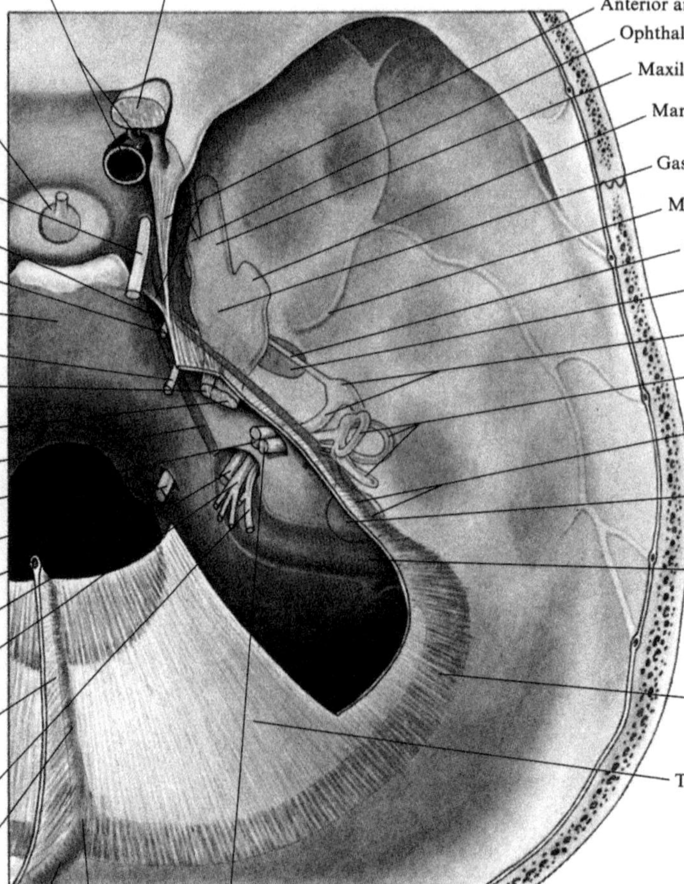

8.23 Spatial Relationships of the Labyrinth and the Nerves of the Internal Auditory Meatus Within the Temporal Bone (Posterosuperior Medial View)

The position of the labyrinth in relationship to the internal and the external acoustic meati and to the superior and posterior surfaces of the temporal bone is schematized in a posterosuperior view. The superior petrosal sinus corresponds to the crista pyramidalis. The endolymphatic sac lies mostly in the dura of the posterior surface of the bone as the *pars intramuralis*. When resecting the posterior wall of the internal acoustic meatus through the suboccipital approach, the dura of the posterior surface of the temporal bone should be removed medial to the endolymphatic sac in order to avoid injury to the cochlea. In actually removing this bone, it should be gradually thinned with a surgical drill to allow identification of the "blue line" of the posterior semicircular canal.

Surgical Importance

Suboccipital exposure of the cerebellopontine angle (see Sect. 8.22)

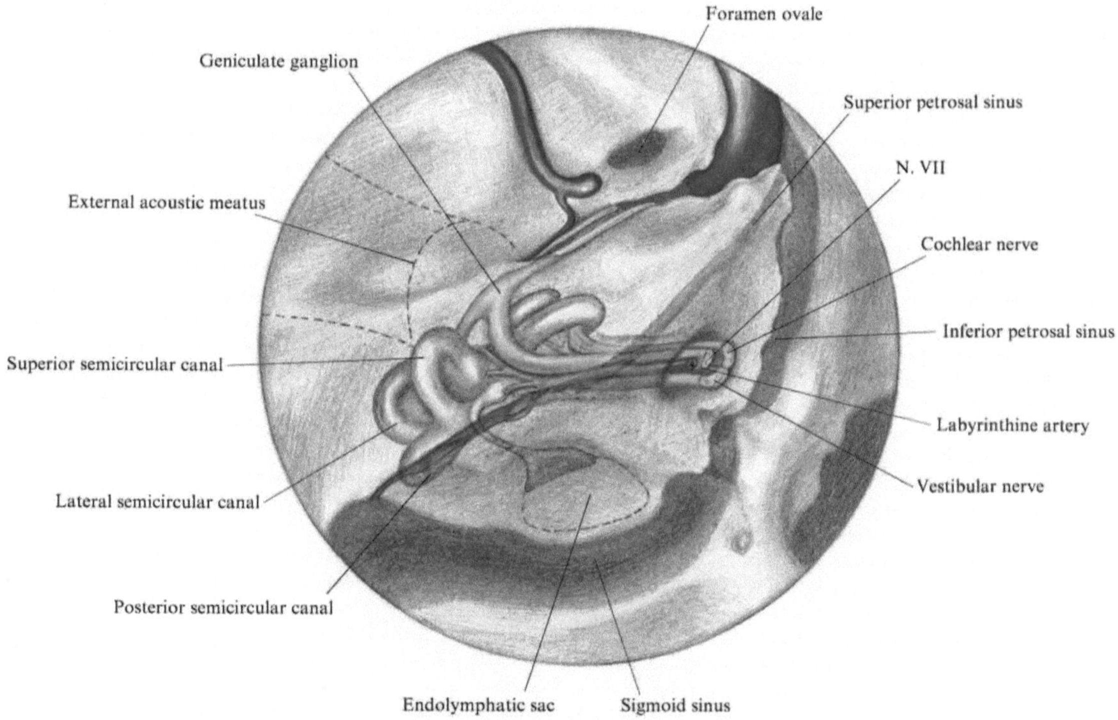

8.24 (Plates 1–2) The Posterior Surface of the Temporal Bone and the Fundus of the Internal Auditory Meatus

As shown in Plate 1, the *crista transversa* is a horizontal ridge which divides the opening of the internal acoustic meatus into a superior and an inferior part [1]. In the superior part, the orifice of the facial canal lies anterosuperiorly in front of the *crista verticalis* (*Bill's bar*). Behind the crista lies the orifice for the superior vestibular nerve. Below the transverse crest and in front is the orifice for the cochlear nerve and is represented in Plate 1 by a series of minute holes, the cribriform plate. Two openings are located posteriorly and are the orifices for the inferior vestibular nerve's branches to the saccule and to the ampulla of the posterior semicircular canal (via the foramen singulare).
In addition to the porus of the internal auditory meatus (Plate 2), the sulcus of the inferior petrosal sinus and the orifice of the vestibular aqueduct (for the endolymphatic duct) and the saccular vein from the endolymphatic sac are located on the posterior surface of the temporal bone. The orifice of the cochlear aqueduct (perilymphatic or cochlear duct) lies in front of the jugular bulb, but is usually obliterated in the adult.

Surgical Importance

Surgery of the internal auditory meatus: vestibular neurectomy, resection of acoustic neurinoma

Reference

1. Helms J (1978)

Plates 1 and 2 from Helms [1]

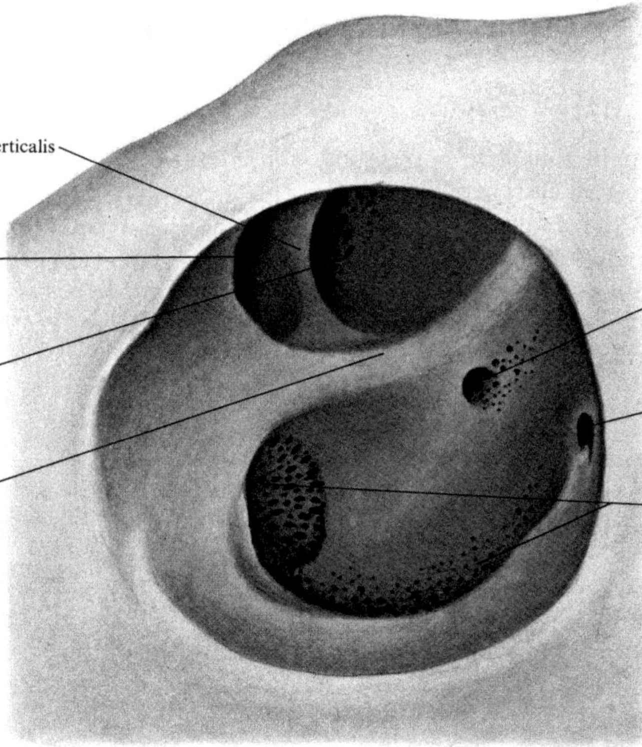
Crista verticalis
Hiatus of the facial canal
Utriculoampullary area of the vestibule
Crista transversa (Transverse crest)
Area of the vestibular saccule
Singular foramen to the posterior ampulla
Area of the cochlear tractus spiralis foraminosus

1

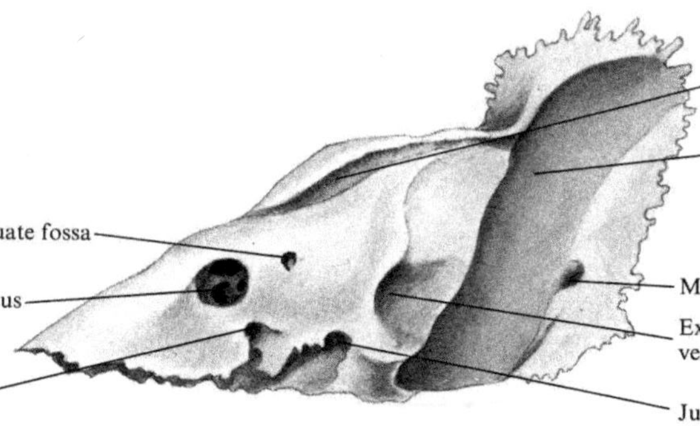
Superior petrosal sinus
Sigmoid sinus
Subarcuate fossa
Internal acoustic meatus
External orifice of the perilymphatic duct
Mastoid emissary vein
External orifice of the vestibular aqueduct
Jugular fossa

2

8.25 Dimensions of the Internal Auditory Meatus and the Fallopian Canal

The dimensions shown in the plate correspond with measurements reported elsewhere [2, 3, 5, 6]. Šercer and Krmpotić [4] have found a progressive apposition of bone at the fundus of the internal auditory meatus in temporal bones of older individuals and have attributed this occurrence with the development of presbyacusis.

Surgical Importance

Surgery of the internal auditory meatus, vestibular neurectomy, surgery of acoustic neurinomas

References

1. Lang J (1981)
2. Papangelou L (1972, 1974)
3. Portmann M et al. (1975)
4. Šercer A and Krmpotić J (1958)
5. Stern-Padovan R (1977)
6. Thomsen JS et al. (1981)

Plates from Lang [1]

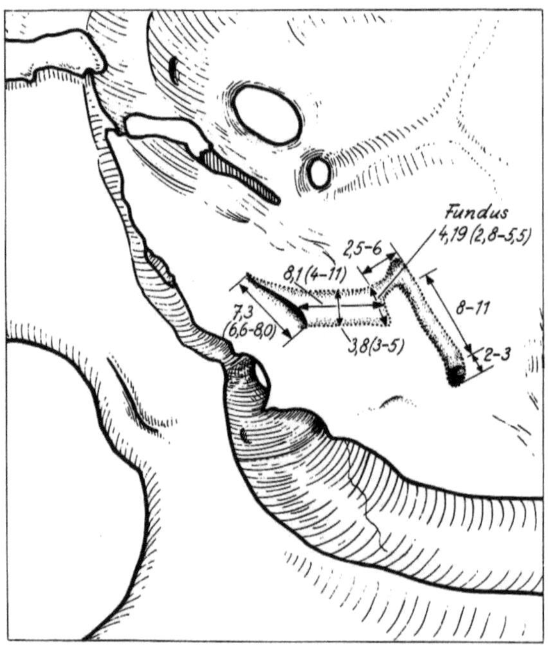

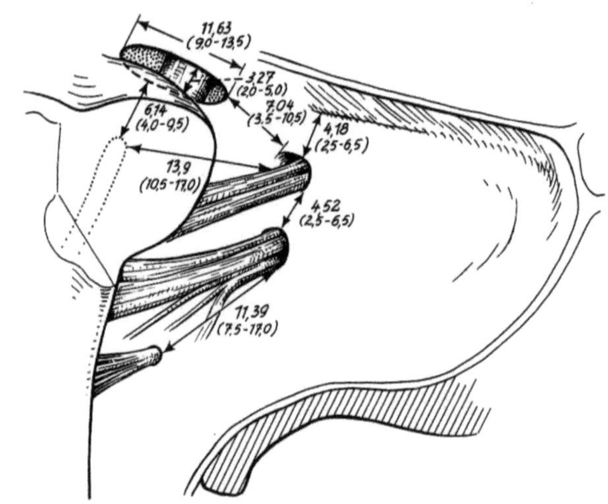

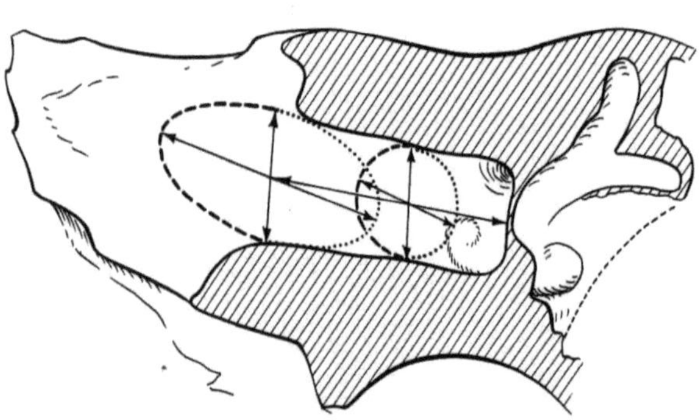

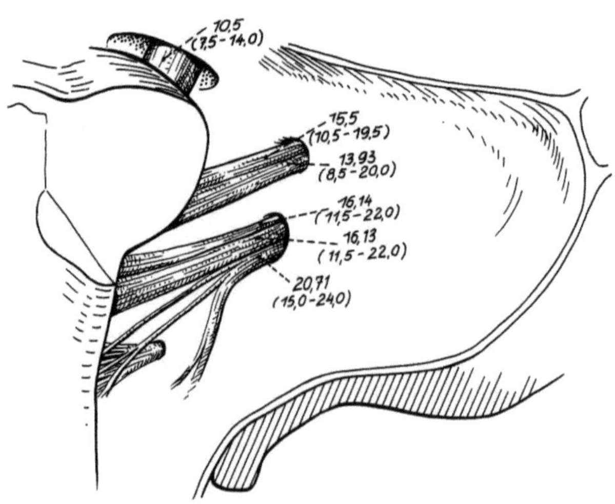

Age	Porus width	Porus height	Meatal length	Meatal height	Meatal width
Adult		3.10–5.80	11.15	3.0–1.0	6.46
15–17 yrs.	9.63	4.75	11.10	6.0	5.50
2 yrs.	7.04	4.08	7.23	4.6	4.30
Newborn	5.90	3.06	5.00	3.0	2.88

8.26 Posterior Spatial Anatomy of the Temporal Bone

The Plate is a schematized drawing of the spatial relationships of the important structures of the temporal bone. The posterior projection of the labyrinth, internal auditory meatus, and endolymphatic duct and sac are shown. The temporal bone is bounded laterally by the sigmoid sinus, inferiorly by the jugular bulb and the inferior petrosal sinus, and superiorly by the superior petrosal sinus. From the labyrinth, the endolymphatic duct passes toward the posterior surface of the temporal bone and opens within the dura with the external orifice of the vestibular aqueduct. The intradural extent of the endolymphatic duct usually extends as far as the sigmoid sinus or even deeper.

The measurements indicated in the plate are based on Lang's studies [1].

Surgical Importance

Transmastoid, translabyrinthine, and suboccipital surgery of the labyrinth, the internal auditory meatus, and the cerebellopontine angle

Reference

1. Lang J (1981)

Plate modified from Lang [1]

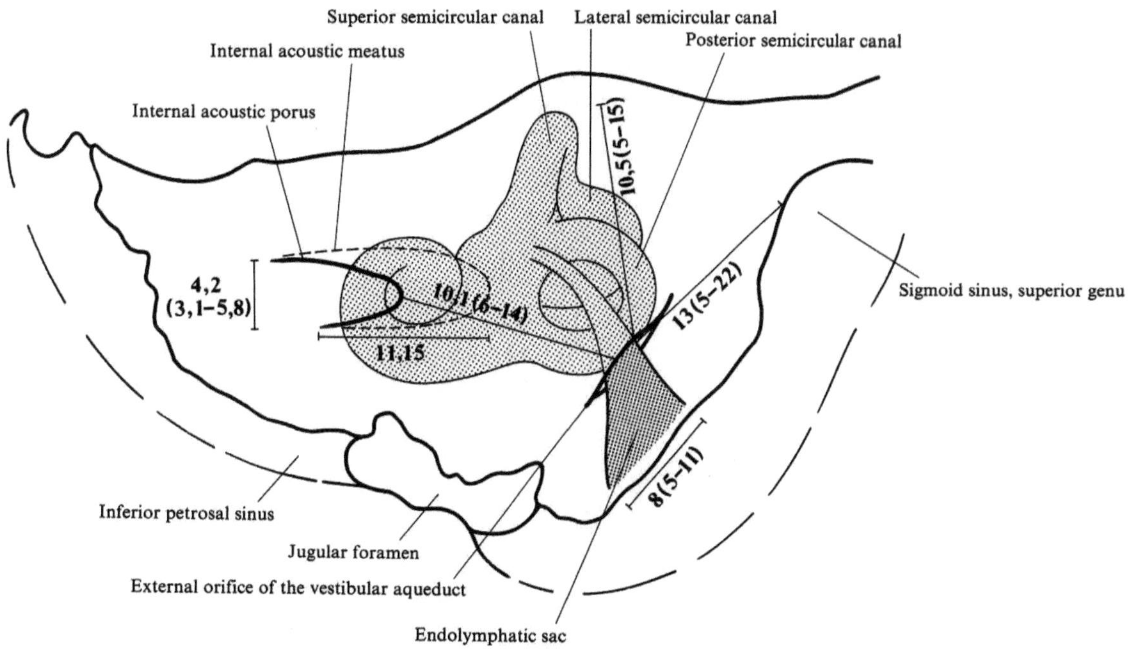

8.27 The Vestibular Aqueduct (Endolymphatic Duct)

The *pars rugosa* of the endolymphatic sac was measured by Skerlev and Mačković [6] from casts of bony specimens. The sac lies between the vestibular aqueduct and the external orifice of the vestibular canal [4, 7]. Depending on the degree of pneumatization of the temporal bone, the distance of this vertically placed structure from the posterior semicircular canal varies. The height of the sac can also vary, as so does its horizontal extension [5, 6].

Casts of the endolymphatic sac containing the pars rugosa show a steplike formation of the adjacent bone, dividing it into two parts: the superior part of the bone containing the sac is continuous with the vestibular aqueduct. The actual sac contains connective tissue as well as a space for the endolymph. The endolymphatic duct is a membranous formation, while the vestibular aqueduct is a bony canal.

Individual variations in the development of the endolymphatic duct can be seen in the neonate. Kodama and Sando [3] describe hypoplastic, normoplastic as well as hyperplastic shapes. Variations in size and positional relationships to the labyrinth and the sigmoid sinus become more evident in advanced age [1, 2, 7].

Surgical Importance

Surgery of the endolymphatic sac, saccotomy, vestibular nerve section.

References

1. Jahrsdorfer RA et al. (1981)
2. Kaufmann Arenberg I (1982)
3. Kodama A and Sando J (1982a, b)
4. Lang J (1979)
5. Ogura YJ and Clemis JD (1971)
6. Skerlev M and Mačković M (1981)
7. Wilbrand HF (1974)

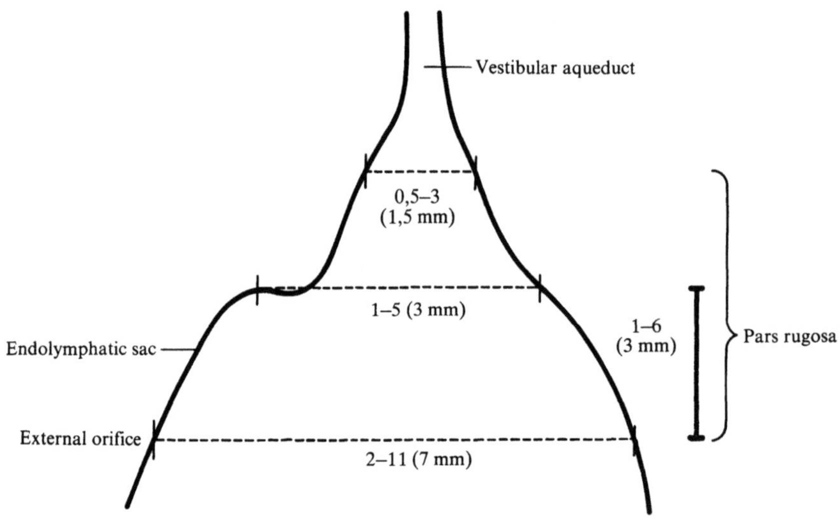

8.28 The Labyrinth and the Nerves in the Internal Auditory Meatus (Mediodorsal View, Right Side)

The membranous labyrinth and its nerves, including the facial nerve, are represented in the Plate. The *crista transversa* facilitates identification of these nerves and is indicated by a dotted line. The size of the endolymphatic sac varies [1], as illustrated schematically.

The vestibular aqueduct runs from the vestibule parallel and medial to the crus commune of the labyrinth, exiting beneath the point where it divides into the superior and posterior semicircular canals [3]. The vestibular aqueduct then runs laterally and downwards to the posterior wall of the temporal bone where it reaches its external orifice approximately 10 mm lateral to the internal auditory meatus.

The endolymphatic duct is surrounded within the vestibular aqueduct by connective tissue and small blood vessels, which become more numerous in the pars rugosa. The paravestibular aqueduct runs parallel to the endolymphatic duct and, according to Ogura and Clemis [4], contains an artery and a vein. The duct displays many variations, in one case being obliterated by an ectatic sigmoid sinus [2] (not represented).

Surgical Importance

Saccotomy, vestibulocochlear neurectomy [5], surgery of the facial nerve, suboccipital approach to the internal auditory meatus.

References

1. Caix M and Outrequin G (1979)
2. Jahrsdorfer RA et al. (1981)
3. Kodama A and Sando J (1983)
4. Ogura Y and Clemis JD (1971)
5. Okano Y et al. (1980)
6. Wilbrand HF et al. (1974)

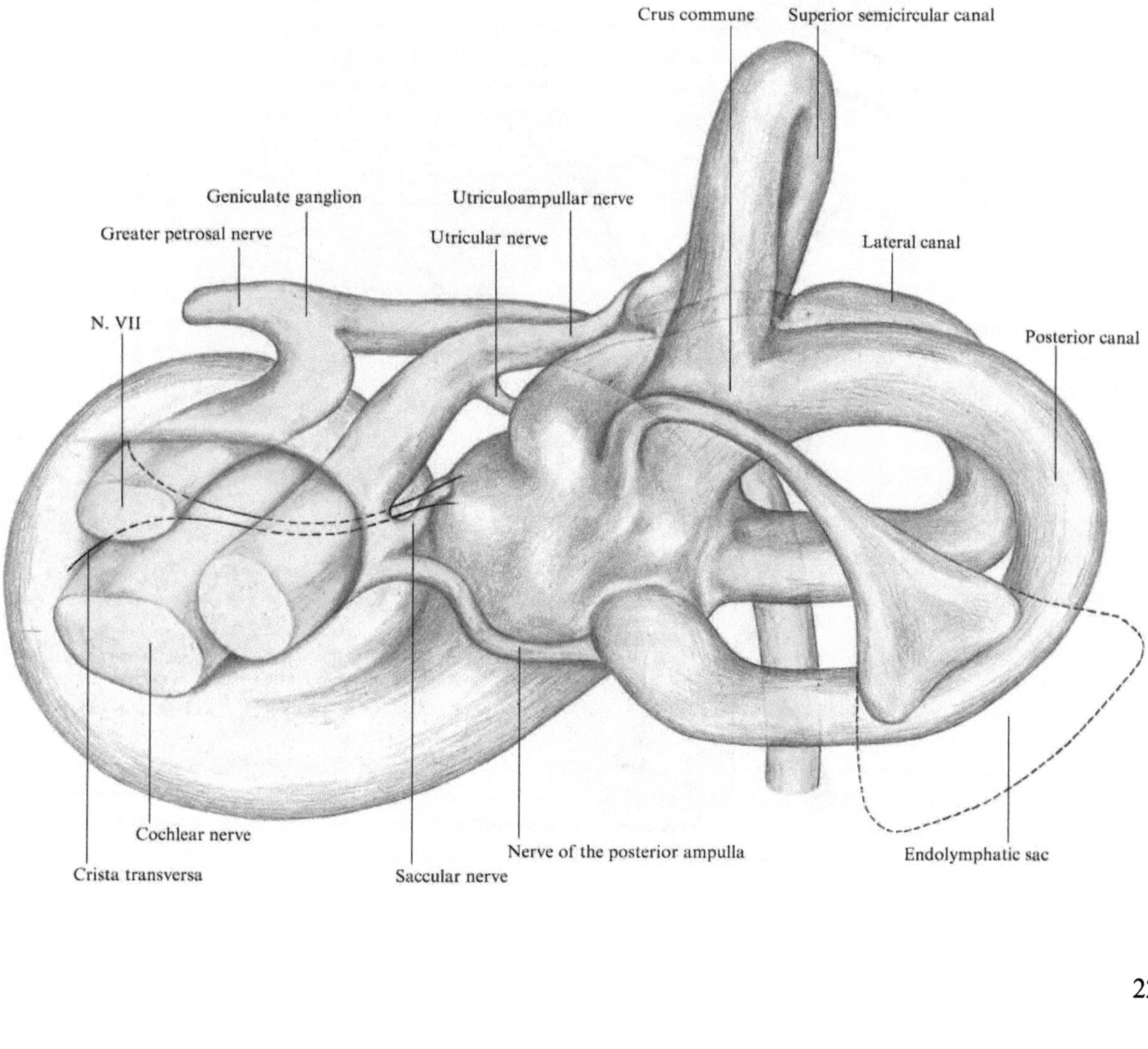

As illustrated by Šimunić [6], the arterial supply of the inner ear consists of three components: the cochlear, the vestibulocochlear, and the vestibular arteries [2, 3–5]. Variations in the number of supplying arteries exist, as listed below [6]:

Plate I

- Ant. inferior cerebellar a.
  - Labyrinthine a.
    - Common cochlear a.
      - Vestibulocochlear a.
        - Saccular br.
        - Cochlear br.
      - Spiral modiolar a.
        - Radiate arts.
    - Ant. vestibular a. (Utricular a.).

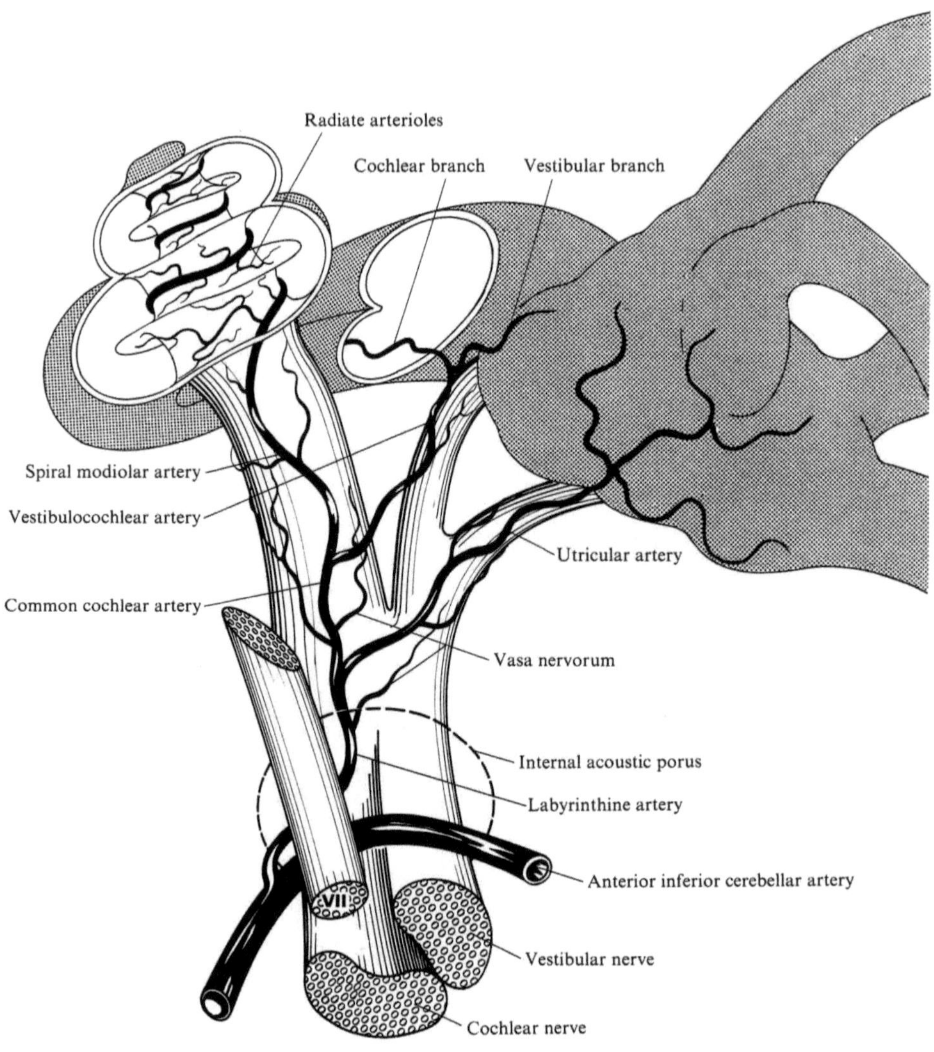

1

The typical blood supply is shown in Plate 1, which depicts a single labyrinthine artery, and in Plate 2 in which there are two arteries [1, 6]. The section of the artery lying beneath the vestibular nerve is shown as a dotted line. The alternate derivation of the labyrinthine vessels is as follows:

Plate II
Ant. inferior cerebellar a.
  Labyrinthine a. I
    Spiral modiolar a. (*cochlearis propria a.*)
      Radiate a.
    Ant. vestibular a. (utricular a.)
  Labyrinthine a. II
    Cochleovestibular a.
      Vestibular (saccular) br.
      Cochlear br.

Surgical Importance

Surgery of the internal auditory meatus and cerebellopontine angle, surgery of acoustic tumors.

References

1. Fisch U (1968)
2. Helms J (1978)
3. Konaschko PJ (1927)
4. Mazzoni A (1969)
5. Mazzoni A and Hansen CC (1970)
6. Šimunić V (1974)

Plates 1 and 2 from Šimunić [6]

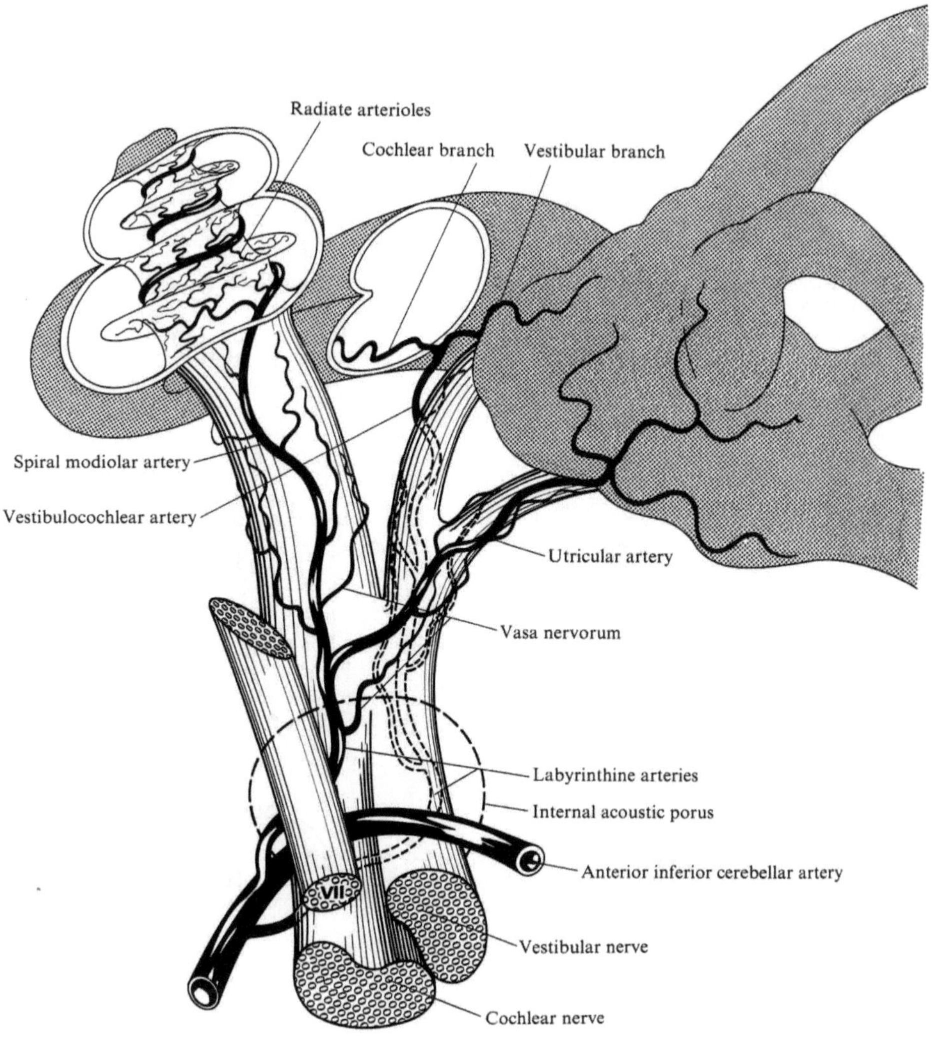

2

8.30 (Plates 1–8) Variations in the Vessels of the Internal Auditory Meatus

The different vascular patterns in the internal auditory meatus have been schematized from the right side in a superior view. In surgery of the internal auditory meatus, these vessels should be preserved in order to avoid functional disturbances.

Surgical Importance

Surgery of the facial nerve and the internal auditory meatus, vestibular neurectomy.

Reference

1. Helms J (1980)

Plates 1–8 from Helms [1]

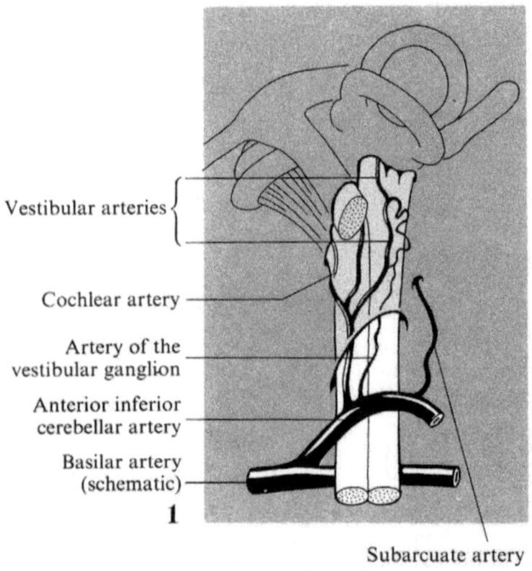

1

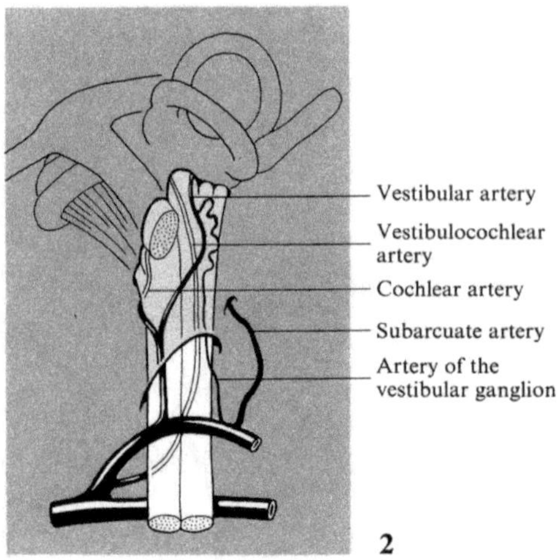

2

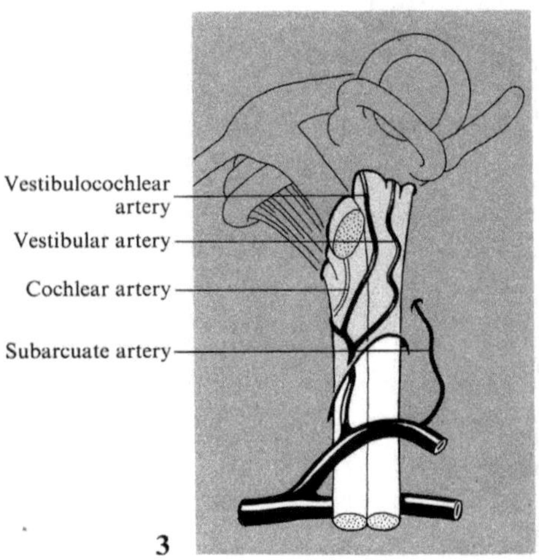

3

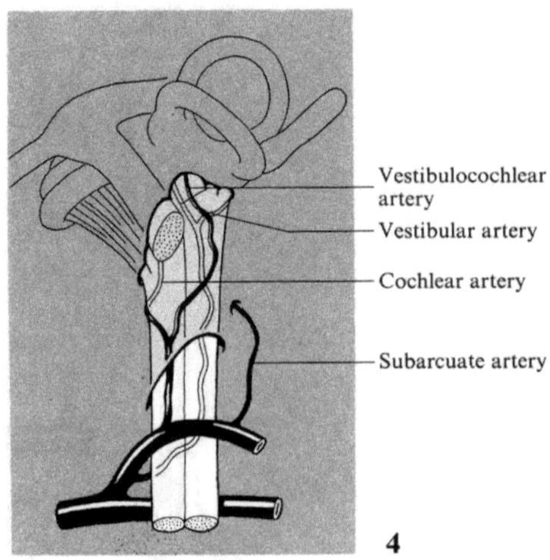

4

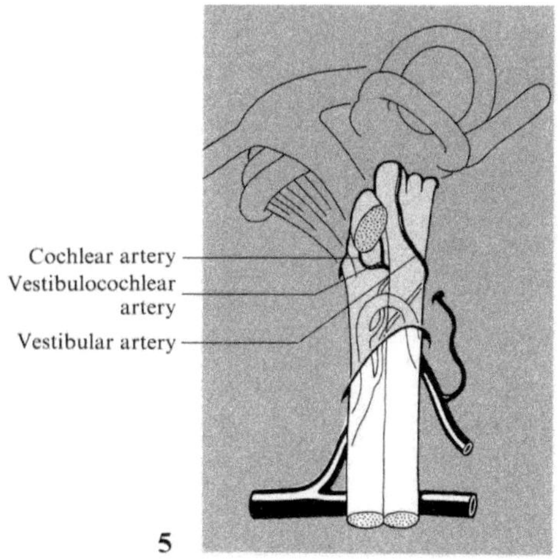

Cochlear artery
Vestibulocochlear artery
Vestibular artery

5

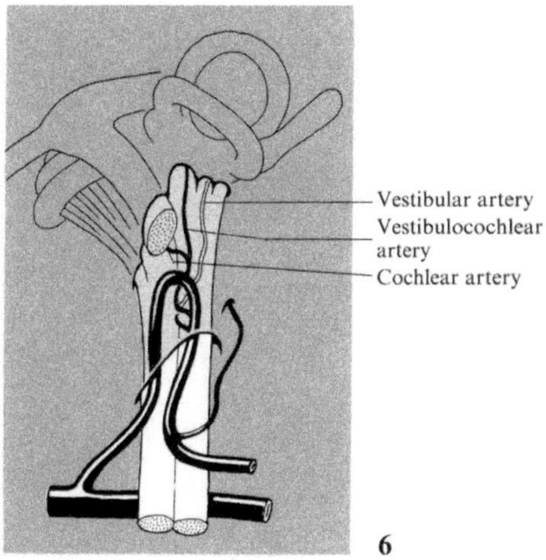

Vestibular artery
Vestibulocochlear artery
Cochlear artery

6

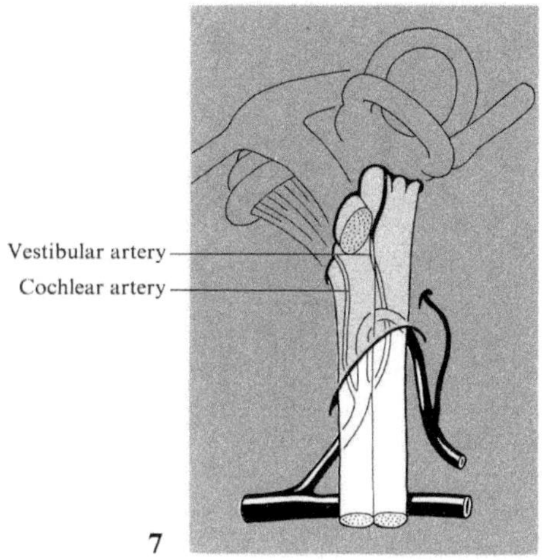

Vestibular artery
Cochlear artery

7

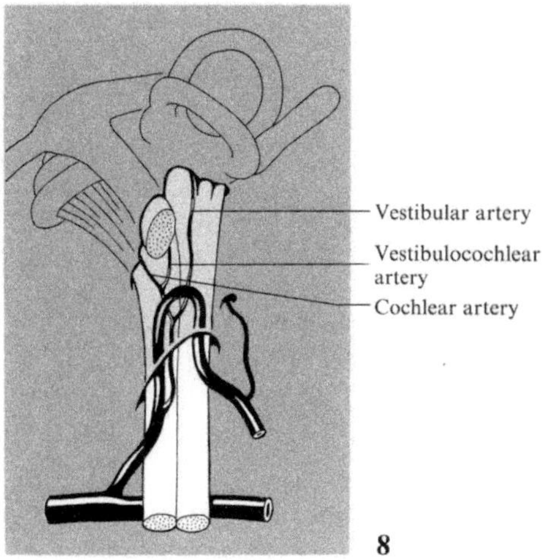

Vestibular artery
Vestibulocochlear artery
Cochlear artery

8

9 The Middle and Posterior Cranial Fossae

In the illustration, the right hemicranium and vital cortical tissues have been removed to the level of the falx cerebri. The left side of the tentorium has been preserved while that on the right has been removed. The sella turcica and the cavernous sinus are shown. The dura of the middle cranial fossa has been elevated and displaced laterally in order to expose the Gasserian (trigeminal) ganglion, dura propria and the middle meningeal artery. By removing a portion of the dorsal temporal bone, the geniculate ganglion, the facial nerve, and the greater petrosal nerve are exposed.

In rare cases, the greater petrosal nerve is not located in its corresponding sulcus, but rather in a canal which can extend as far as the apex of the temporal bone, such that the hiatus of the facial canal is missing [1]. The lesser petrosal nerve, a branch of the tympanic nerve (from the glossopharyngeal nerve) is not represented. The hiatus of the facial canal serves as a surgical landmark for finding the internal acoustic meatus via the transtemporal extradural approach (see Sect. 9.7.2). In 50 percent of cases, the hiatus lies at the same distance from the temporal squama as does the porus of the internal auditory meatus.

Surgical Importance

Transtemporal intracranial surgery of the floor of the middle cranial fossa, exposure of the internal auditory meatus and its porus, resection of the geniculate ganglion and the greater petrosal nerve, surgery of the Gasserian ganglion and hypophysis.

References

1. Denecke HJ (1966)
2. Kempe LG (1968)

Plate from Kempe [2]

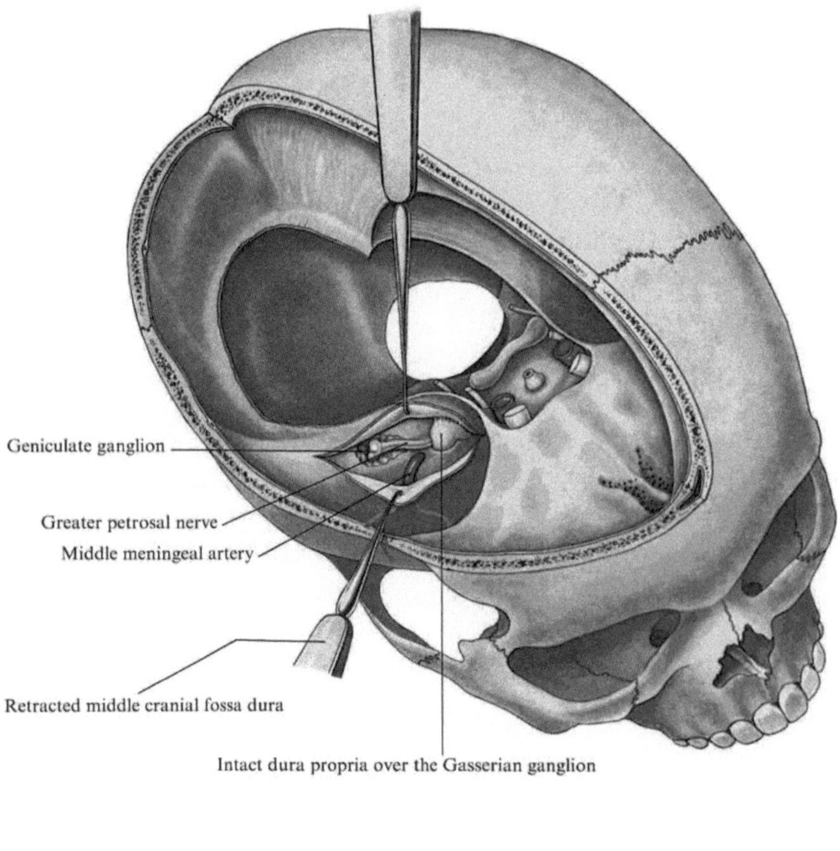

9.2 Schematic Representation of the Nerves and Vessels of the Lateral and of the Middle Part of the Cranial Base

The extracranial and intracranial courses of surgically important nerves and arteries in the middle part of the cranial base and cranial fossa are represented in the Plate. The styloid process lies lateral to the internal carotid artery and is an important surgical landmark.

Surgical Importance

Inferior skull base surgery (e.g., glomus surgery), surgery of the infratemporal fossa, approach to the sphenopalatine fossa, surgery of the facial and vidian nerves (see Sects. 6.15, 22, 23, 24.1).

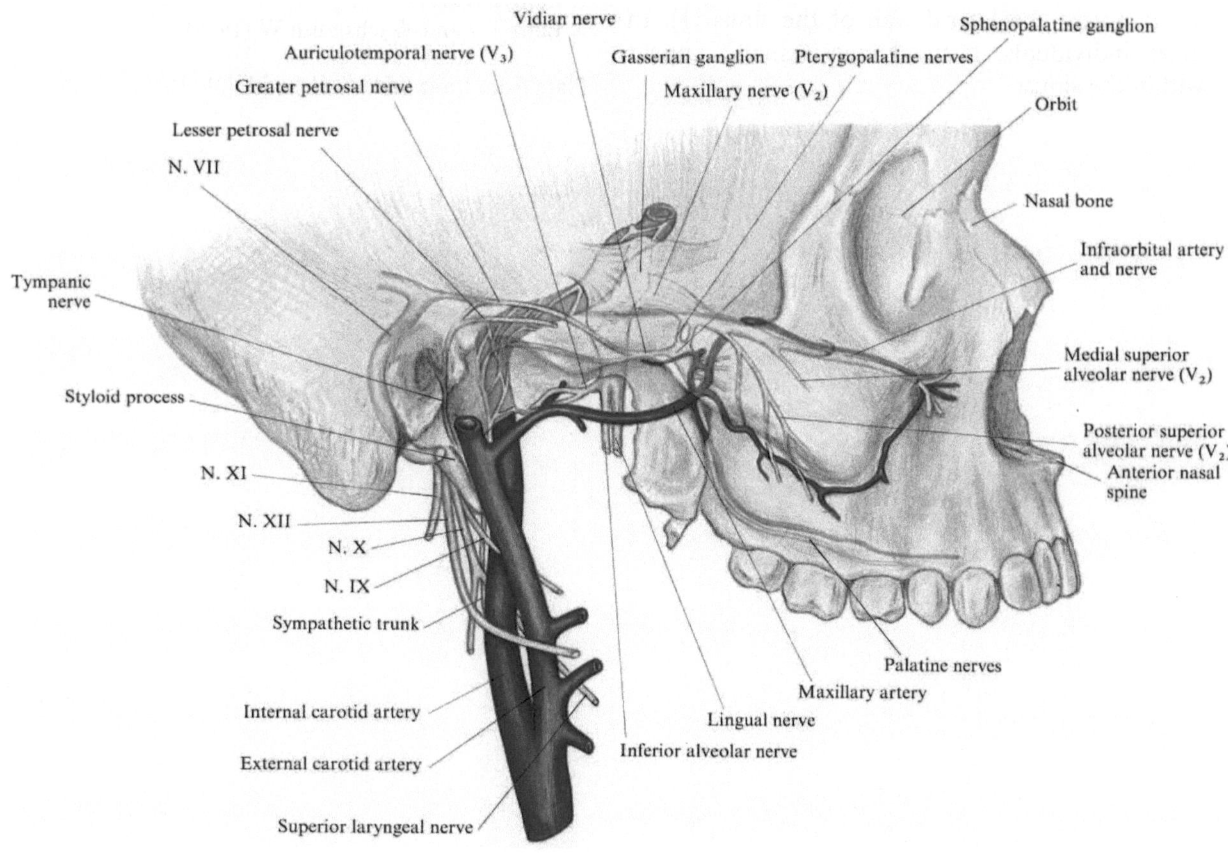

9.3 (Plates 1–2) Frontal Section Through the Right Cavernous Sinus

The lateral wall of the cavernous sinus is formed by a sagittal dural plate from the middle cranial fossa. The medial wall is formed superiorly by a connective tissue layer from the dura covering the hypophysis and inferiorly by the body of the sphenoid and the roots of the greater wing of the sphenoid. The right and left cavernous sinuses are connected by the intercavernous sinuses and the basilar sinus. The anterior portion of the sinus within the venous blood spaces is crossed by numerous trabeculae [2, 3].

The internal carotid artery and the abducens nerve lie external to the sinus cavity in 77% of cases, such that the medial side of the internal carotid artery forms the lateral wall of the sinus [1]. In older individuals, fatty changes usually appear within the sinus.

The cavernous and the intercavernous sinuses encircle the sella turcica and the hypophysis.

Surgical Importance

Surgery of the sphenoid sinus and hypophysis, fistulae between the internal carotid artery and the cavernous sinus, traumatology and tumors of the skull base, intracranial surgery of trigeminal nerve.

References

1. Bedford MA (1966)
2. Lang J (1979)
3. Lanz T v and Wachsmuth W (1979)

Plate from Lanz and Wachsmuth [3]

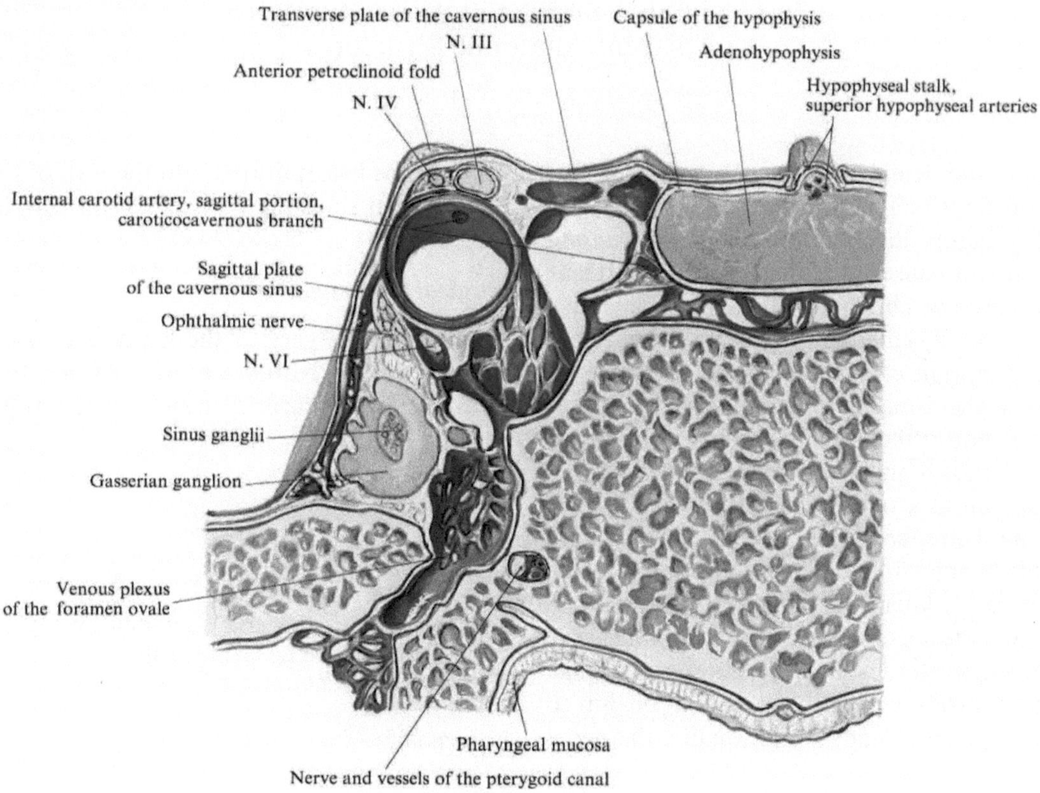
Transverse plate of the cavernous sinus
Capsule of the hypophysis
N. III
Adenohypophysis
Anterior petroclinoid fold
Hypophyseal stalk,
superior hypophyseal arteries
N. IV
Internal carotid artery, sagittal portion,
caroticocavernous branch
Sagittal plate
of the cavernous sinus
Ophthalmic nerve
N. VI
Sinus ganglii
Gasserian ganglion
Venous plexus
of the foramen ovale
Pharyngeal mucosa
Nerve and vessels of the pterygoid canal

1

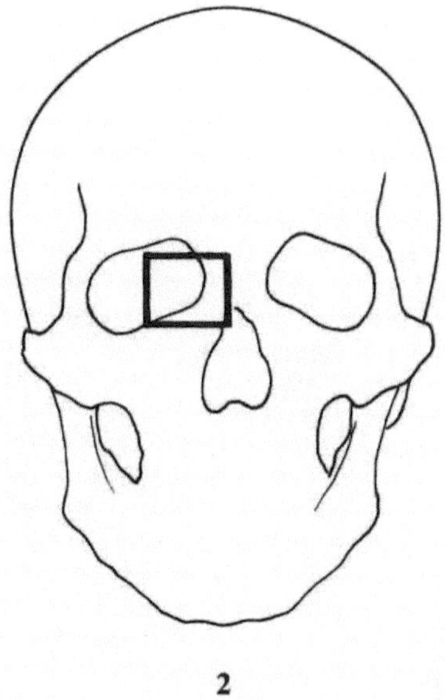

2

9.4 The Optic Chiasm and its Relationships to Adjacent Anatomic Structures

The shape and length of the optic chiasm frequently displays individual variations [1–4]. If the chiasm lies against the anterior sellar wall (*chiasma praefix*), it can cause difficulties in the treatment of frontobasal cerebrospinal fistulae. The abducens nerve usually enters into Dorello's canal near the lateral margin of the clivus (see Sect. 9.4, 5). Its porus in the dura lies 2–7 mm caudal and medial to the root of the trigeminal nerve. The abducens nerve passes under the superior sphenopetrous ligament in a sulcus on the apex of the temporal bone. Infrequently, the nerve can also pass above this ligament. The nerve then runs into the cavernous sinus lateral to the internal carotid artery to reach the superior orbital fissure. The nerve next enters the orbit through the tendinous anulus. This course varies, especially in its relationship to the inferior petrosal sinus. Additionally, the nerve may also be incorporated into the wall of the inferior petrosal sinus [5].

Surgical Importance

Transfrontal surgery of the hypophysis and sella, surgery on the hiatus tentorii and the branches of the internal carotid artery or the trigeminal nerve.

References

1. Gibo et al. (1981)
2. Grisoli F et al. (1982)
3. Lang J in: Lanz T v and Wachsmuth W (1979)
4. Laude M and De Le Gars (1980)
5. Piffer CR and Zorzetto NL (1980)

Plate modified from Lang [3]

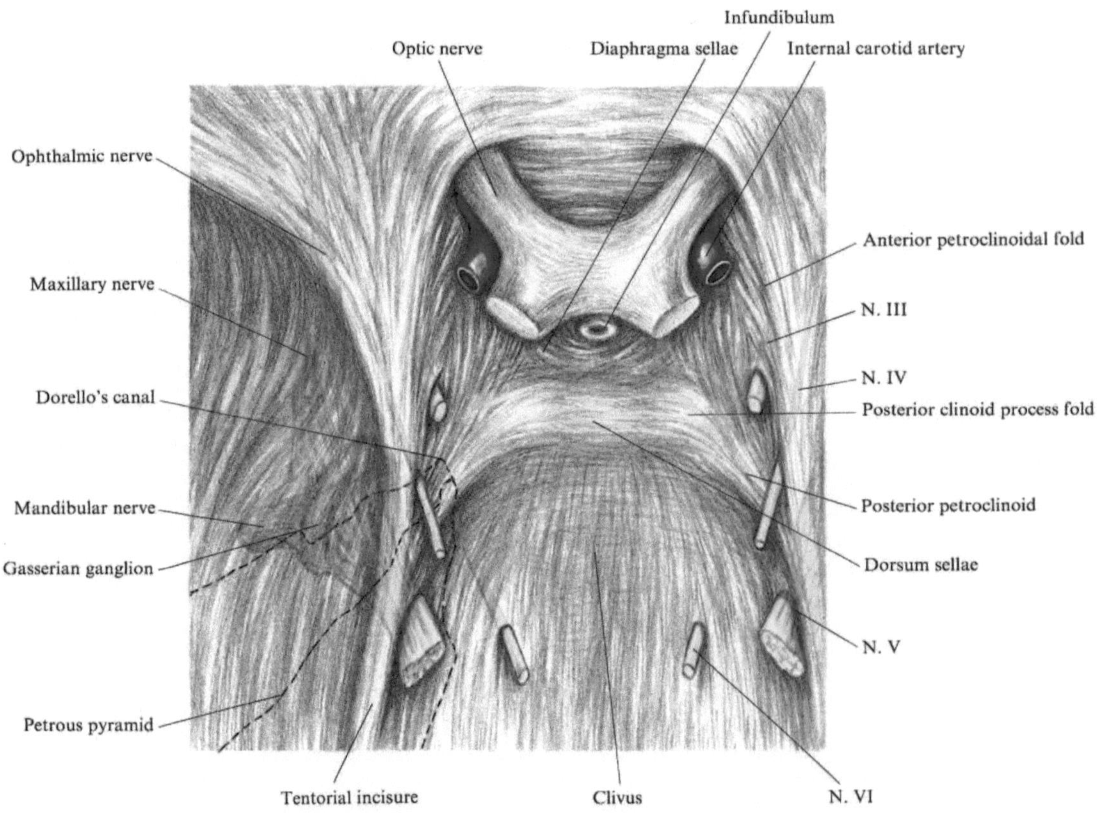

9.5 The Sphenoid Bone and Cavernous Sinus and the Region of the Petrous Pyramid (Left Side, Medial View)

The internal carotid artery, the optic nerve, the cavernous sinus, and the three divisions of the trigeminal nerve can protrude into the lateral wall of the sphenoid sinus in the presence of extensive pneumatization (see Sects. 6.10, 11, 12, 13).
The internal carotid artery runs below and medial to the anterior clinoid process, and then passes retrogradely to form a syphon before passing through the diaphragm of the sella to reach the subarachnoid space.
The abducens nerve passes under the sphenopetrous ligament (*Dorello's canal*), and then enters the region of the cavernous sinus (see Sect. 9.4).
The cavernous sinus can display many variations in shape, [3] while cranial nerves III, IV, VI are located on its lateral surface (see Sect. 9.3.1).
The oculomotor nerve passes lateral to the posterior clinoid process and medial to the petroclinoid fold (see Sect. 9.4). It then passes along the lateral wall of the cavernous sinus to continue into the superior orbital fissure.
The trochlear nerve reaches the cavernous sinus at different points of the tentorial insertion, but is usually medial to the anterior petroclinoid fold. It is situated in the cavernous sinus below or lateral to the oculomotor nerve, but also occasionally occurs at the upper border of the ophthalmic nerve, extending as far as the superior orbital fissure (see Sects. 9.4, 6).

Surgical Importance

See Sect. 9.4

References

1. Corlieu P et al. (1981)
2. Grisoli F et al. (1982)
3. Krayenbühl H and Yasargil MG (1965)

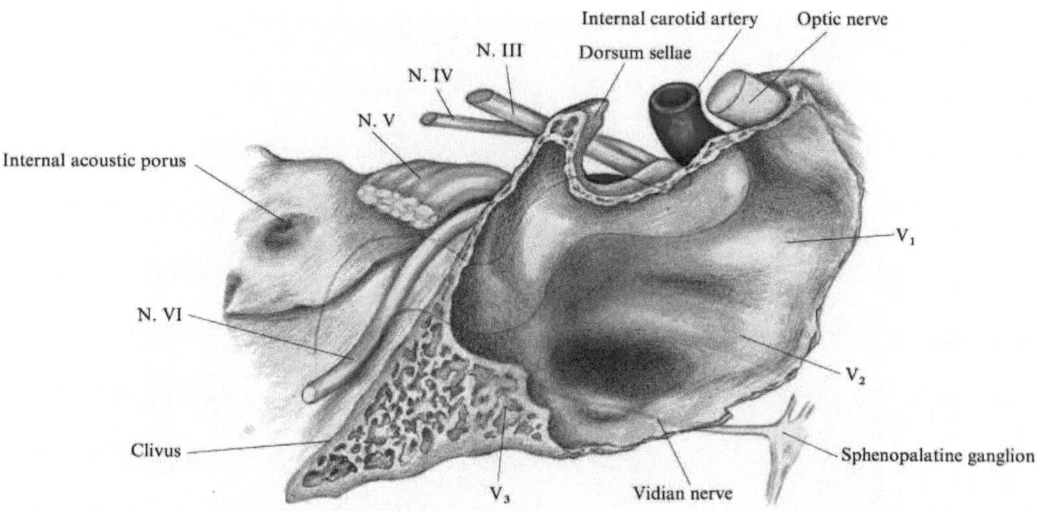

9.6 The Petrous Pyramid After Removal of the Dura and a Section of the Superior Petrosal Sinus

The trigeminal or *Meckel's* cave resembles a sac which is formed by thin connective tissue except for its thicker lateral and rostral walls. This connective tissue is continuous with the sagittal dural plate of the middle cranial fossa. The thin inferior periosteal layer joins the inferior sphenopetrous ligament. The medial part of the ligament is situated sagittally while the lateral part has a transverse location. Both parts cover the internal carotid artery, the accompanying venous plexus, and the carotid nerve. Above the sphenopetrous ligament, only loose connective tissue is present between the Gasserian ganglion, the internal carotid artery, and the abducens nerve. The ganglion is surrounded by the trigeminal cistern, which is connected through the porus of the trigeminal nerve with the lateral pontine cistern (see Sect. 10.8). The inferior sphenopetrous ligament covers the greater petrosal nerve, which passes through the sphenopetrous synchondrosis below and lateral to the internal carotid artery when leaving the cranial cavity (see Sects. 9.3(1–2), 9.4, 9.5, 9.7(1–3).

Surgical Importance

Surgery of the apex of the petrous pyramid (including surgery for *Gradenigo's syndrome* [1]), lateral approach to the sellar region, surgery of the trigeminal nerve.

References

1. Gradenigo G (1907)
2. Kempe LG (1968)

Plate from Kempe [2]

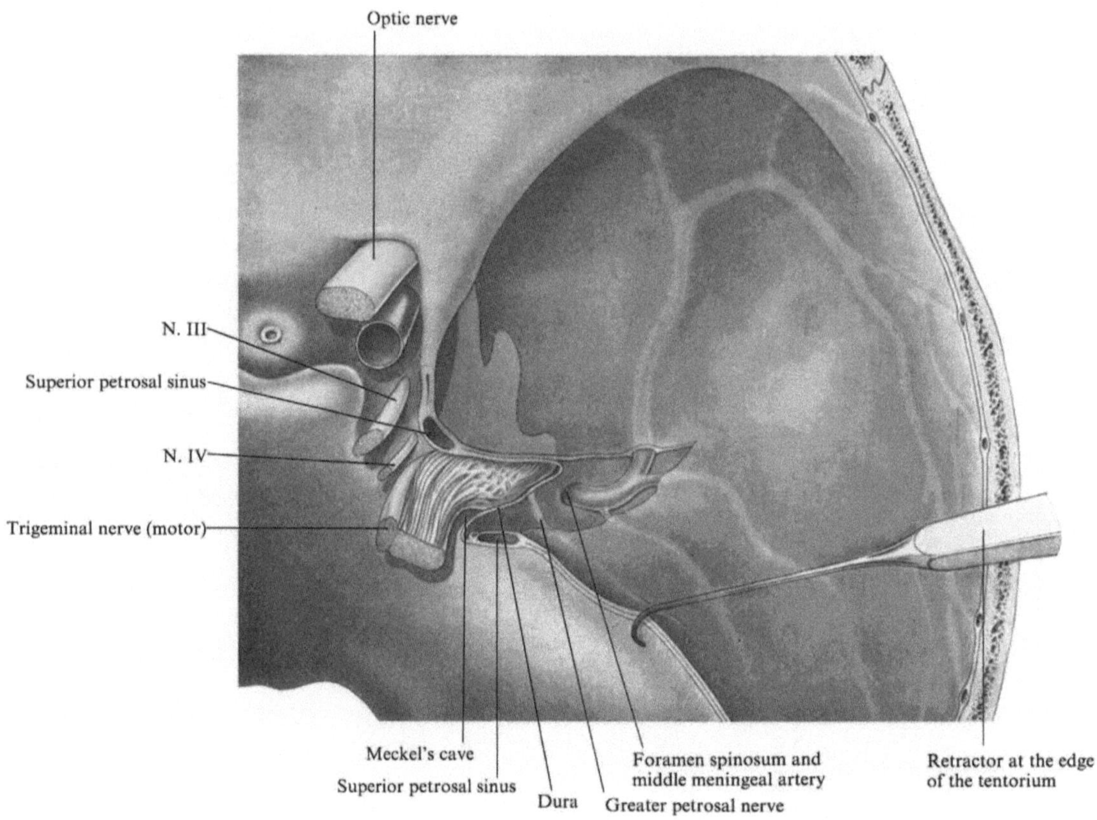

9.7 (Plates 1–3) Surgery of the Greater Petrosal Nerve

9.7 (Plate 1) Topography of the Greater Petrosal Nerve (see Sect. 9.2)

The greater petrosal nerve is the parasympathetic preganglionic component of the vidian nerve [1–6]. Its secretory fibers innervate the lacrimal gland, the glands of the palate, and the nasal mucosa. The nerve passes from the geniculate ganglion through a canal to the hiatus of the greater petrosal nerve. From there it passes into its corresponding sulcus to the foramen lacerum where it enters the Vidian canal, lateral to the internal carotid artery. It is joined by the deep petrosal nerve (postganglionic sympathetic fibers from the carotid plexus) to form the nerve of the pterygoid canal. The latter reaches the sphenopalatine ganglion, from which emanate postganglionic parasympathetic fibers. The nerve is then joined by sympathetic and sensory fibers (which pass only through the ganglion), which proceed as the posterior lateral nasal and septal nerves to supply the nasal cavity. These nerves also supply the palate as the palatine nerves (see Sect. 6.26). The secretory fibers for the lacrimal gland reach the gland by the zygomaticolacrimal anastomosis. The zygomatic nerve is a branch of the second division of the trigeminal nerve, while the lacrimal nerve is a branch of its first division (ophthalmic nerve).

Surgical Importance

Diagnosis and treatment of facial and petrosal neuralgias, surgery of the petrosal nerve and geniculate ganglion, transtemporal exposure of the lateral part of the skull base.

References

1. Fisch U (1968)
2. Guerrier Y (1981)
3. House WF (1964)
4. Krmpotić J and Sunarić R (1963)
5. Parry RH (1904)
6. Portmann M et al. (1975)

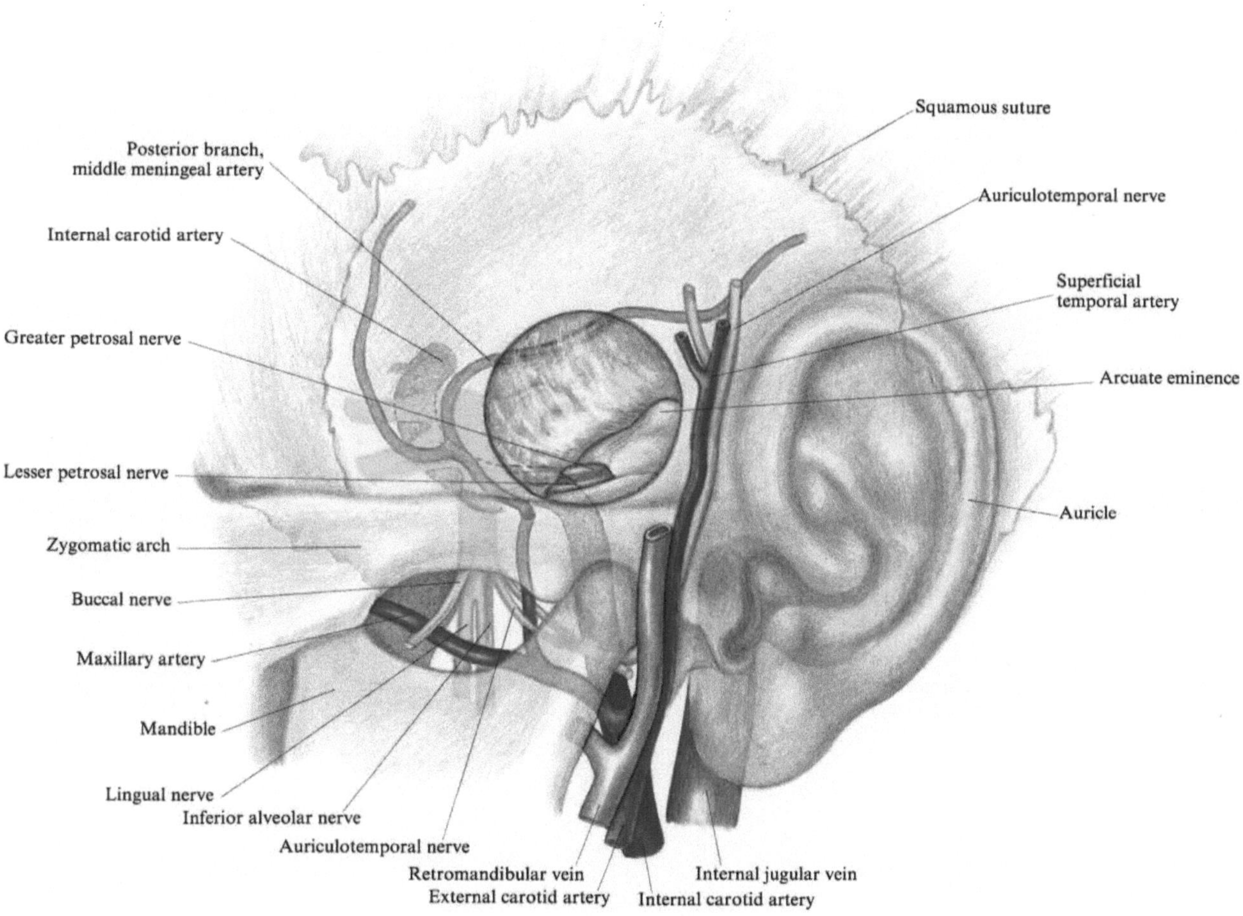

9.7 (Plate 2) Useful Topographical Distances for Finding the Greater Petrosal Nerve Intracranially

The distances from the landmarks depicted in the Plate in the middle cranial fossa to the trepanation opening in the temporal squama are specified [1].

Surgical Importance

See Sects. 9.1, 7.1.

Reference

1. Krmpotić J and Nikolić V (1964)

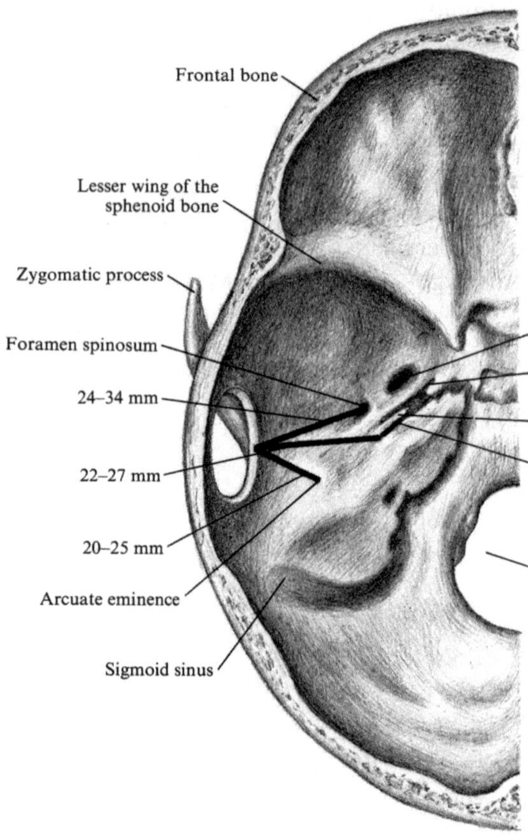

9.7 (Plate 3) Anatomic Landmarks for Locating the Internal Auditory Meatus via the Transtemporal-Extradural (Middle Cranial Fossa) Approach

The internal auditory meatus can be identified on the floor of the middle cranial fossa by using the following landmarks [1–3]:

1. The greater petrosal nerve is found (see Sects. 8.23, 9.7.2) and followed to the geniculate ganglion, after which the labyrinthine course of the facial nerve is exposed to the internal auditory meatus (see Sect. 8.23).

2. The "blue line" of the superior semicircular canal is identified in the arcuate eminence. By defining a 60° angle from the anterior end of this line, the roof of the internal meatus can usually be found and exteriorized [1].

3. The roof of the internal meatus can be located at a distance of 28 mm from the external surface of the temporal squama. The greater petrosal nerve is an additional anatomic landmark for proper orientation [3].

4. The line dividing the angle between the greater petrosal nerve and the blue line of the superior semicircular canal leads to the roof of the internal auditory meatus. The uppermost point of the superior semicircular canal may not always correspond to the arcuate eminence, especially in temporal bones with excessive pneumatizations.
The consistency and the lighter color of the labyrinthine bone in comparison to the bone of the pyramid are the first signs allowing the two structures to be differentiated.

Surgical Importance

See Sects. 9.1 and 9.7.1, 2; transtemporal extradural approach to the internal auditory meatus.

References

1. Fisch U (1968)
2. House WF and Belal A (1980)
3. Portmann M et al. (1975)

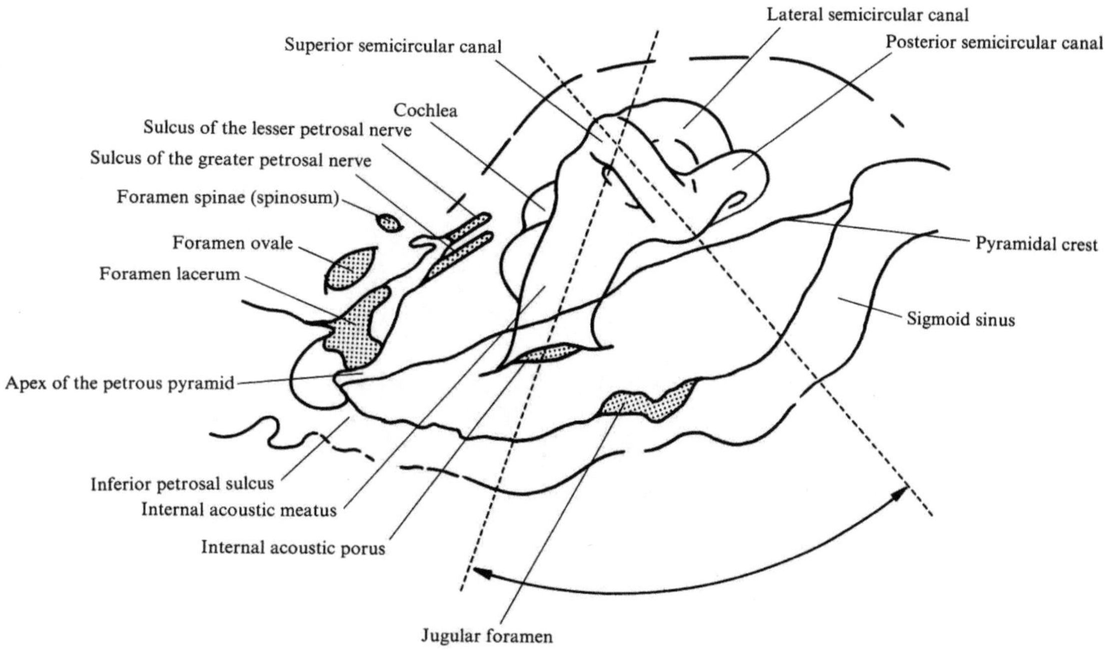

9.8 The Brain Stem and the Inferior Cerebellar Surface With its Nerves and Vessels

The Plate has been adopted from Kempe [1]. The vertebral and the basilar arteries are represented in relation to the cranial nerves. The basilar artery gives off terminal arteries to the pons (*rami ad pontem*), the superior cerebellar artery, and the posterior cerebral artery. The oculomotor nerve passes between these two arteries. The arteries of the labyrinth originate from the anterior inferior cerebellar artery in the majority of cases and, less commonly, from the basilar artery [2]. The anterior inferior cerebellar artery forms variably sized loops at or in the internal auditory meatus (see Sect. 11.15). The branches of this artery supply the antero-inferior cerebellar surface. In certain cases, vascular loops from this vessel may enter the internal auditory meatus and compress the nerves present or pass between them.
The venous blood from the surface of the cerebellum is drained by the superior cerebellar veins into the sinus rectus, the transverse sinus, or the superior petrosal sinus. The venous blood from the inferior cerebellar surface is drained by the inferior cerebellar veins to the occipital, sigmoid, or superior petrosal sinuses. The veins of the posterior cranial fossa can display many variations. The petrosal vein drains the blood from the superior and middle cerebellar peduncles, the anterior part of the inferior cerebellar surface, the pons, the flocculus, the medulla oblongata, and the lateral recess of the fourth ventricle. This vessel should be preserved if possible in surgery of the cerebellopontine angle. The petrosal vein is formed by small veins and – in 40% of cases – lies dorsal to the entrance of the sensory root of the trigeminal nerve in the pons. It can, however, lie lateral to the flocculus or at the level of the horizontal cerebellar fissure. It drains mostly laterally, but it can also drain in a laterocaudal direction to the trigeminal porus and then into the inferior petrosal sinus. Uncommonly, there may be two or even three petrosal veins present.

Surgical Importance

Surgery of the cerebellopontine angle and the brain stem.

References

1. Kempe LG (1968)
2. Perneczky A and Tschabitscher M (1976)

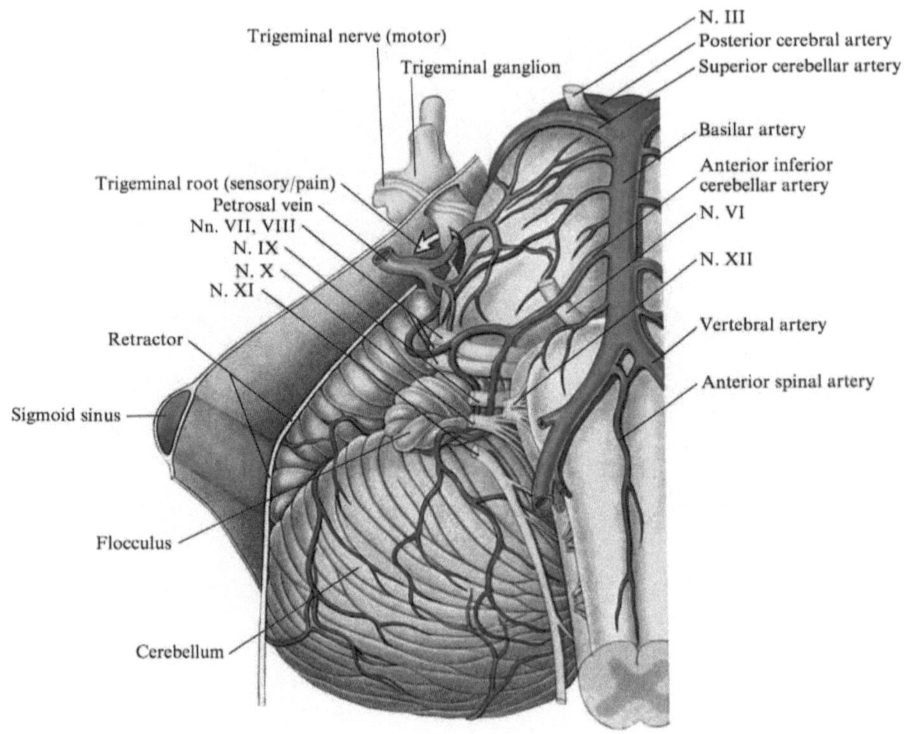

9.9 Topography of the Cerebral Structures in the Posterior Cranial Fossa

The Plate shows the rhomboid fossa which is delimited by the structures of the superior and inferior cerebellar peduncles i.e. the restiform bodies and the brachia conjunctiva [1]. The corresponding cranial nerves and their dural orifices are also represented. The projection of the occipital bone is indicated. The *black dotted line* corresponds to the hiatus tentorii and the falx cerebri, while the red line corresponds to the cerebellum.

Surgical Importance

Surgery of the posterior cranial fossa (including the suboccipital approach to the cerebellopontine angle).

Reference

1. Hermann F (1908)

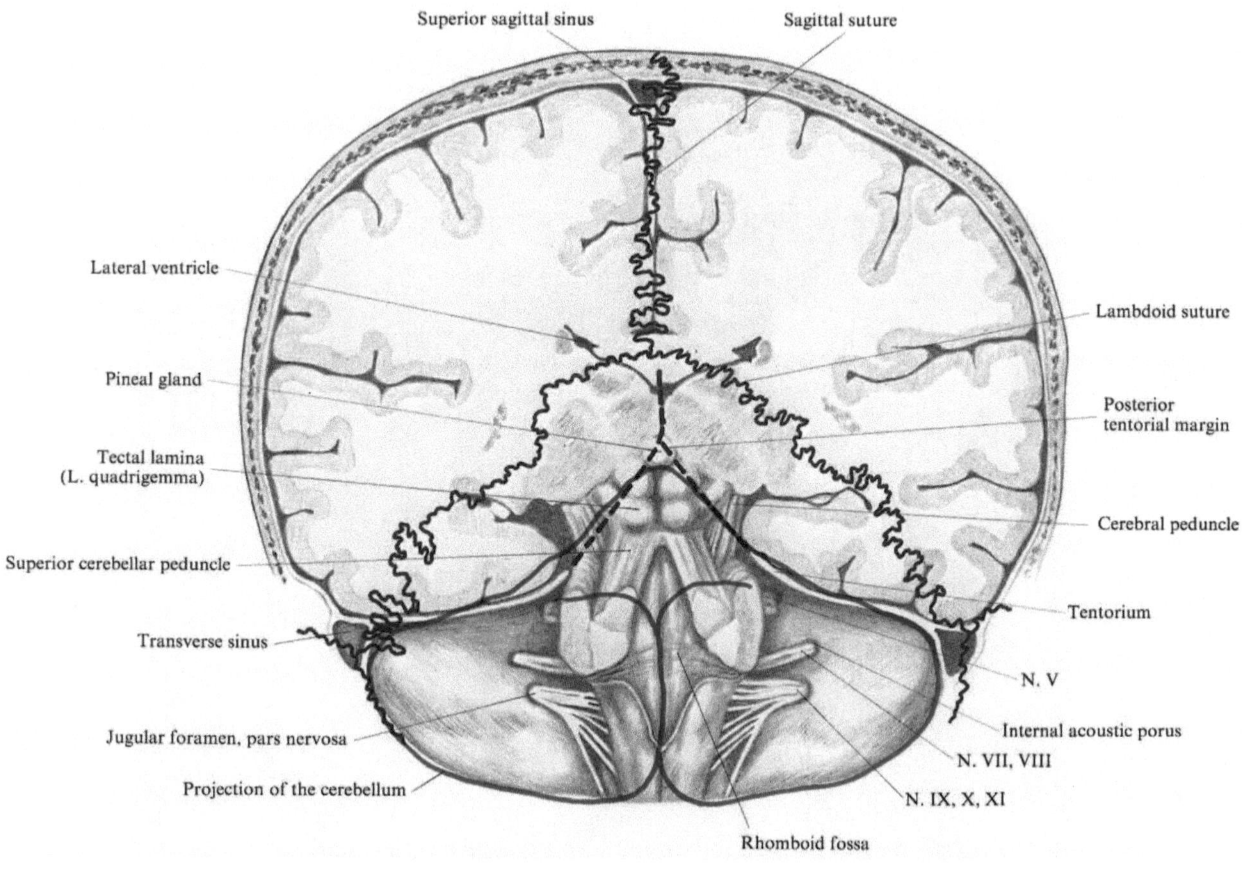

The Plate is a mediolateral view of a paramedian section of the posterior cranial fossa at the level of the vertebral artery in the region of the atlas. The various distances of cranial nerves V–XII to important landmarks are given. The vertebral artery leaves the atlas to enter the intradural space and gives off the posterior inferior cerebellar artery which (in this specimen) makes a loop through the foramen magnum into the vertebral canal. The anterior inferior cerebellar artery makes a loop in the region of the internal auditory meatus and gives off the labyrinthine artery. The cerebral and spinal roots of the accessory nerve are represented.

Surgical Importance

Ligature of the vertebral artery (see Sects. 9.9, 11.11–13), vascular decompression of the cranial nerves.

Reference

1. Lang J in: Lanz T v and Wachsmuth W (1979)

Plate from Lang [1]

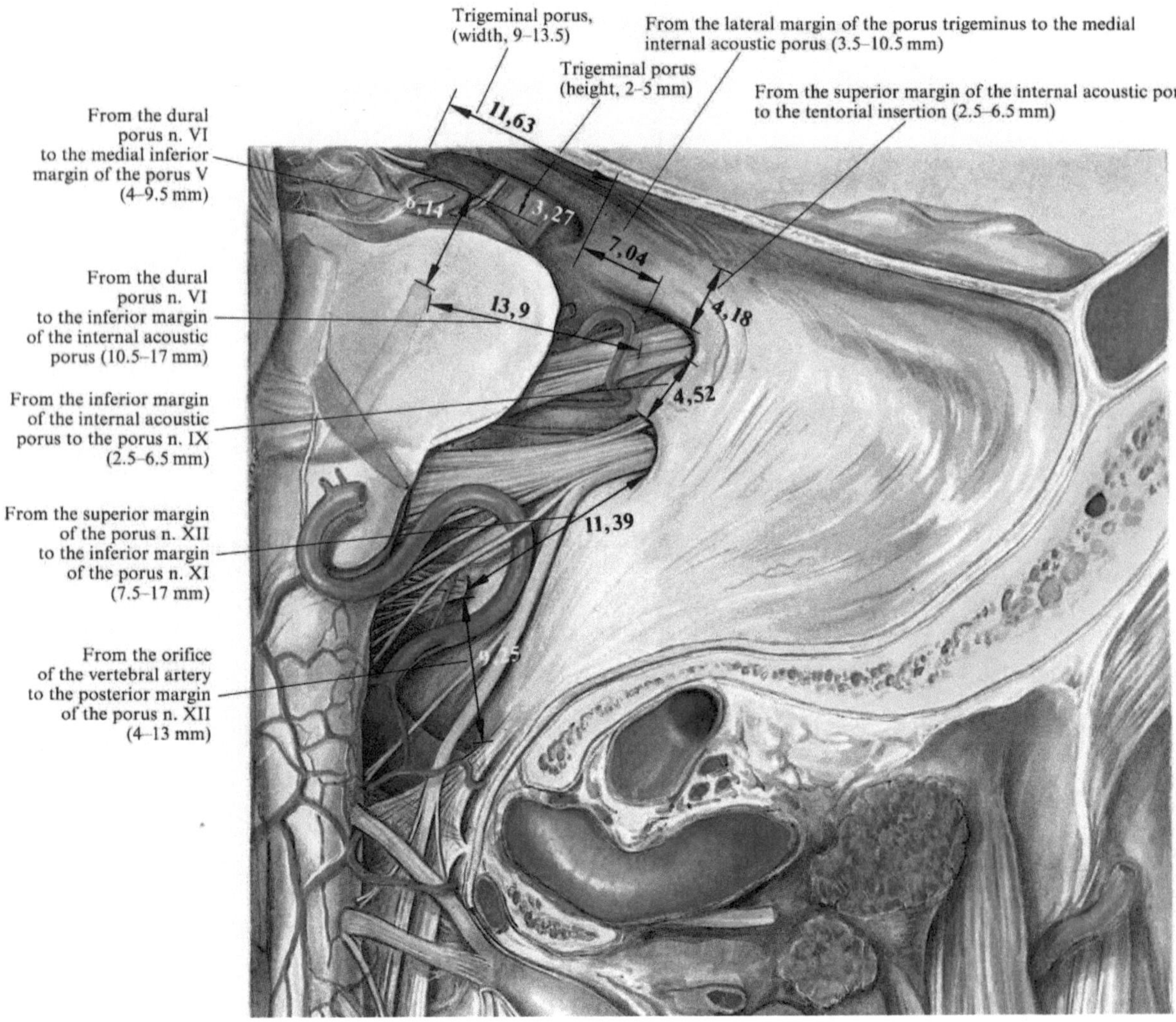

9.11 Topography of the Cerebellopontine Angle in the Suboccipital Approach

The nerves are schematically displayed and the cerebellum is retracted medially. The insert shows the direction of the visual axis after suboccipital craniotomy. The size of the defect depends on the extent of surgery required.

Surgical Importance

Suboccipital approach to the cerebellopontine angle.

Reference

1. Helms J (1978)

Plate from Helms [1]

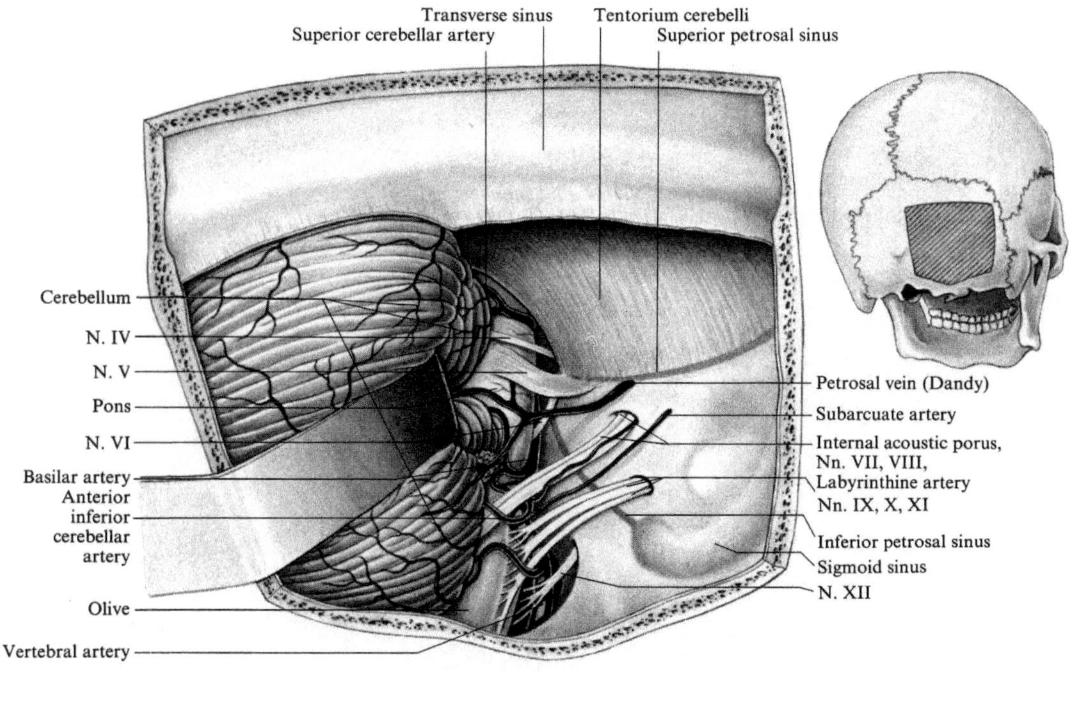

9.12 Topography in the Extradural-Transmastoid Approach to the Internal Auditory Meatus

The Plate shows the surgical approach to the posterior surface of the temporal bone through the superior part of the mastoid process. The dura and the sigmoid sinus have been elevated and displaced medially. The porus of the internal auditory meatus is visible and the meatal dura has been sectioned. Below and medial to the porus lies the jugular foramen and the glossopharyngeal, vagus, and spinal-accessory nerves. The subarcuate artery (not represented) enters the subarcuate fossa, which is located lateral to the porus beneath the crista pyramidis. The vestibular aqueduct for the endolymphatic sac is located more laterally. In the extradural-transmastoid approach to the internal auditory meatus, both the intradural part of the sac and the subarcuate artery are cut through at their point of egress from the temporal bone.

Surgical Importance

Surgery of the posterior surface of the temporal bone extending to the internal auditory meatus, endoscopy of the cerebellopontine angle.

Reference

1. Šercer A (1960)

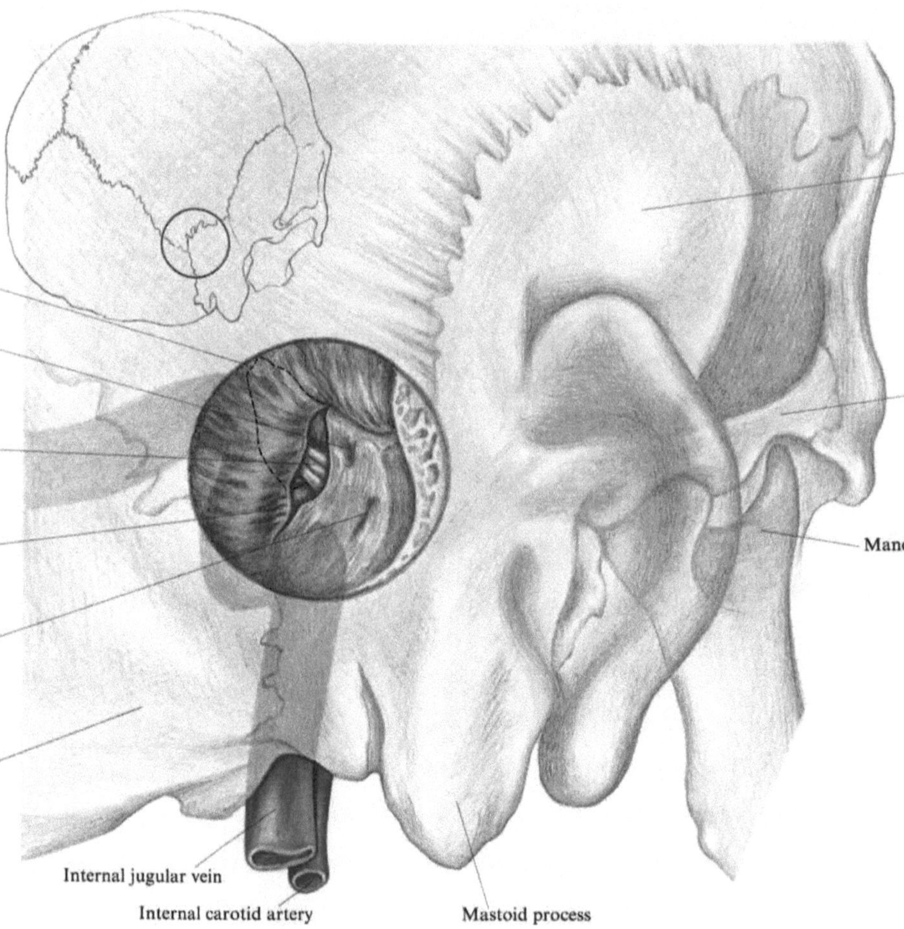

10 The Skull and the Gross Anatomy of the Central Nervous System

10.1 The Projections of the Cerebral Cortex to the Lateral Part of the Skull in the Neonate

The bones of the calvaria are joined by fontanelles in the neonate [1]. The largest fontanelle is the anterior fontanelle, and is situated between the paired anlage of the frontal bone and the anterior part of the parietal bones. The small posterior fontanelle is located between the occipital bone and the posterior parts of the parietal bones. The space between the greater wing of the sphenoid, the frontal, and the parietal bones contains the sphenoid fontanelle. These fontanelles consist of membranous connective tissue. The mastoid fontanelle is situated in the region of the future asterion between the parietal, the occipital, and the temporal bones, and consists of hyaline cartilage.

The posterior and sphenoid fontanelles close very soon after birth, while the anterior fontanelle is closed by bone during the second year of life. In this region, a bony elevation called the bregma can appear.

The cranial sutures normally ossify between 40 and 50 years of age. This ossification commences as a senile synostosis on the inner side of the skull while the sutures remain visible on the external surface. Premature synostoses can begin at different parts of the sutures, and involve their entire thickness. Such ossifications then spread from the primary ossification center to both sides of the suture. Craniofacial malformations are a consequence of premature synostoses. As the result of isolated ossification centers, *ossa suturarum* (such as bregma bone) can be formed within the affected sutures.

Surgical Importance

Surgery of Craniofacial Malformations (e.g. Crouzons desease).

Reference

1. Heiderich F (1934)

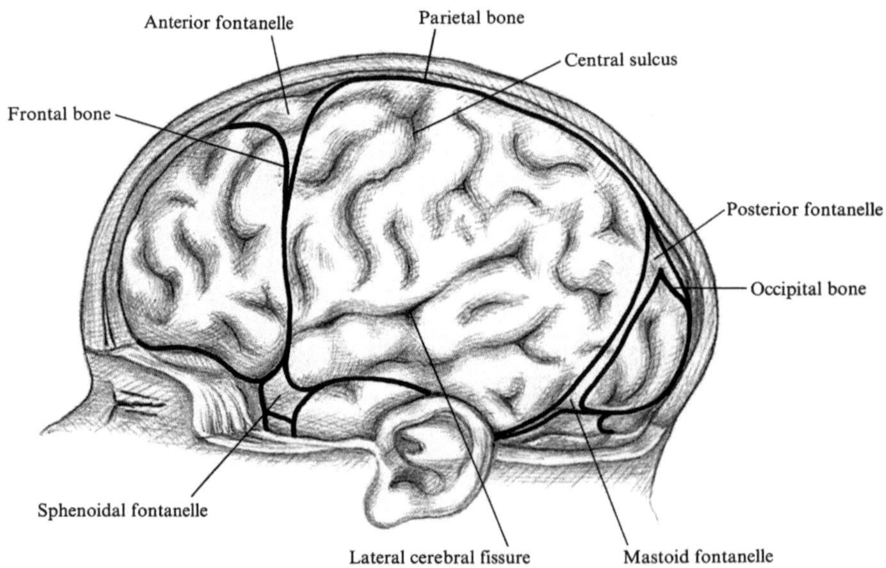

10.2 Projections of the Cerebral Cortex, the Ventricles, and the Meningeal Vessels to the Lateral Surface of the Adult Skull

The sutures are closed but are not ossified in the adult skull (see Sect. 10.1). The dimensions and the directions of the venous sinuses are usually not bilaterally symmetrical. This is especially true of the transverse sinus, which is usually larger on the right side. Both transverse sinuses can also be absent, in which case they are replaced by the occipital sinus (see Sects. 8.8.1, 2; 11.18). Additionally, the size of the middle meningeal artery can increase if the vessel supplies a tumor [1].

Surgical Importance

Ventricular puncture, craniotomy, surgery of the lateral cerebral surface.

Reference

1. Hermann F (1908)

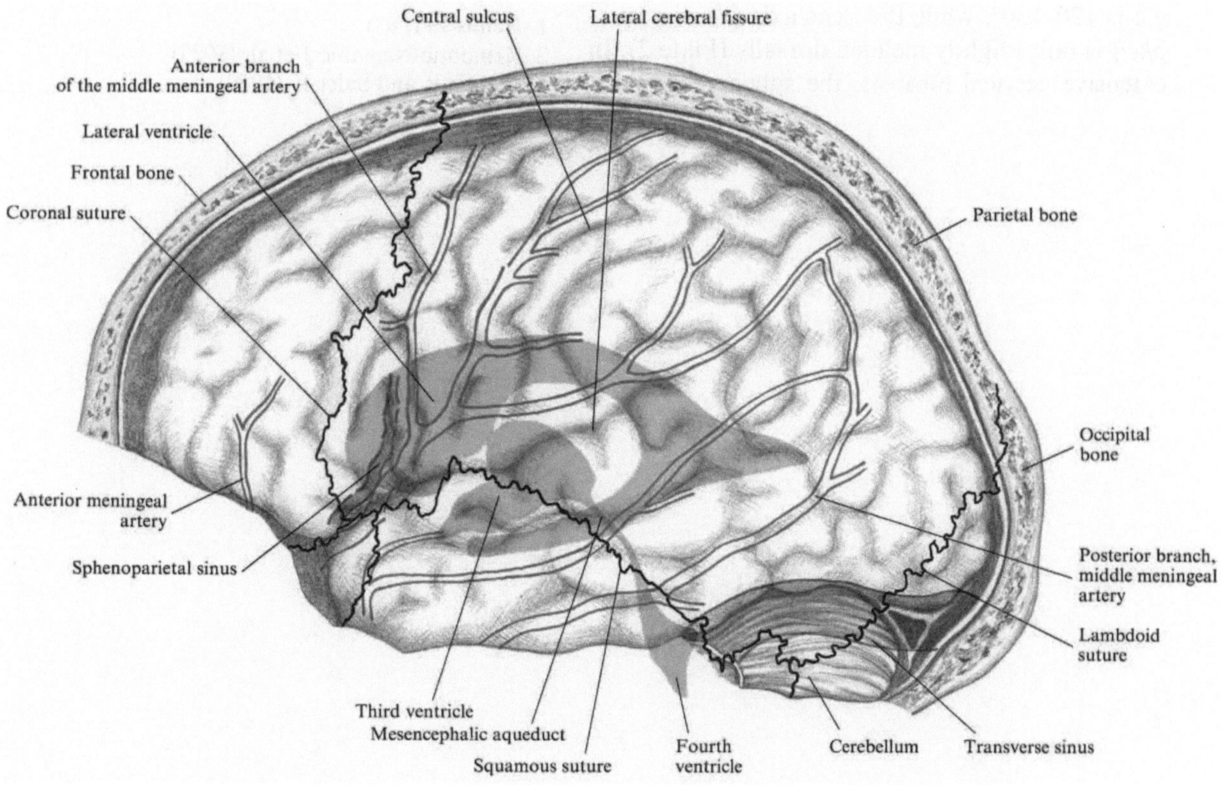

10.3 (Plates 1–4) Variations in the Sphenoid Angle and in the Bend of the Dens Axis

Kyphosis of the skull base as a compensatory curvature is a consequence of the upright human posture [2]. The degree of kyphosis present is expressed by the sphenoid angle. In Welcker's measurements, this angle is located between the line connecting the *nasion* with the anterior edge of the sella turcica and a straight line connecting the latter point with the *basion* (the anterior rim of the foramen magnum) (see Plate 1, *dotted line*). In contrast, Landzert has indicated that the sphenoid angle is bounded by a line running across the planum sphenoidale and a line running from the posterior clinoid process (Dorsum sellae) to the anterior rim of foramen magnum (see Plate 1, *solid line*) [1, 3].

The degree of kyphosis present at the skull base depends on the degree of curvature of the spine. In a normal spinal configuration, the sphenoid angle is 120–130°, while the dens axis (*dens epistrophei*) is only slightly inclined dorsally (Plate 2). In extensive cervical lordosis, the sphenoid angle is small and the dens inclined posteriorly (Plate 3). In a straighter spine, the sphenoid angle is 130° or more and the dens is inclined forwards (Plate 4) [2]. The spine and any kyphosis effecting the skull base are subjected to new static and dynamic forces throughout life, and are influenced by such factors as age and profession, as well as the presence or absence of spinal disease.

Surgical Importance

Surgery of the dens, clivus and sphenoid sinus, diagnosis of basilar impression.

References

1. Helms J (1963)
2. Krmpotić-Nemanić J et al. (1973)
3. Martin R and Saler K (1959)

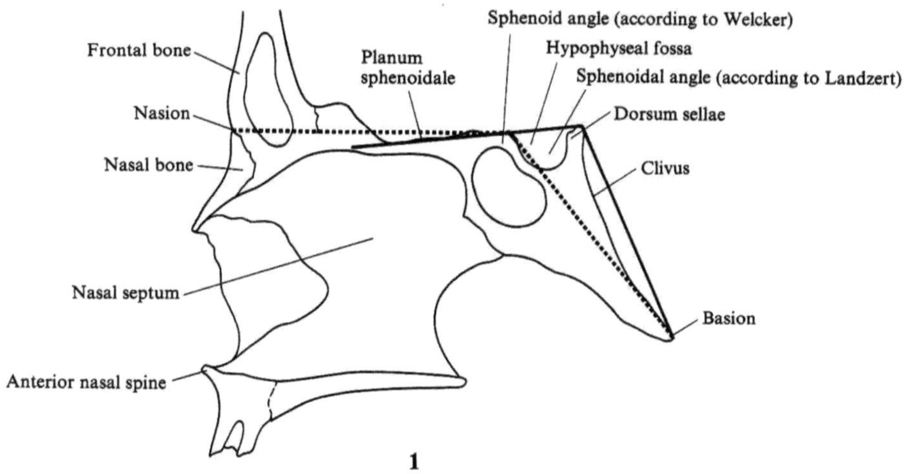
Sphenoid angle (according to Welcker)
Hypophyseal fossa
Sphenoidal angle (according to Landzert)
Frontal bone
Planum sphenoidale
Nasion
Dorsum sellae
Nasal bone
Clivus
Nasal septum
Basion
Anterior nasal spine

1

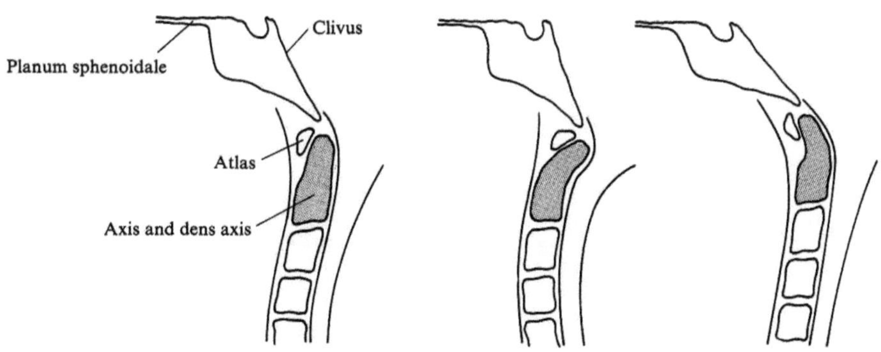
Clivus
Planum sphenoidale
Atlas
Axis and dens axis

2 3 4

10.4 The Superficial Cerebral Veins (Lateral View)

The position of the superficial cerebral veins varies. The anastomoses of the superficial middle cerebral vein with the superior sagittal and transverse sinuses are clinically important. The superior anastomotic vein of Trolard connects the superficial middle cerebral vein with the superior sagittal sinus, while the inferior anastomotic vein of Labbé links the superficial middle cerebral vein with the transverse sinus. Both veins can be of hemodynamic importance.

Surgical Importance

Craniotomy.

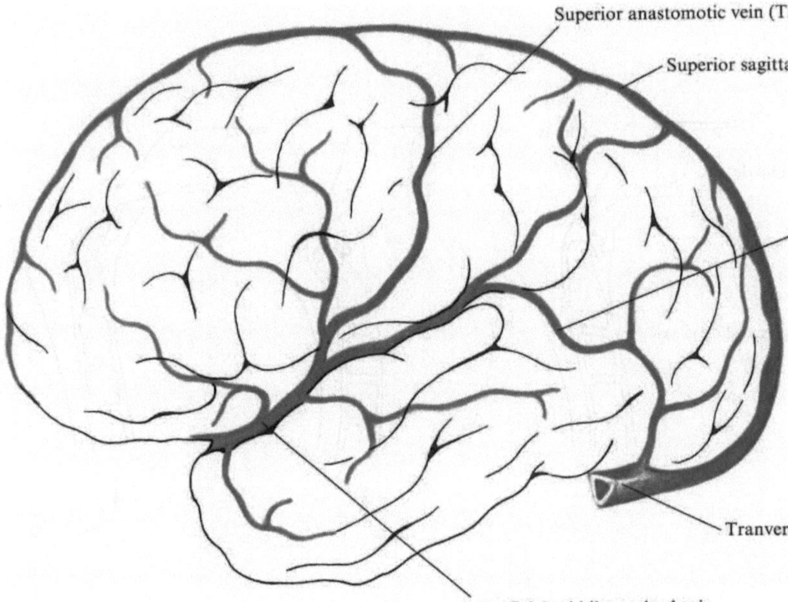

10.5 Spatial Relationships of Anatomic Structures at the Skull Base

The dura, the cranial nerves, and the internal carotid and vertebral arteries are shown in the Plate. The contours of the sphenoid bone and the foramen magnum facilitate orientation. A schematized section through the mesencephalon lies caudal to the root of the oculomotor nerve. The trigeminal nerve is visible in the middle cerebellar peduncle. The exposed fourth ventricle is bounded by the superior (brachium conjunctivum) and inferior cerebellar peduncles (the restiform bodies) [1].

Surgical Importance

Endocranial surgery of the skull base and the hypophysis, surgery of the brain stem and cerebellopontine angle, otoneurosurgery, ligature of the sigmoid sinus, traumatology of the skull base.

Reference

1. Hermann F (1908)

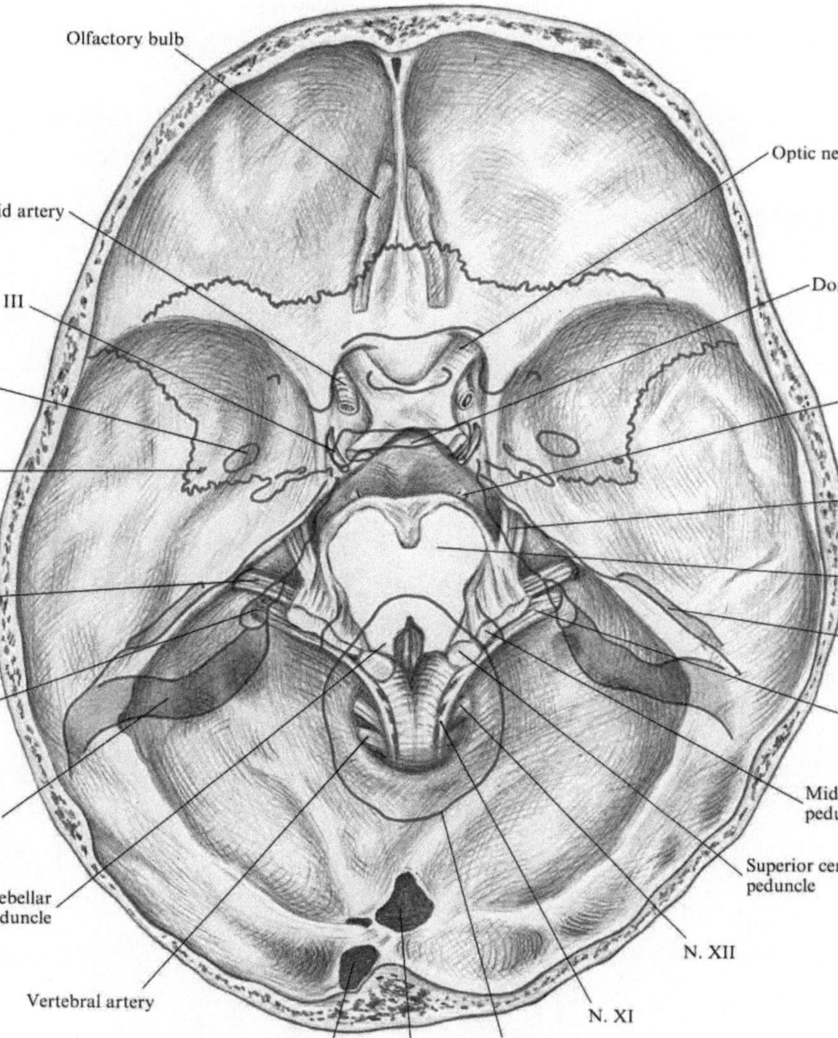

10.6 Spatial Relationships of the Cerebral Ganglia and the Ventricular System to the Calvaria

The cranial sutures and the ventricles (*outlined*) render orientation easier. The choroid plexus is located in the occipital horn of the lateral ventricle. The section shows the cerebral ganglia at mid level of the insula.

Surgical Importance

CT diagnosis, traumatology, ventriculo-jugular shunts.

Reference

1. Hermann F (1908)

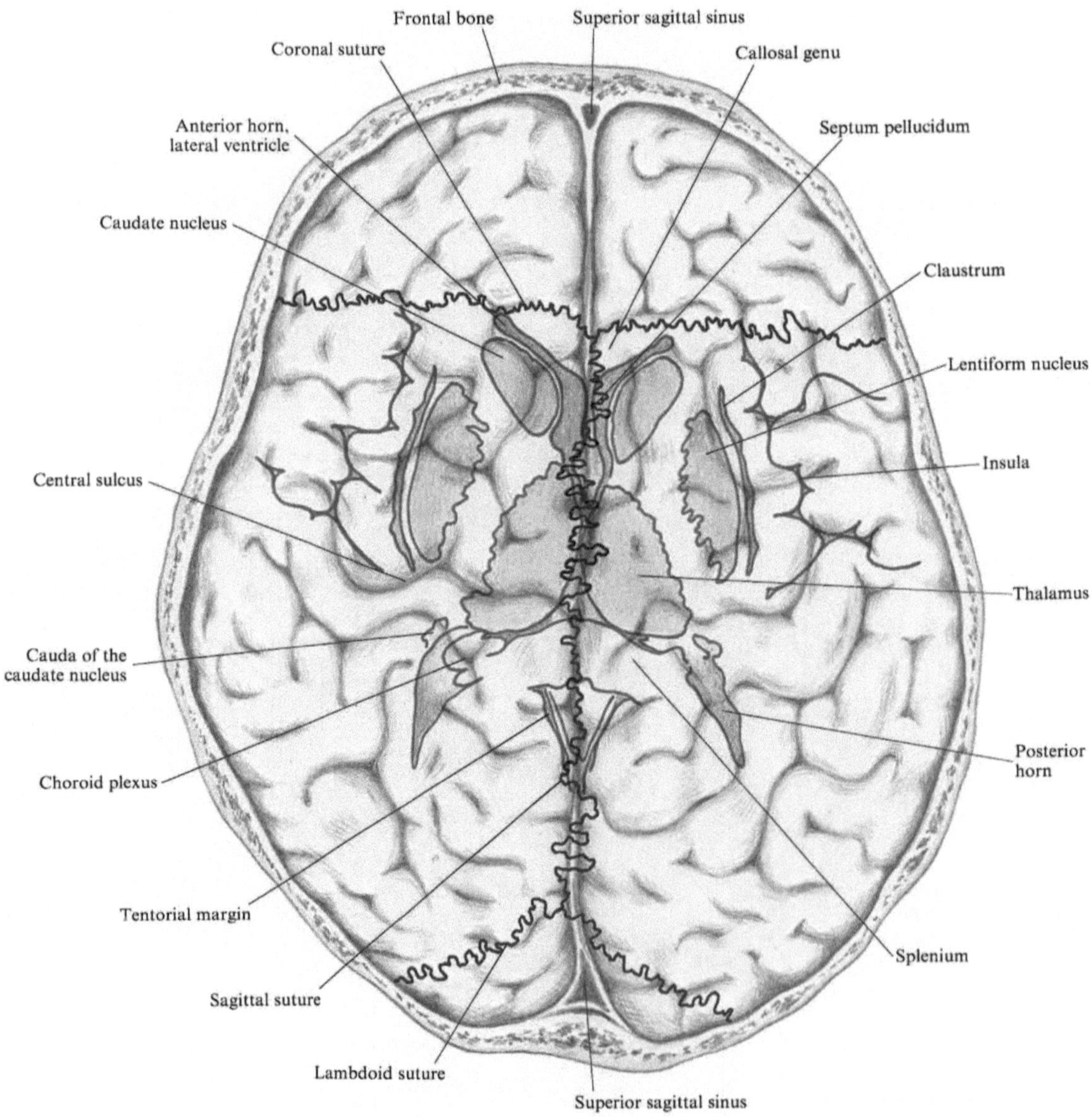

10.7 Schematic Relationships of the Inferior Cerebral Surface to the External Surface of the Skull Base

The cerebral arterial circle of Willis and the most important arteries and cranial nerves of the skull base are represented in the plate. The structures of the base of the skull and the hard palate are outlined in *black*. The positional relationships of the cerebellar tonsils, medulla oblongata and foramen magnum are also shown.

Surgical Importance

Diagnosis, diseases, trauma, intracranial and extracranial surgery of the skull base.

Reference

1. Hermann F (1908)

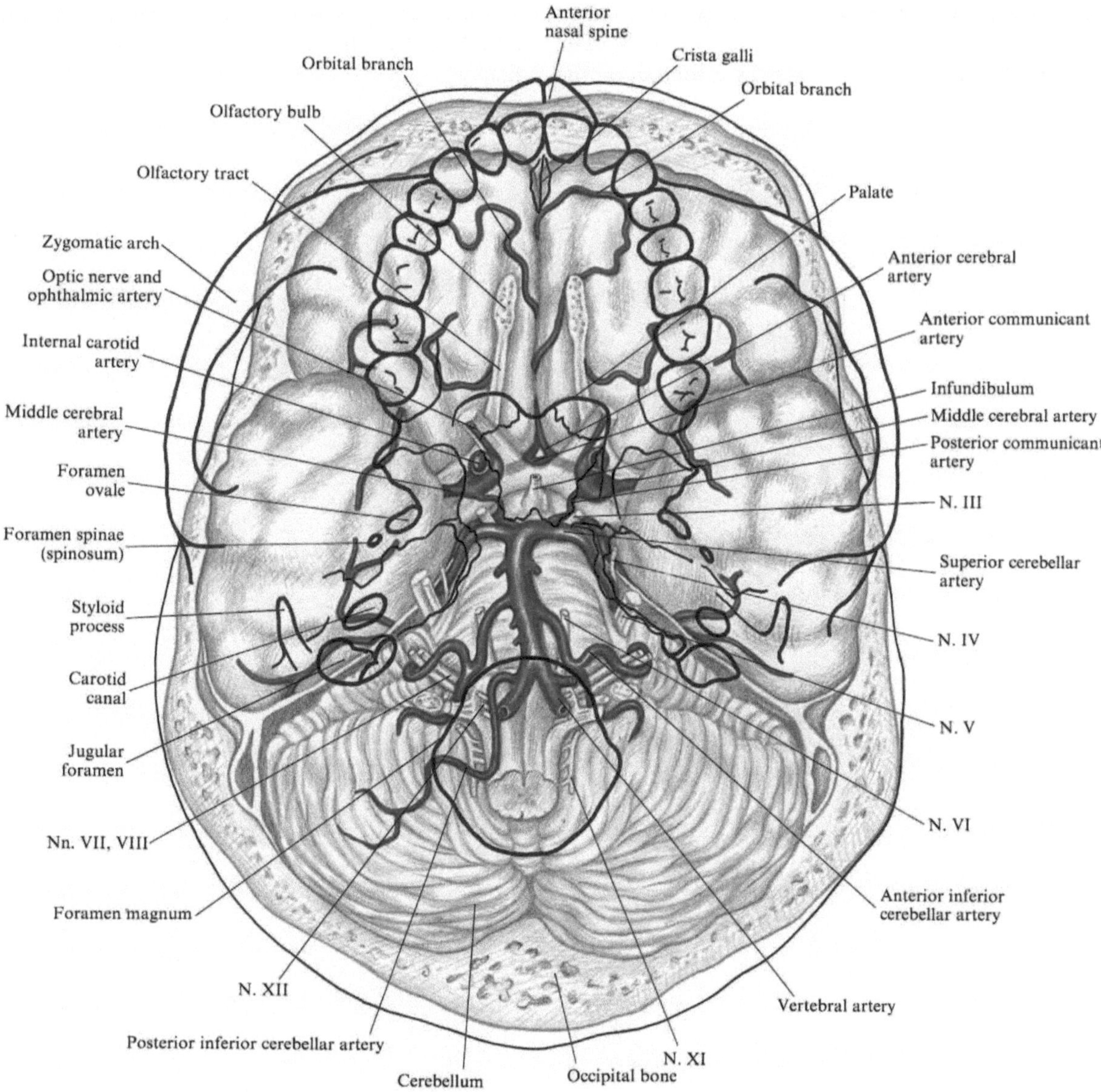

10.8 (Plates 1–2) The Cerebral Cisterns

Plate 1 is a schematized lateral view of the cerebral cisterns, while Plate 2 is a postero-superior view of the same structures.

The largest cistern is the *cisterna magna*, or the *cisterna cerebellomedullaris*. The cerebrospinal fluid flows into this cistern from the middle rhombencephalic aperture in the roof of the fourth ventricle (*foramen of Magendi* or the median aperture of the rhombencephalon. The cisterna magna continues anteriorly into the *cisternae pontis, interpeduncularis* and *chiasmatis*. These cisterns communicate in front of the lamina terminalis with the *cisterna corporis callosi*, and further on with the *cisterna vermis*, which joins the cisterna magna. The mesencephalon is surrounded by the *cisterna ambiens* at the level of the hiatus tentorii. The *cisterna fossae lateralis cerebri* is situated in the lateral cerebral fissure. Protrusions of the cisterns run parallel to the cranial nerves for a certain distance. All cisterns communicate with each other and with the subarachnoid space, lying between the pia mater and the arachnoid membranes. The cerebrospinal fluid flows into the cisterna pontis through the (lateral) foramina of Luschka. The choroid plexus appears on the cerebellar surface dorsal to the IX, X, and XI cranial nerves in the region of the foramina of Luschka, and should not be mistaken for the cerebellar flocculus.

Surgical Importance

Endocranial diseases and their treatment, impaired circulation of the cerebrospinal fluid, hydrocephalus.

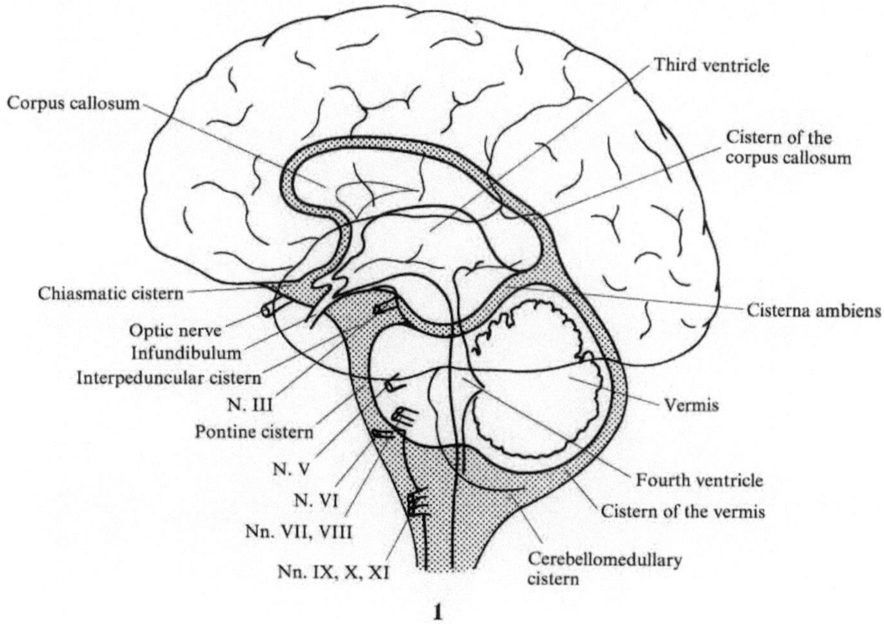

Corpus callosum
Third ventricle
Cistern of the
corpus callosum
Chiasmatic cistern
Cisterna ambiens
Optic nerve
Infundibulum
Interpeduncular cistern
N. III
Vermis
Pontine cistern
N. V
Fourth ventricle
N. VI
Cistern of the vermis
Nn. VII, VIII
Nn. IX, X, XI
Cerebellomedullary
cistern

1

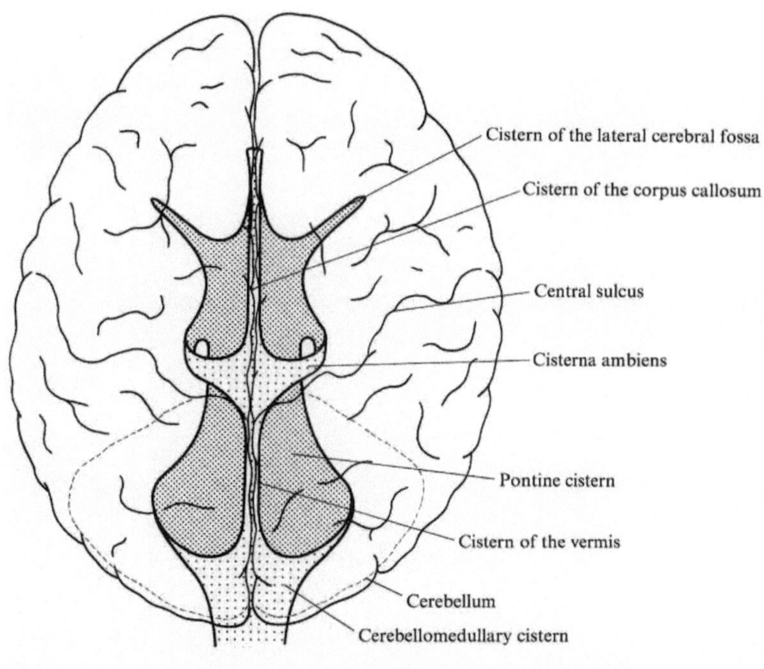

Cistern of the lateral cerebral fossa
Cistern of the corpus callosum
Central sulcus
Cisterna ambiens
Pontine cistern
Cistern of the vermis
Cerebellum
Cerebellomedullary cistern

2

11 The Topography and Variations of the Extra- and Intracranial Vessels

11.1 (Plates 1–8) The Course and Possible Locations of the Internal Carotid Artery in the Neck, Anterior View

The position of the internal carotid artery in relation to the external carotid artery can have many surgically important variations [2]. The internal carotid artery can lie lateral to the external carotid artery in 50% of cases (Plate 1), dorsal in 21% (Plate 2), dorsomedial in 18% (Plate 3), medial in 3% (Plate 4) or ventromedial in 9% (Plate 5). The internal carotid artery does not normally give off branches in the neck and only exceptionally gives rise to the ascending pharyngeal and the occipital arteries. Bryce et al. [1] have reported a case in which the left vertebral artery originated from the internal carotid artery 3 cm above the carotid bifurcation. Additionally, Kwak et al. [3] have described a case in which the posterior meningeal artery was found to originate from the extracranial part of the internal carotid artery (not represented).

Uncommonly, the internal carotid artery may display prominent tortuosities (Plates 6, 7, 8; see Sects. 4.7, 11.2). On rare occasions, the vessel may also be absent.

Surgical Importance

Surgery of the lateral neck and the pharynx.

References

1. Bryce S et al. (1979)
2. Krmpotić-Nemanić J (1978)
3. Kwak S et al. (1980)

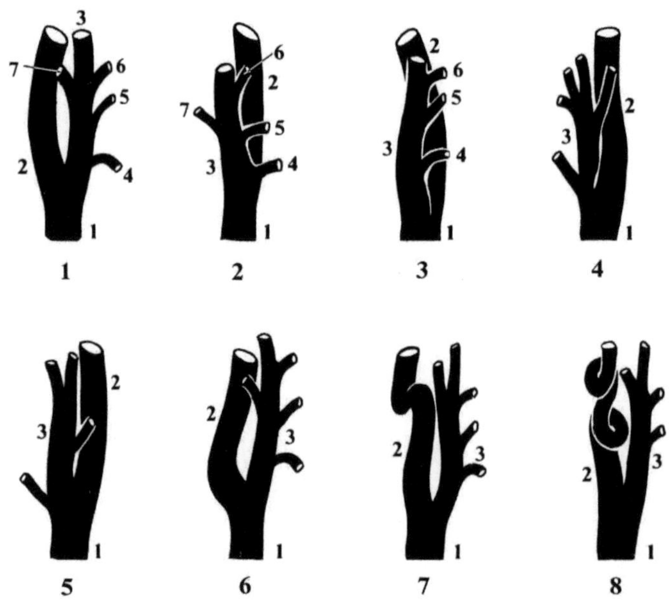

1 Common carotid artery
2 Internal carotid artery
3 External carotid artery
4 Superior thyroid artery
5 Lingual artery
6 Facial artery
7 Occipital artery

11.2 (Plates 1–2) Formation of Loops of the Internal Carotid Artery

The internal carotid artery can display loops in 30% of cases. Angulation can occur in 6% of cases and flat curves in 7% [1–5]. However, these variations do not necessarily cause hemodynamic disturbances.
On rare occasions, the internal carotid artery forms a loop around the transverse process of one of the cervical vertebra (see Sects. 4.7, 11.1).

Surgical Importance

Surgery of the internal carotid and vertebral arteries and cervical spine.

References

1. Asamrah DK and Foy PM (1979)
2. Bouchet A (1978)
3. Francke JP et al. (1982)
4. Goodman RS and Cohen NL (1981)
5. Lazorthes G et al. (1979)

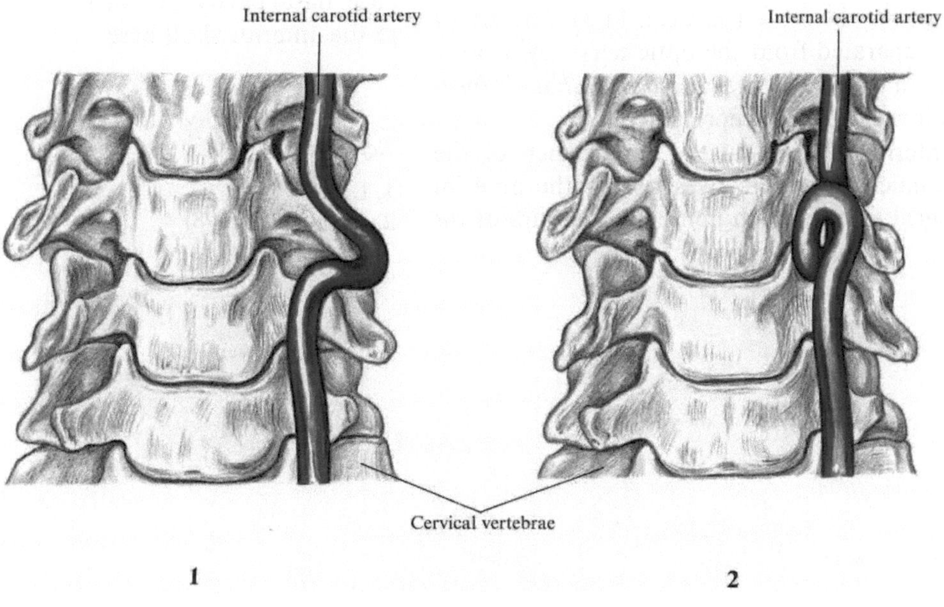

11.3 (Plates 1–2) Branches of the Internal Carotid Artery

11.3 (Plate 1) Branches of the Internal Carotid Artery in the Anterior Cranial Fossa and in the Skull Base

After emerging from the cavernous sinus in the anterior cranial fossa, the internal carotid artery gives off the ophthalmic artery and then branches into the anterior and middle cerebral arteries [1]. This arrangement is shown in Plate 1 while Plate 2 schematizes the segment shown. Both anterior cerebral arteries are connected by the anterior communicant (communicating) artery. In 75% of cases, the ophthalmic artery lies medial and caudal to the optic nerve within the optic canal and leaves the canal lateral to the nerve. If the artery does not cross the nerve (in 25% of cases), its course runs along the medial side of the optic nerve to the orbit. In exceptional cases, the artery crosses the nerve twice to leave the canal medially, or it passes above the nerve (see Sect. 11.3). The artery can be separated from the optic nerve by a bony septum, in which case the *clinoideoophthalmic foramen* is formed [2] (not represented).

The anterior ethmoid artery is a branch of the ophthalmic artery and gives rise to the anterior meningeal artery, which nourishes the dura of the anterior cranial fossa. The anterior cerebral artery gives rise to the frontobasal artery (not represented), which supplies the mediobasal part of the frontal lobe. Additional branches include the frontopolar artery to the frontal lobe, the pericallosal artery passing over the corpus callosum, and the callosomarginal artery (see Sect. 11.23). (The nomenclature of these latter vessels has been derived from recent angiographic studies of the cerebral vasculature.)

Surgical Importance

Surgery and traumatology of the anterior cranial base, nasal cavity, and orbits; transcranial surgery to the anterior skull base.

References

1. Lang J (1979)
2. Taguchi K (1898)

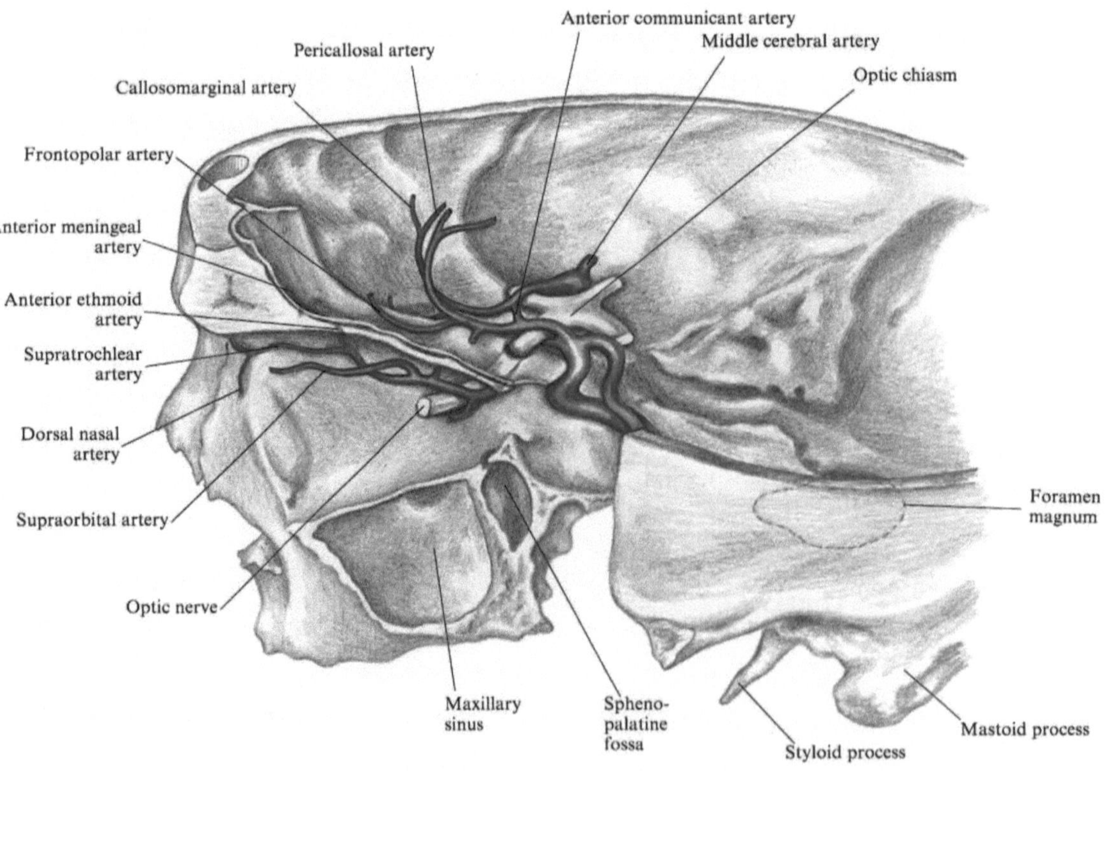

11.3 (Plate 2) Schematic Orientation of the Skull Section
Depicted in 11.3, Plate 1

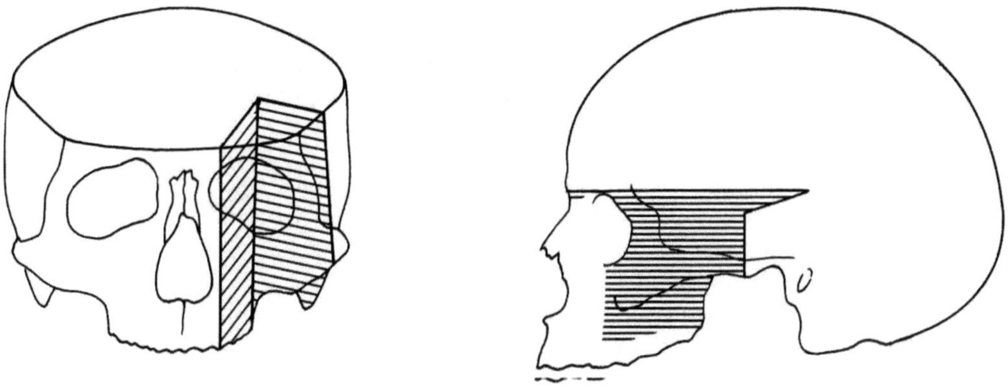

11.4 The Anterior and Posterior Ethmoid Arteries, Superior View

The Plate depicts the structures present under the left frontobasal region of the skull. The anterior ethmoid artery is a branch of the ophthalmic artery and traverses the orbitocranial foramen (the *anterior ethmoid foramen*). This artery passes beneath and within the dura alongside the ethmoid air cells to the falx cerebri, forming the anterior meningeal artery and the *anterior falcine artery*.
The posterior ethmoid artery is also a branch of the opthalmic artery. It usually runs cranial to the superior oblique muscle, passing to the posterior ethmoid air cells and supplying the posterior part of the nasal cavity. In this specimen, the vessel enters the ethmoid bone caudal to the superior oblique muscle.

Surgical Importance

Surgery of the paranasal sinuses and orbits, intracranial and extracranial surgery of the anterior cranial fossa and skull base, ethmoidosphenoidal surgery of the orbits, transseptal and transantral approaches to the hypophysis.

Reference

1. Denecke HJ and Guleke N (1953)

Plate from Denecke and Guleke [1]

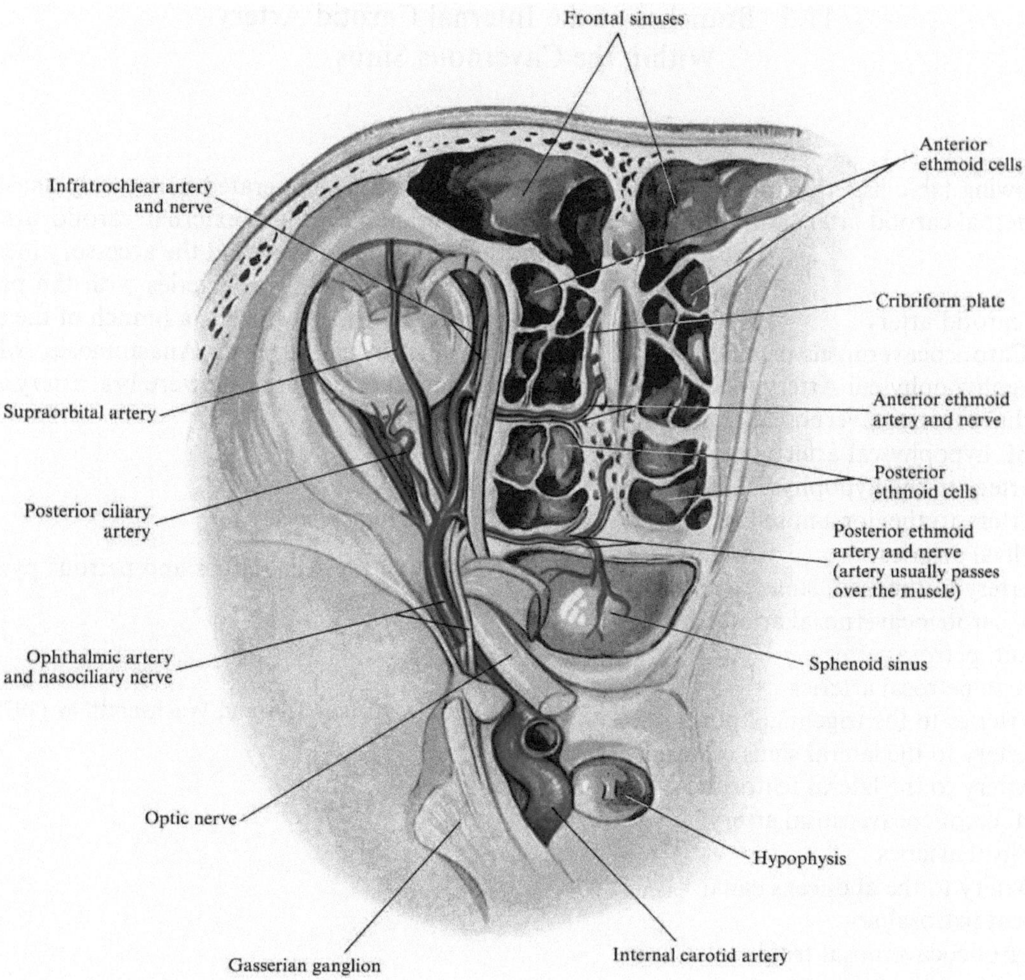
Frontal sinuses
Anterior ethmoid cells
Infratrochlear artery and nerve
Cribriform plate
Supraorbital artery
Anterior ethmoid artery and nerve
Posterior ethmoid cells
Posterior ciliary artery
Posterior ethmoid artery and nerve (artery usually passes over the muscle)
Ophthalmic artery and nasociliary nerve
Sphenoid sinus
Optic nerve
Hypophysis
Gasserian ganglion
Internal carotid artery

11.5 Branches of the Internal Carotid Artery Within the Cavernous Sinus

The following table lists the intracranial branches of the internal carotid artery within the cavernous sinus.

- Internal carotid artery
  - Post. Caroticocavernosus trunk
  - Meningohypophyseal Artery
    - Medial caroticocavernoseal artery
      - Inf. hypophyseal artery
      - Artery to the hypophyseal fossa
      - Artery to the dorsum sellae
      - Clival arteries
      - Artery to the med. sinus cavernosus
    - Sup. caroticocavernosal artery
      - Ant. petrosal artery
      - Post. petrosal arteries
      - Arteries to the trigeminal porus
      - Artery to the lateral sinus cavernosus
      - Artery to the lateral tentorium
    - Post. caroticocavernosal artery
      - Clival arteries
      - Artery to the abducens canal
      - post petrosal art.
  - Lat. caroticocavernosal trunk
    - Central meningeal artery
      - Artery to the oculomotor nerve
      - Artery to the trochlear nerve
      - Clinoid artery
    - Artery to the orbital fissure
    - Artery to the trigeminal nerve
      - Artery to the pars triangularis
      - Artery to the trigeminal ganglion
      - Artery to the ophthalmic nerve
      - Artery to the maxillary nerve
      - Artery to the mandibular nerve
    - Artery to the abducens nerve
    - Ant. petrosal arteries
  - Capsular arteries
  - Inf. med. hypophyseal artery
  - Inf. ant. hypophyseal artery

The branches enumerated here partly anastomose with branches of the external carotid artery via the middle meningeal and the accessory meningeal arteries and the clival arteries with the posterior meningeal artery (which is a branch of the ascending pharyngeal artery). Anastomoses with the meningeal branches of the vertebral artery are also present [1].

Surgical Importance

Surgery of the sella, clivus and petrous pyramid.

Reference

1. Lang J in: Lanz T v and Wachsmuth W (1979)

Plate from Lang [1]

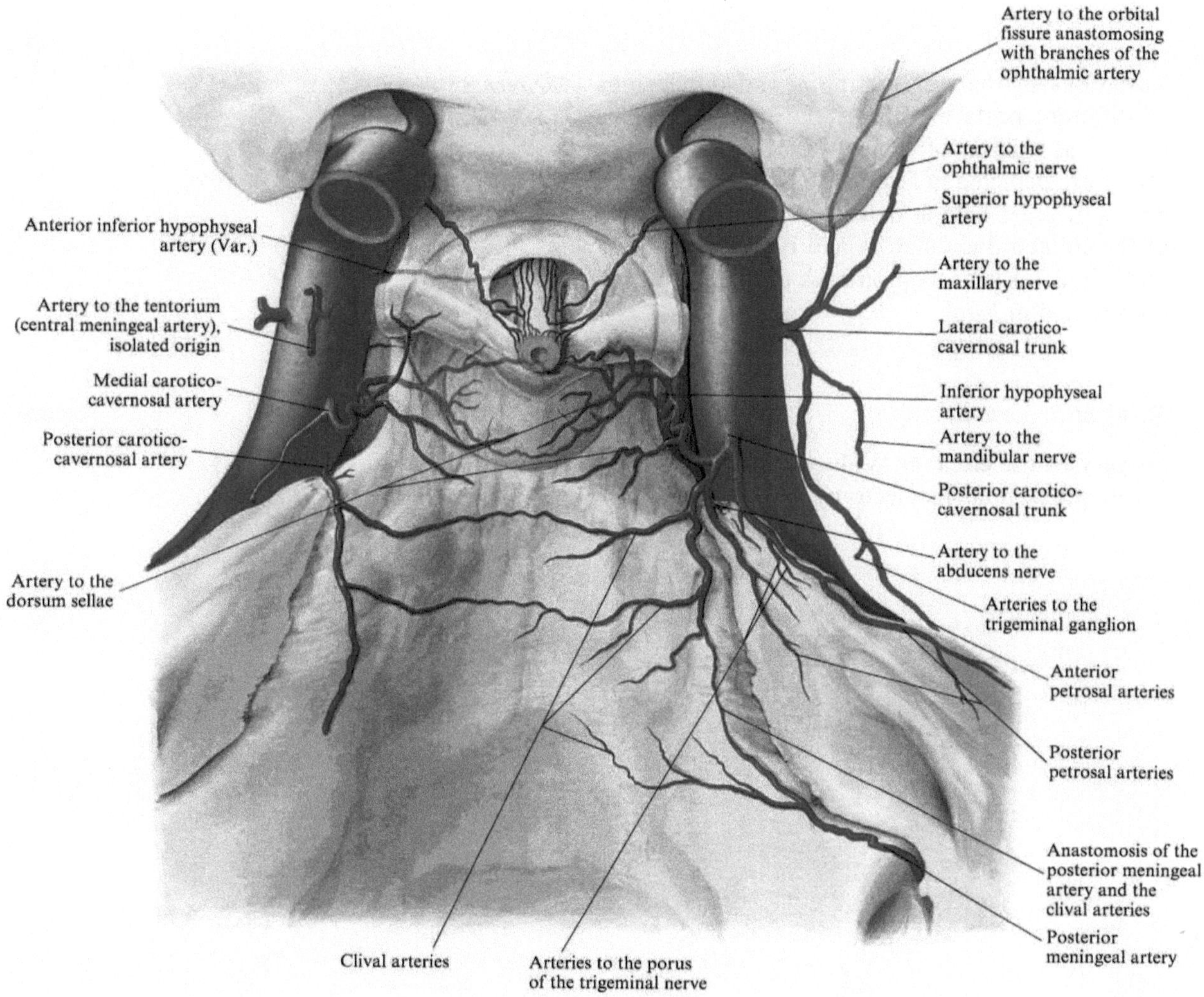
Artery to the orbital fissure anastomosing with branches of the ophthalmic artery
Artery to the ophthalmic nerve
Superior hypophyseal artery
Anterior inferior hypophyseal artery (Var.)
Artery to the maxillary nerve
Artery to the tentorium (central meningeal artery), isolated origin
Lateral carotico-cavernosal trunk
Medial carotico-cavernosal artery
Inferior hypophyseal artery
Posterior carotico-cavernosal artery
Artery to the mandibular nerve
Posterior carotico-cavernosal trunk
Artery to the abducens nerve
Artery to the dorsum sellae
Arteries to the trigeminal ganglion
Anterior petrosal arteries
Posterior petrosal arteries
Anastomosis of the posterior meningeal artery and the clival arteries
Posterior meningeal artery
Clival arteries
Arteries to the porus of the trigeminal nerve

11.6 (Plates 1–12) Variations in the Arterial Circle of Willis

Variations in the configurations and dimensions of different parts of the arterial circle of Willis as well as aplasia of the posterior communicant artery are depicted in Figs. 1–12 [1–14]. Angiography is of great help in determining the formation of the collateral circulation and its variants in actual clinical practice. Anomalies in the figures are represented in *black*.

Surgical Importance

Surgery of the circle of Willis (aneurismectomies, caroticocavernous fistulectomy).

References

1. Adachi B (1928)
2. Askenazy M et al. (1952)
3. Firbas W and Sinzinger H (1972)
4. Fischer E (1938)
5. Guerrier Y and Villaceque G (1949)
6. Hochstetter F (1937, 1938)
7. Iontov A (1962)
8. Krmpotić-Nemanić J (1978)
9. Miller-Fischer C (1965)
10. Mitterwallner F (1955)
11. Müller F (1978)
12. Platzer W (1962)
13. Puchades-Orts A et al. (1976)
14. Vogel G and Bauer G (1978)

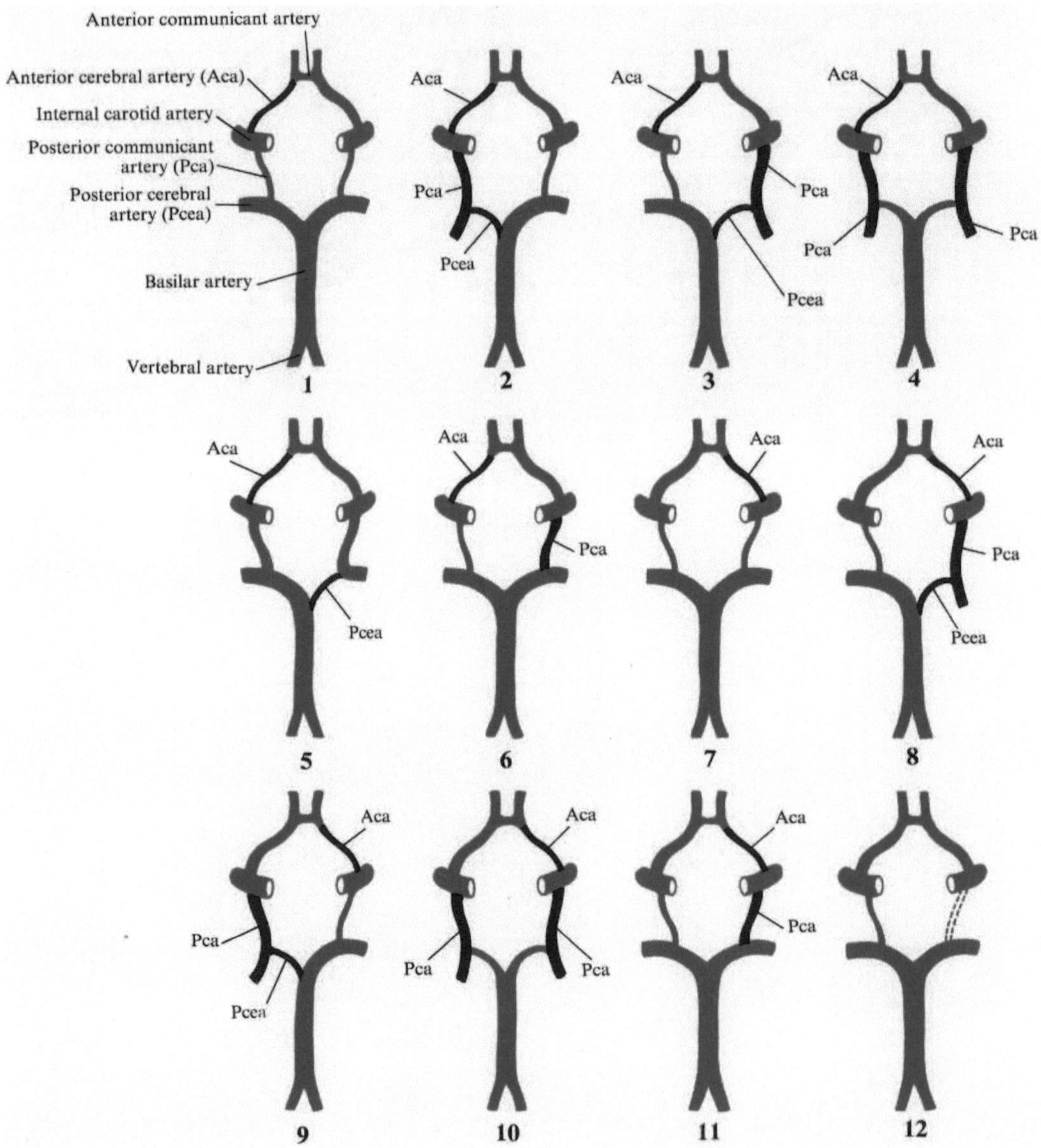

11.7 (Plates 1–73) Variations in the Anterior Communicant Artery

Variations in the flow patterns of the anterior communicant artery are shown. In any surgery performed in this region, especially surgery for aneurisms, the functionally patent communications of the vessels present are very important for possible collateral circulations [1].
Variations in the anterior cerebral artery are often found in combination with variations in the posterior cerebral artery [2].

Surgical Importance

Intracranial surgery of the anterior portion of the arterial circle of Willis and the internal carotid artery, surgery for aneurysms and caroticocavernous fistulae.

References

1. Miller-Fischer C (1965)
2. Krmpotić-Nemanić (1978)

Plates 1–73 modified from Miller-Fischer [1]

11.8 (Plates 1–8) Variations in the Configurations of the Anterior and Posterior Cerebral Arteries

All of the vessels of the circle of Willis have the same dimensions by the 4th month of fetal development. Differences appear at later stages. Anomalous caroticobasilar anatomoses can involve the primitive trigeminal artery or – more rarely – the hypoglossal artery or the acoustic artery. The circle of Willis includes many different variations, as shown in Plates 1–8 [1]. The vessels of an ideal circle have nearly the same caliber. While larger posterior communicant arteries make possible functionally useful anastomoses between both right and left internal carotid and vertebral arteries, small posterior communicant arteries result in a nearly independent circulation in the regions of the brain supplied by each internal carotid and vertebral arteries. The diameter of the anterior communicant artery varies from 0.1 to 2 mm, while that of the posterior communicant artery is 0.1 to 3 mm. This size can also change if a more abundant blood supply is needed and varies in the course of life [2].

Surgical Importance

Surgery for aneuryms, ligature of the common carotid artery or internal carotid artery, subclavian steal syndrome, surgery of the vertebral artery.

References

1. Krmpotić-Nemanić J (1978)
2. Lazorthes G et al. (1979)

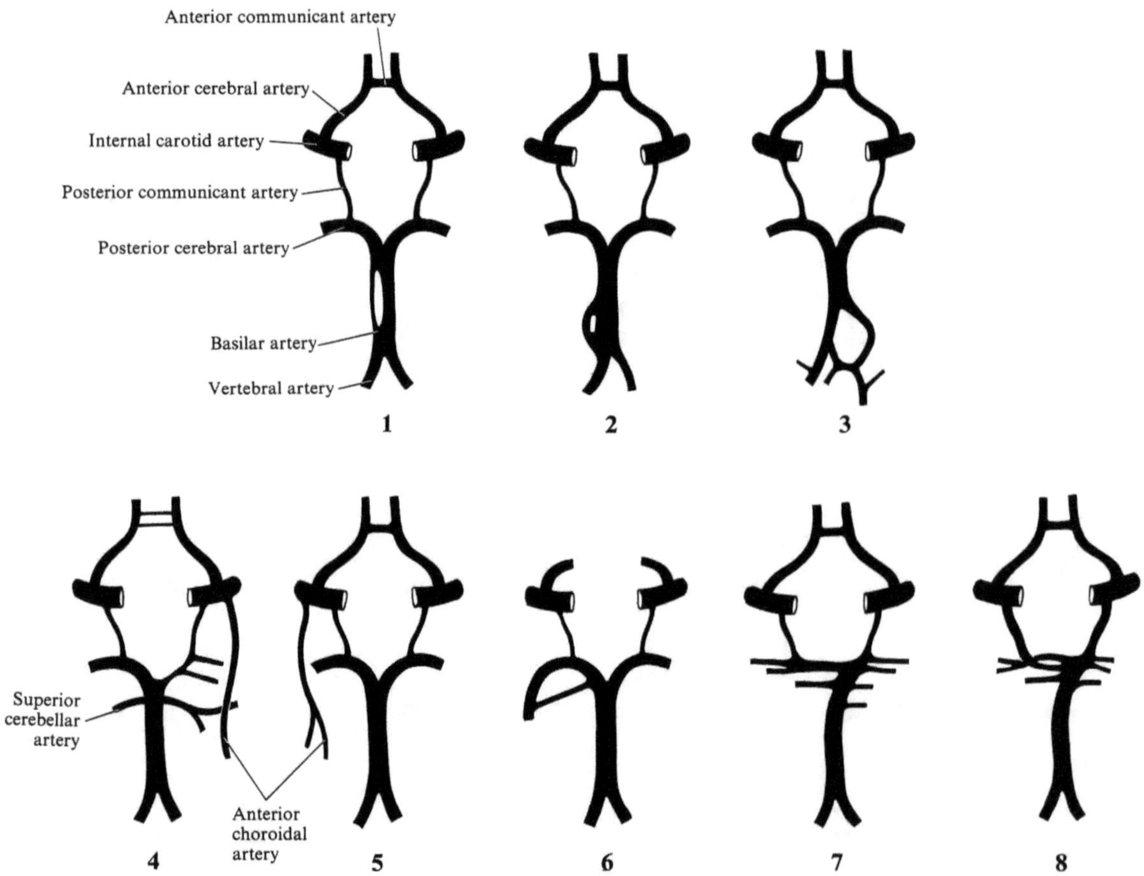

The vertebral artery normally originates from the subclavian artery (Plate 1), entering the transverse foramen of the VIth cervical vertebra [1–4]. It exceptionally enters the foramina of either the Vth or IVth cervical vertebra. In the latter two cases, the ventral tubercle of the vertebra involved is better developed than is that of the VIth vertebra. The vertebral artery can originate from any of the branches of the aortic arch or from the aortic arch itself as a single or a double artery. When the vessel originates as a branch of the aortic arch, it lies in front of and *not* – as it usually does – behind the thoracic duct. In Plates 2–9, the different variations are shown in *black* (see Sect. 11.11).

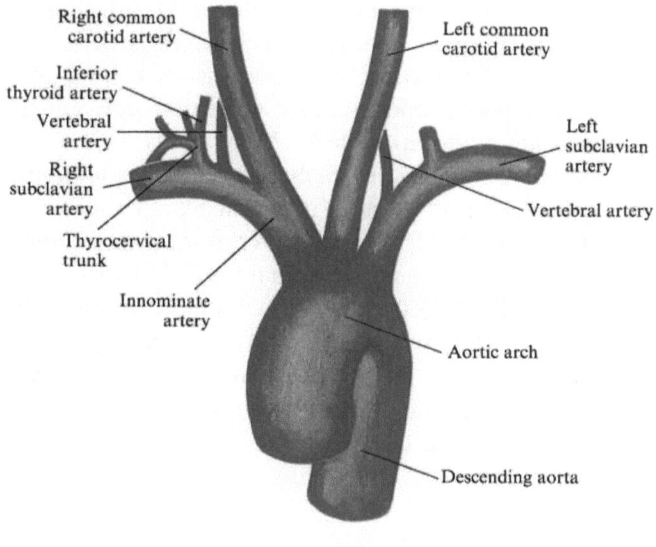

1

Surgical Importance

Surgery of the vertebral artery and cervical spine, neck dissection, ligature of the vertebral artery.

References

1. Cormier JM et al. (1979)
2. Krmpotić-Nemanić J (1978)
3. Lang J (1979)
4. Scotti G et al. (1978)

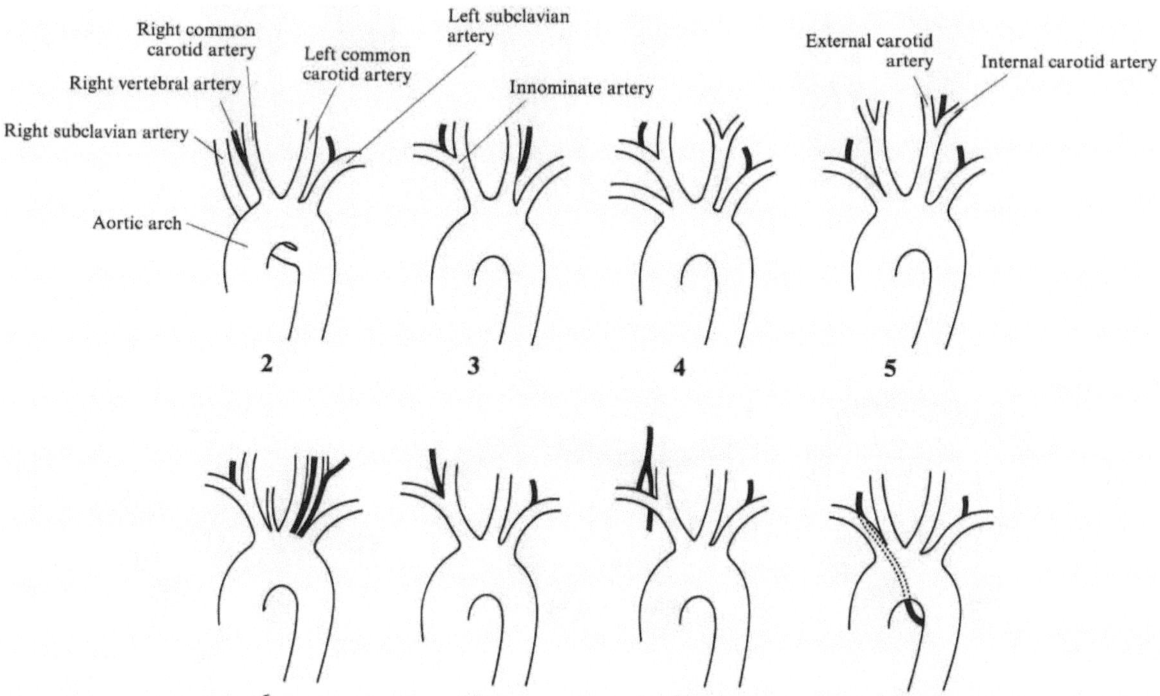

2 3 4 5

6 7 8 9

11.10 (Plates 1–6) Anomalous Origin of the Right Subclavian Artery From the Aorta and Variations in the Relationships of the Right Subclavian Artery to the Esophagus and the Trachea

In an anomalous origin of the right subclavian artery (Plate 1), the vessel originates as the last branch of the aortic arch and passes behind the esophagus. Clinically, this condition can cause *dysphagia lusoria* (Plate 4). More rarely, the vessel passes between the trachea and the esophagus (Plate 5) or in front of the trachea to the right side (Plate 6) [2]. The right vertebral artery (*blue*) can originate from the anomalous right subclavian artery (Plate 1) or from the common or internal carotid arteries (Plates 2, 3) [1, 3]. When the subclavian artery passes between the trachea and esophagus (Plate 5), stridor can result due to compression of the trachea. Plate 6 shows the subclavian artery passing in front of the trachea. Clinically, such vessels can be injured during low tracheotomies if not carefully done.

Surgical Importance

Dysphagia lusoria, esophagoscopy, tracheotomy, stomal surgery, exploration of the recurrent laryngeal nerve, neck dissection.

References

1. Bernardi L and Dettori P (1975)
2. Locatelli E et al. (1977)
3. Sutton D (1950)

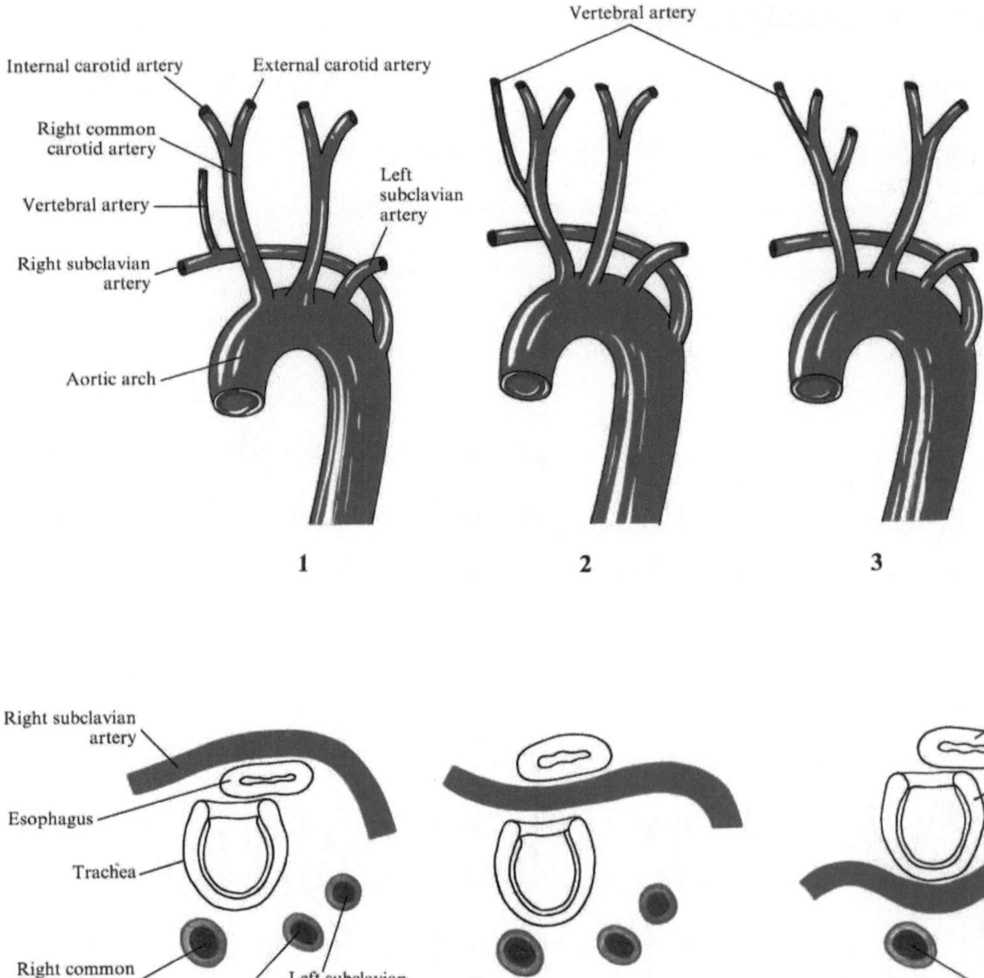

11.11 Location of the Vertebral Artery in the Suboccipital Triangle of the Neck

After leaving the transverse foramen of the atlas, the vertebral artery enters the corresponding sulcus of the posterior arch of the atlas and enters the skull through the atlanto-occipital membrane and the foramen magnum [1]. The vessel is best found in the suboccipital triangle of the neck, which is located in the lateral part of the arch of the atlas. This triangle is delimited by the rectus capitis major muscle, as well as the obliquus capitis superior and inferior (oblique atlantal) muscles (see Sect. 4.1).

The *punctum nervosum* of Erb is a zone on the posterior border of the sternocleidomastoid muscle in the region where it crosses the inferior border of the splenius capitis muscle. It is usually the site of pain produced by neuralgias due to the lesser occipital nerve. The accessory nerve is relatively superficial in its location here, and can be easily injured during extirpation of any cervical lymph nodes. The greater occipital nerve can usually be palpated lateral to the external occipital protuberance.

Surgical Importance

Surgery of the vertebral artery and atlas, posterior cervical tumors and metastases, intra- and extracranial glomus tumors, prophylactic ligature of the vertebral artery, neck dissection.

Reference

1. Lang J (1979)

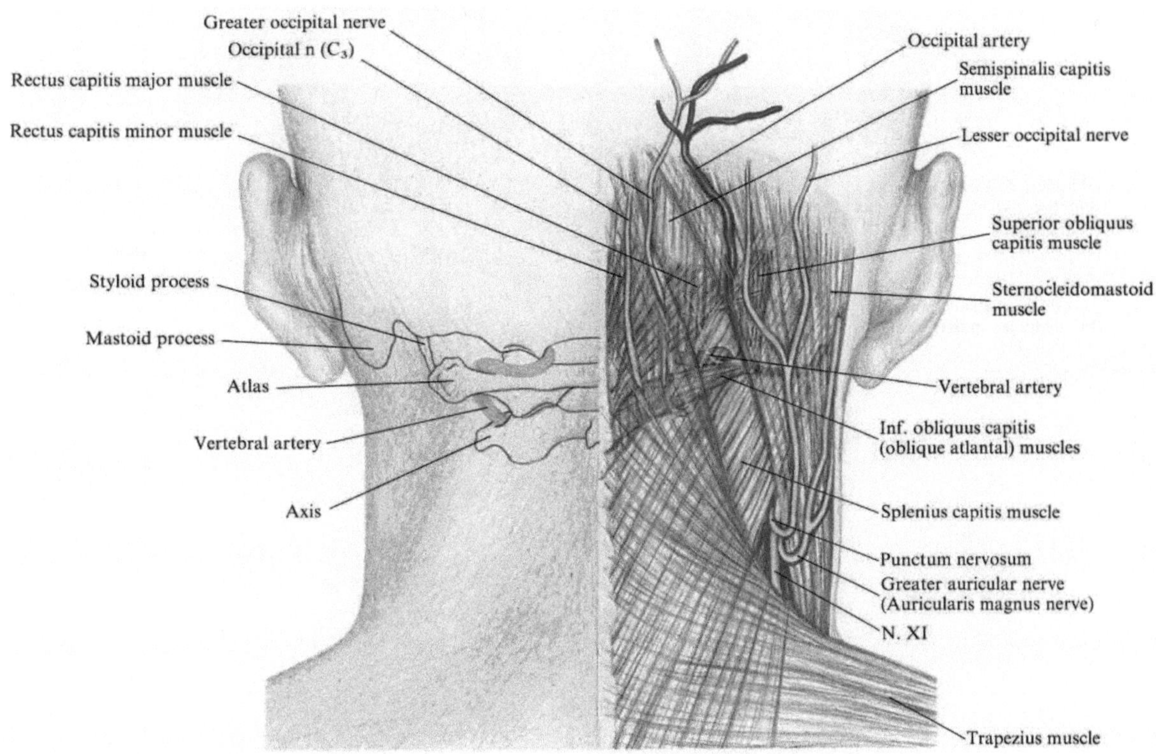

11.12 Anatomic Relationships of the Vertebral Artery, Spinal Column, and Dens Axis at the Level of the Atlas (Superior View)

The vertebral artery lies in a groove located on the arch of the atlas, and is enclosed by a fibrous ligament (the posterior atlanto-occipital ligament) or a bony bridge (the arcuate foramen of the atlas) in 6.2–10% of cases. This cover can occur unilaterally or bilaterally. In the presence of an arcuate foramen, retroflexion, rotation, and inclination of the head can compress the artery between the atlas and the posterior margin of the foramen magnum. This leads to an impaired blood flow and possible irritation of the periarterial sympathetic plexus, a situation that can contribute to the development of vertebrobasilar insufficiency (as in the Lieou-Barré syndrome) [1–4].

Surgical Importance

Surgery in the region of the foramen magnum and atlas, surgery of the deep neck, diagnosis of vertebrobasilar insufficiency.

References

1. Argenson C et al. (1980)
2. Bernasconi M and Marasciuolo S (1972)
3. Lang J (1979)
4. Saraniti G et al. (1976)

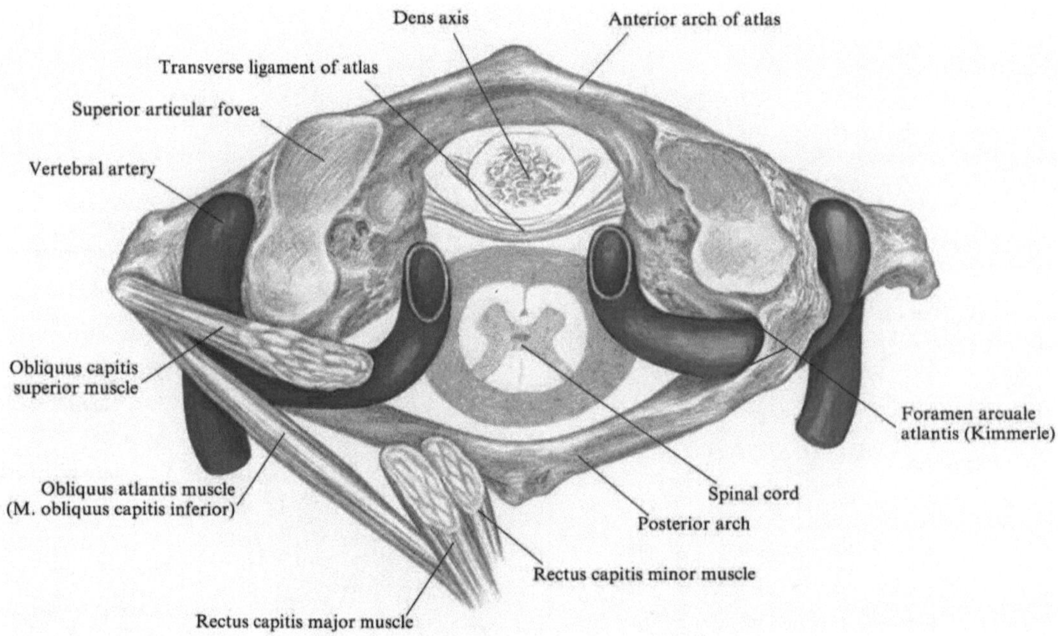

11.13 The Vertebro-Occipital Course of the Vertebral Artery (Posterior View)

In the occipital-vertebral region of the neck, the vertebral artery runs from the lateral part of the arch of the atlas, crosses medially from behind the first cervical root, and pierces the dura between the arch of the atlas and the occipital bone. The cervical root of the accessory nerve lies dorsal to the intradural course of the vertebral artery. The posterior inferior cerebellar artery branches from the vertebral artery immediately after the latter vessel enters the intradural space of the posterior cranial fossa [1].

Surgical Importance

Vertebrobasilar insufficiency, ligature of the vertebral artery in the region of the foramen magnum.

Reference

1. Lang J (1979)

Plate modified from Lang [1]

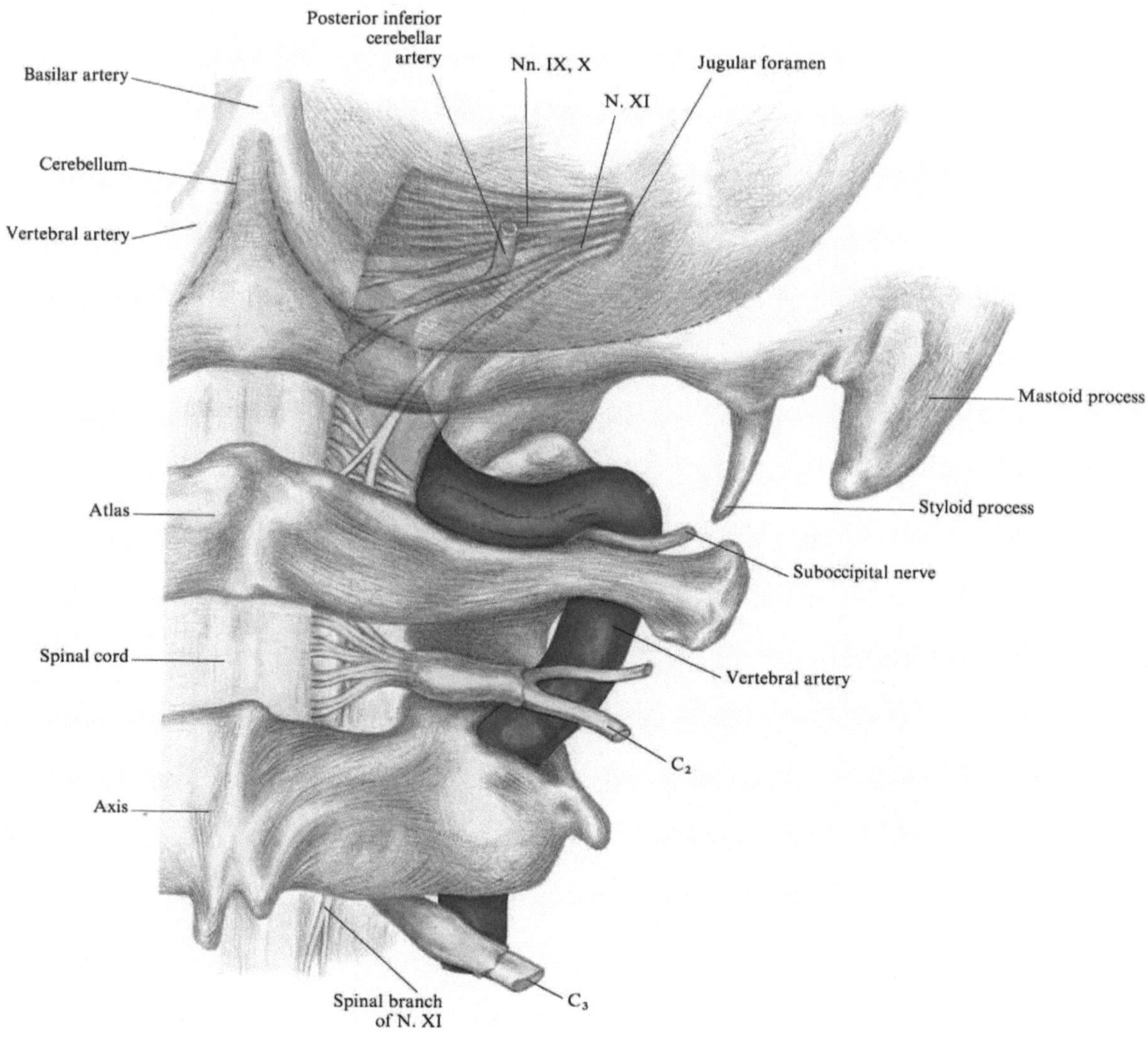

11.14 (Plates 1–2) Exposure of the Vertebral Artery in the Arch of the Atlas and its Course Through the Transverse Foramina

In cases of arteriovenous fistulae or bleeding after injuries, the vertebral artery can be surgically controlled by one of two methods. In *Drüner's technique* [1], the artery is exposed by dissecting away the muscles covering the arch of the atlas (Plate 1). In order to expose the artery in the region of the transverse foramina of the cervical vertebra, the carotid sheath is displaced posteriorly. In *Herlyn's method* [2], the transverse processes are approached anterolaterally, after which the vertebral artery is exposed, taking care not to injure the nerve roots (Plate 2).

Surgical Importance

Ligature of the vertebral artery.

References

1. Drüner (cit. Guleke N 1953)
2. Herlyn H (1943) and cit. Guleke N (1953)

Plates 1 and 2 from Herlyn [2]

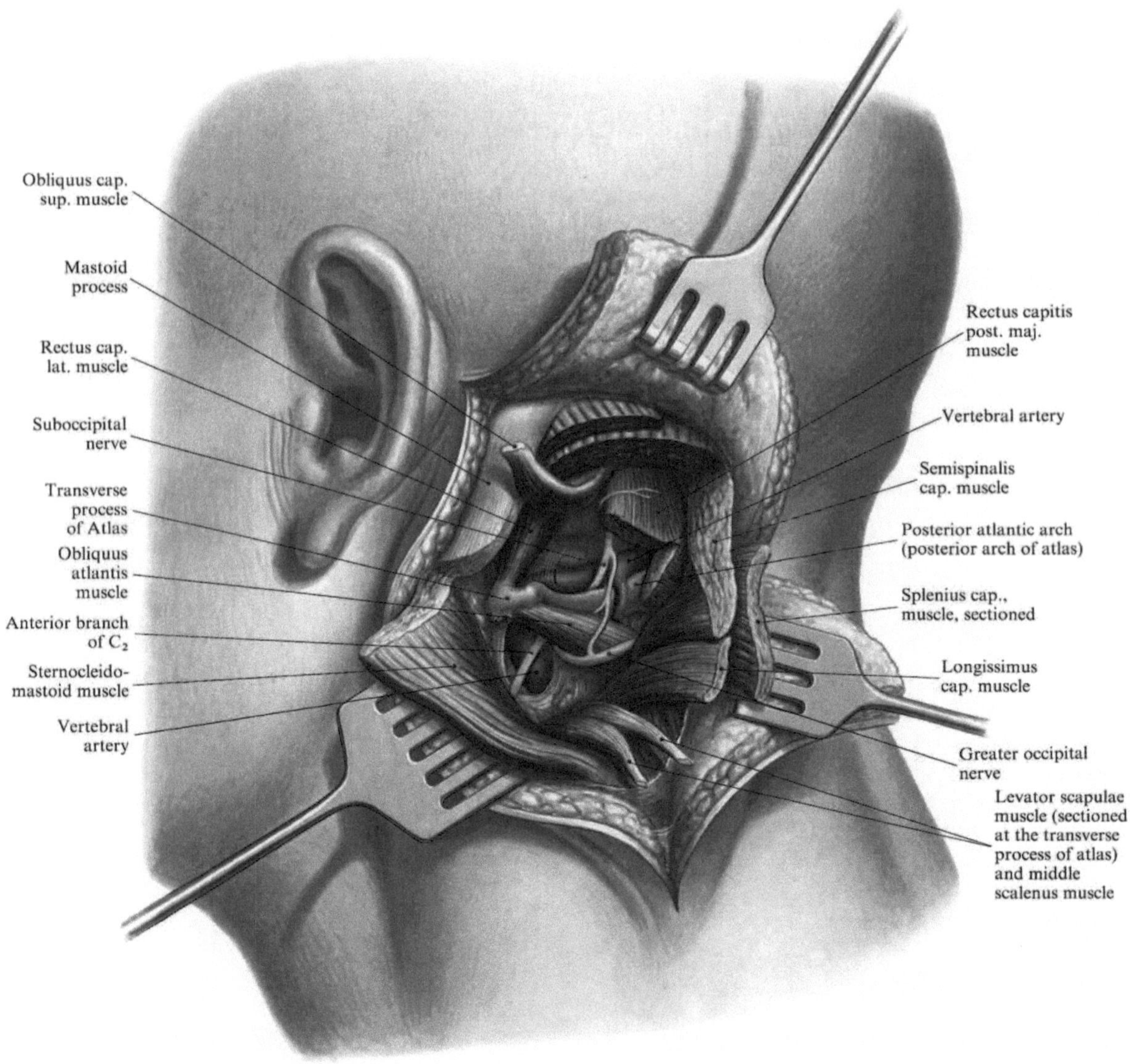

1

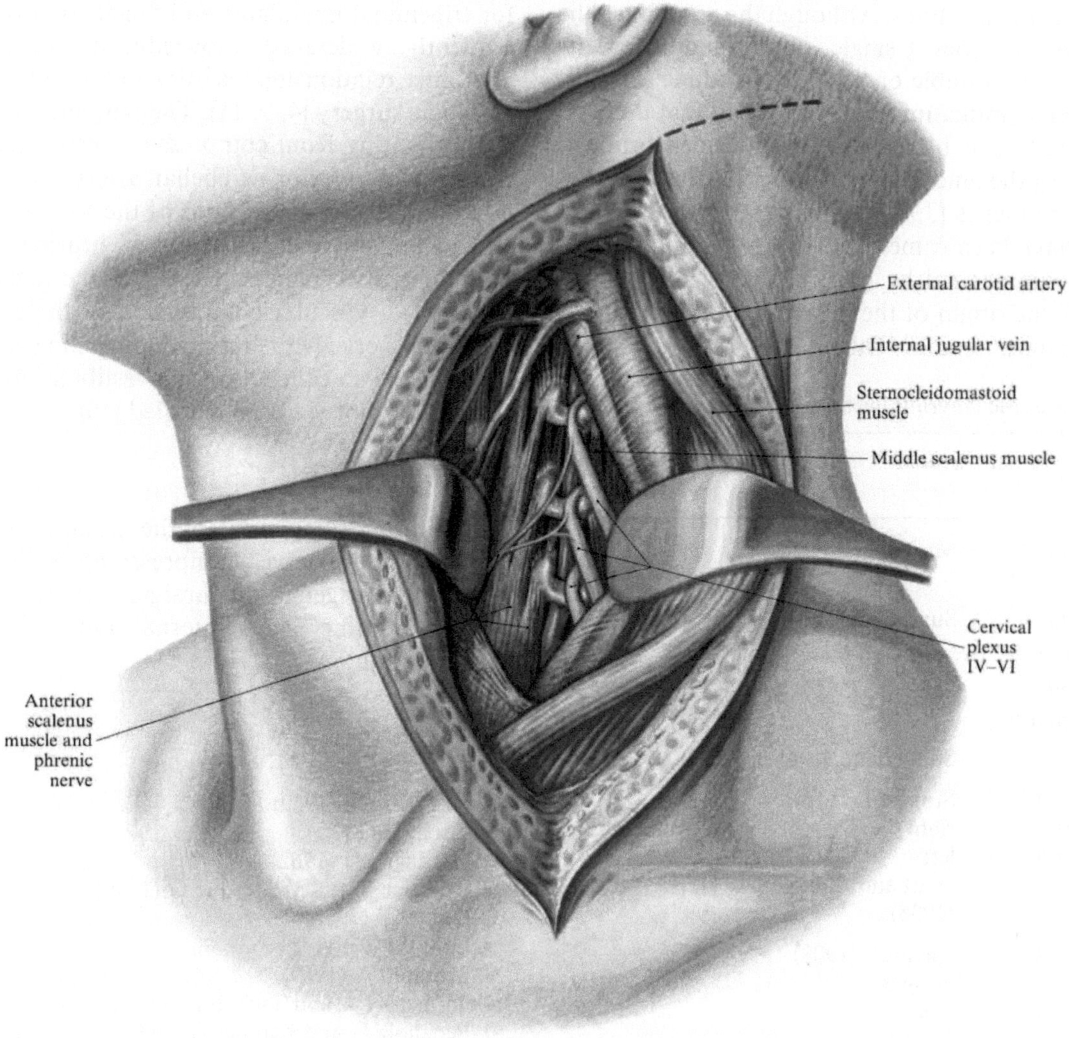
External carotid artery
Internal jugular vein
Sternocleidomastoid muscle
Middle scalenus muscle
Cervical plexus IV–VI
Anterior scalenus muscle and phrenic nerve

2

The origin of the labyrinthine artery can have a number of variations. Although the anatomical literature describes a single labyrinthine artery (as well as rare double or triple labyrinthine arteries), clinical investigations have suggested that there occasionally can be as many as fourteen branches reaching the inner ear from either the basilar artery or its branches [3, 8, 9]. The data from the literature have been somewhat controversial, and have been summarized by Lang [7]. This latter review shows the origin of the labyrinthine artery in various clinical studies (Table).

Origins of the labyrinthine artery from brainstem vessels

Source Vessel	*Author*	*References*	*Percent cases*
Ant. inf. cerebellar a.	Mazzoni (1969)	[8]	80%
Ant. inf. cerebellar a.	Sunderland (1945)	[13]	83%
Ant. inf. cerebellar a.	Stopford (1916)	[12]	63%
Basilar a.	Sunderland (1945)	[13]	17%
Basilar a.	Stopford (1916)	[12]	
Basilar a.	right		36%
Basilar a.	left		28%
Basilar a.	Watt and McKillop (1935)	[15]	5%
Basilar a.	Cavatori (1908)		
Basilar a.	right		70%
Basilar a.	left		67%
Basilar a.	Pernetzky and Tschabitscher (1976)	[9]	5%
Ant. inf. cerebellar and vertebral a.	Mazzoni (1969)	[8]	3%
Ant. inf. cerebellar and vertebral a.	Walker (1965)	[14]	seldom
Ant. inf. cerebellar a.	Ferrari-Lelli (1939)	[2]	15%
Ant. inf. cerebellar a.	Guerrier and Villaceque (1949)	[5]	23%

As described by Ferrari-Lelli [2] and Guerrier and Villaceque [5], the labyrinthine artery may originate at several sites from the anterior inferior cerebellar artery. According to Mazzoni [8], the vessel can also occur doubly in 45% of cases or triply in 4% of cases.

Compression of the nerves in the region of the brain stem is considered to be a significant cause for trigeminal neuralgias and facial spasms. Consequently, a detailed knowledge of the various anatomic relationships is very important in neurotologic surgery [4, 9, 11]. The labyrinthine arteries originating from compressing vessels (such as the anterior inferior cerebellar artery) can be injured during decompressions of the Vth and VIIth cranial nerves, and can result in hearing and/or vestibular disturbances. However, the actual importance of vascular compression of the cochleovestibular nerve with corresponding clinical symptoms (as the so-called "cochleovestibular dysfunctions") has not yet been clarified [16].

Surgical Importance

Angiographic diagnosis in the posterior cranial fossa (cranial nerve compressions with facial spasms and trigeminal neuralgias), surgery of the cerebellopontine angle, internal auditory meatus and labyrinth.

References

1. Cavatori (1908)
2. Ferrari-Lelli (1939)
3. Fisch (1968)
4. Gardner WJ (1962)
5. Guerrier Y and Villaceque G (1949)
6. Janetta P (1982)
7. Lang J (1979)
8. Mazzoni A (1969)
9. Perneczky A and Tschabitscher M (1976)
10. Penzholz H and Kühner A (1981)
11. Samii M and Janetta PJ (1981)
12. Stopford (1916)
13. Sunderland (1945)
14. Walker (1965)
15. Watt and McKillop (1935)
16. Wigand ME et al. (1982)

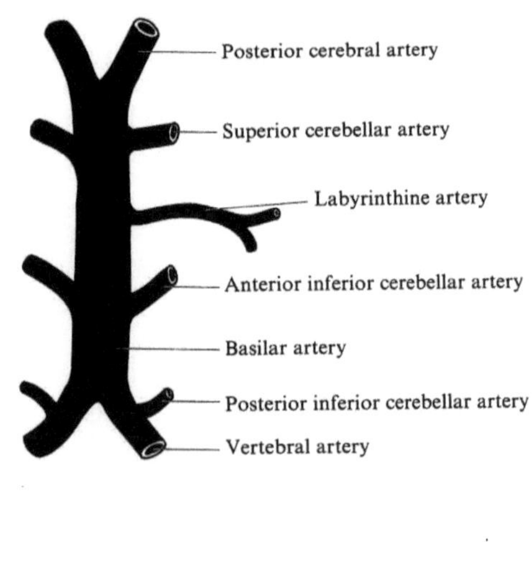

11.16 (Plates 1–9) Loop Formation of the Anterior Inferior Cerebellar Artery at the Internal Acoustic Porus (Posterior View)

Vascular loops of the internal acoustic porus are of clinical importance because they can lead to dysfunctions of the facial nerve and – most probably – of the cochlear and vestibular nerves [1, 2]. Attachment of the cerebellar arteries to the dura covering the petrous pyramid or to the nerves of the internal auditory meatus can cause technical difficulties during surgical procedures involving the petrous pyramid and the cerebellopontine angle. Since the labyrinthine arteries can originate at different sites from the looped cerebellar arteries, displacement of the main vessel can create disturbances in the middle ear and vestibular organ.

Mazzoni [3] has found a loop formation at the internal acoustic porus in 33% of cases, while Pernetzky and Tschabitscher [4] have found loops in 64% (Plates 1–3, 7, 8). Actual displacement of the artery into the porus was found in 27% of cases by Mazzoni and in 22% of cases by Pernetzky and Tschabitscher (Plates 4, 5, 9).

The loop of the cerebellar artery is located between the facial and cochlear nerves in most cases (Plates 1, 2, 5, 7, 8). One, two or more labyrinthine and/or subarcuate arteries can originate from the anterior inferior cerebellar artery at the porus (Plates 1, 3–5, 7, 8). The anterior inferior cerebellar artery and the subarcuate artery may be fixed to the dura above the porus [3] and cause difficulties during surgery of this site. The former vessel lies mostly basal to the abducens nerve (81%–86% of cases) and only in rare cases is dorsal to the nerve [2] (not represented).

Surgical Importance

Surgery of the labyrinth, internal auditory meatus, and cerebellopontine angle.

References

1. Fisch U (1968)
2. Lang J (1979)
3. Mazzoni A (1969)
4. Perneczky A and Tschabitscher M (1976)

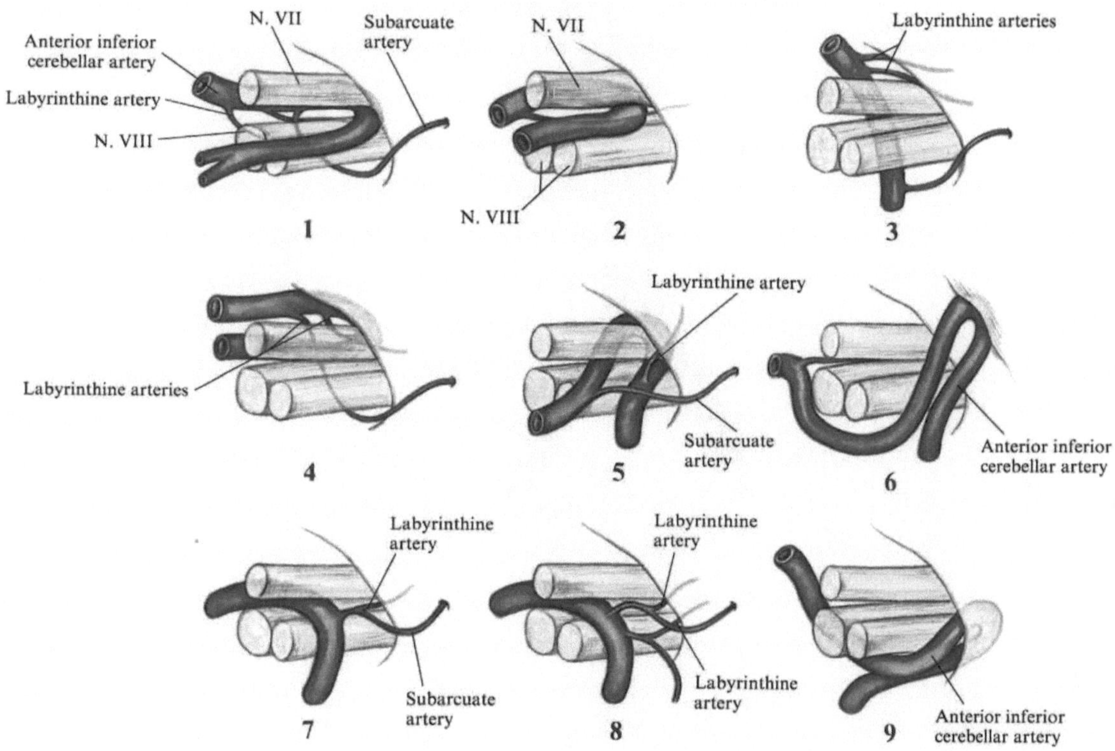

11.17 Data Concerning the Dimensions of the Arteries of the Base of the Brain

The Plate is a schematic drawing of the basilar artery and the arteries of the circle of Willis. The diameters of the major vessels are shown.

Reference

1. Lang J in: Lanz T v and Wachsmuth W (1979)

Plate modified from Lang [1]

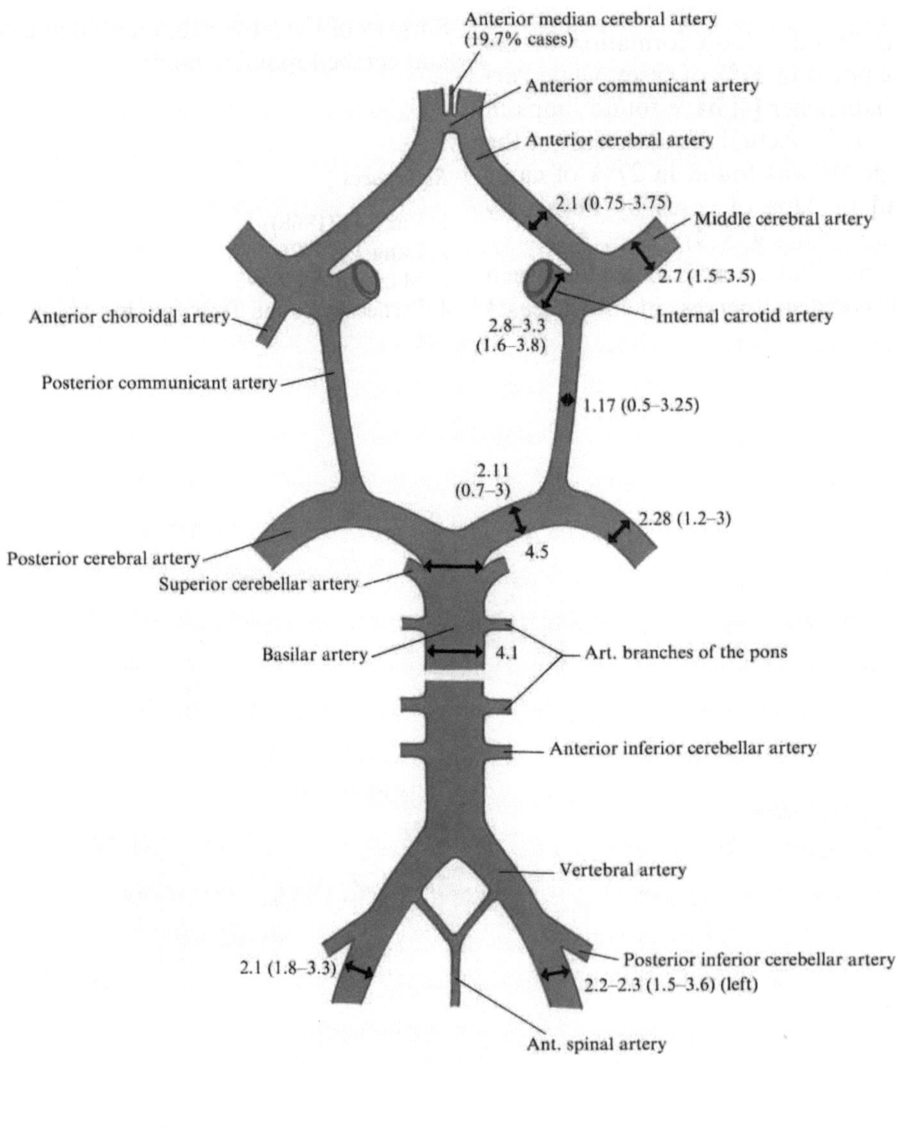

11.18 (Plates 1–6) Some Clinically Important Anomalies of the Dural Sinuses (Posterior View)

In most cases, the superior sagittal sinus divides into a larger right and a smaller left transverse sinus (Plate 1). The bifurcation of the sinuses lies at the level of the internal occipital protuberance. This site is called the *confluens sinuum* because the sagittal sinus joins here with both transverse sinuses and the occipital sinus. The transverse sinus then continues into the sigmoid sinus. If the occipital sinus is well-developed (Plates 3, 4, 6), it gives off the right and left marginal sinus at the occipital foramen (Plates 5, 6). These latter sinuses flow into the corresponding jugular bulbs together with the sigmoid sinus. On occasion, the right or left transverse sinus can be missing (Plates 2, 4). In such case the remaining occipital sinus shows compensatory enlargement (Plate 5) [1, 2].

Surgical Importance

Surgery for sinus thrombosis, neck dissection.

References

1. Helms J (1978)
2. Krmpotić J and Šercer A (1960)

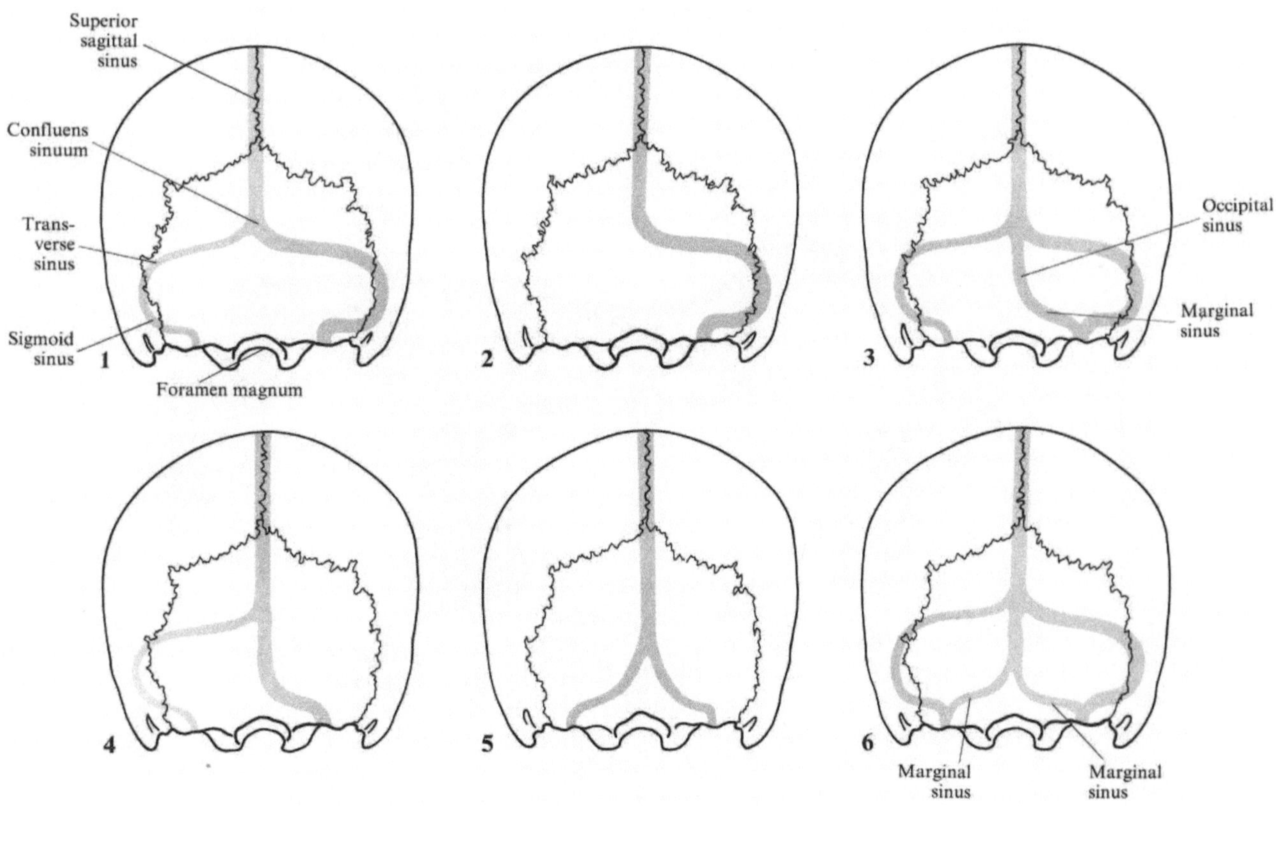

11.19 (Plates 1–2) The Transbasal Venous Channels (Superior and Lateral Views)

The Plates illustrate the venous channels connecting the transverse sinus, the jugular bulb and jugular vein, and the cavernous sinus. The large sizes of these channels show why any bleeding after division of the sigmoid sinus or of the jugular bulb will not be stopped completely by isolated ligature of the transverse sinus or the internal jugular vein [2]. Chemodectomas or glomus tumors tend to grow along these vascular channels to become non-resectable.

Surgical Importance

Surgery of the lateral skull base and the petrous pyramid, ligature of the sigmoid sinus, *Grunert's* jugular bulb operation, translabyrinthine and retrosigmoid approaches to the cerebellopontine angle and internal acoustic meatus.

References

1. Denecke HJ (1966)
2. Teufel J (1964)

Plates 1 and 2 from Denecke [1]

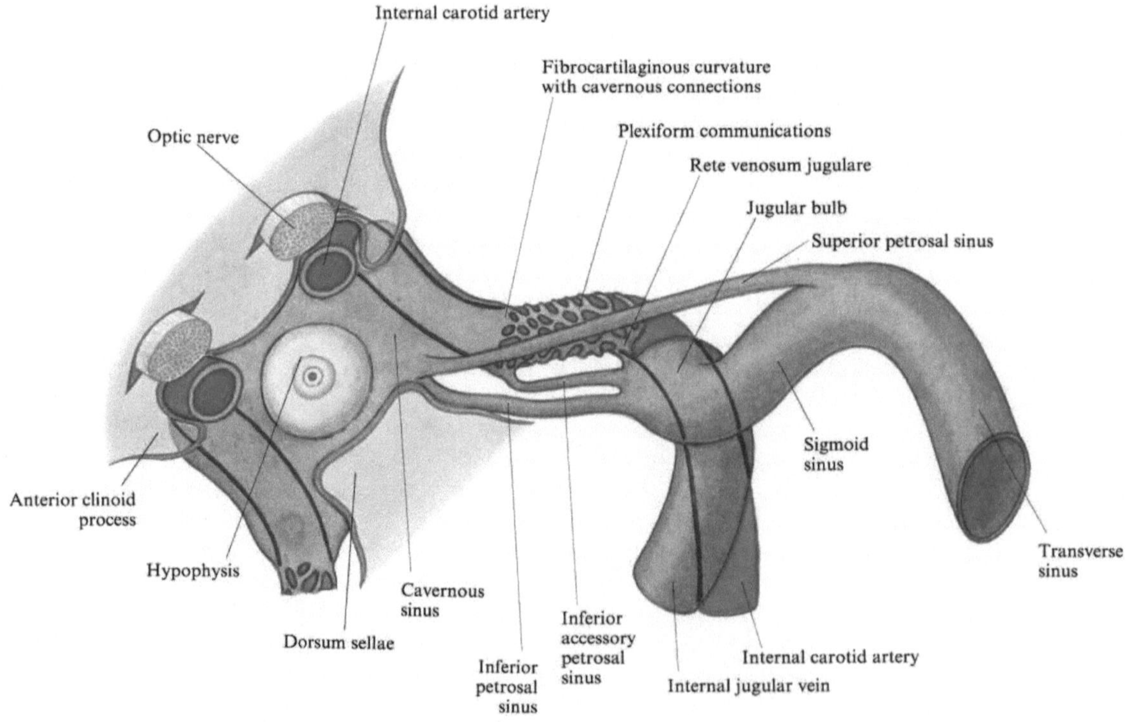

1

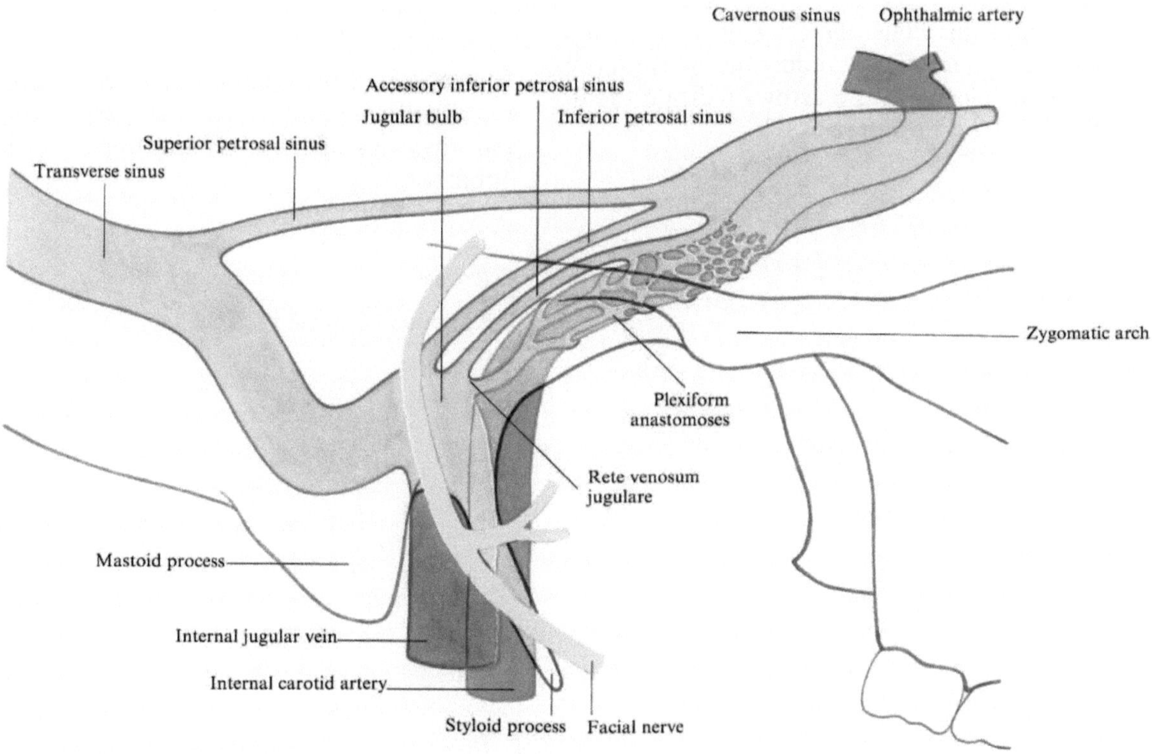
Cavernous sinus
Ophthalmic artery
Accessory inferior petrosal sinus
Jugular bulb
Inferior petrosal sinus
Superior petrosal sinus
Transverse sinus
Zygomatic arch
Plexiform anastomoses
Rete venosum jugulare
Mastoid process
Internal jugular vein
Internal carotid artery
Styloid process
Facial nerve

2

11.20 Vascular Communications Between the Intracranial and Extracranial Venous Channels

As shown in this Plate, the veins of the head and brain are connected by numerous anastomoses and venous plexuses. The arrows indicate the direction of blood flow present [1].

Surgical Importance

Thrombosis of the sigmoid, transverse, cavernous, and superior sagittal sinuses; surgery of furuncles and abscesses of the nose and upper lip, jugular thrombectomy.

Reference

1. Krmpotić-Nemanić J (1978)

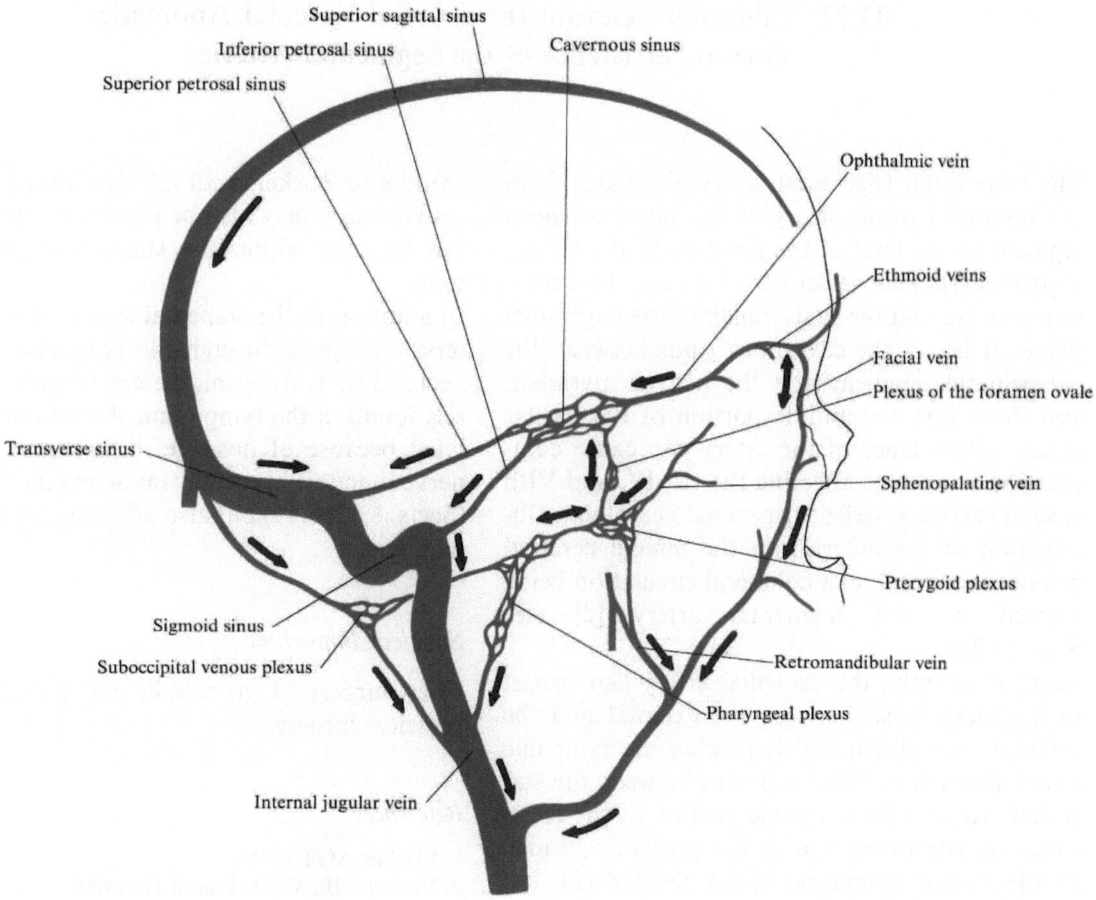
Superior sagittal sinus
Inferior petrosal sinus
Cavernous sinus
Superior petrosal sinus
Ophthalmic vein
Ethmoid veins
Facial vein
Plexus of the foramen ovale
Transverse sinus
Sphenopalatine vein
Pterygoid plexus
Sigmoid sinus
Suboccipital venous plexus
Retromandibular vein
Pharyngeal plexus
Internal jugular vein

11.21 Clinically Relevant Intracranial Vascular Anomalies Due to Persistence of the Segmental Arteries

The primordial trigeminal artery originates from the internal carotid artery in the intracavernous segment at the level of the junction of the C_4–C_5 segments [1]. This vessel runs between the oculomotor nerve and the first branch of the trigeminal nerve. It leaves the cavernous sinus between the petroclinoid ligament and the petrous pyramid, and flows into the middle portion of the basilar artery. (Persistence of the artery can cause compression symptoms affecting the III, IV, and VIth cranial nerves, as well as trigeminal neuralgia). Obstruction of the anterior or the middle cerebral arteries may result in a collateral circulation being formed by this segmental artery [2] (see Sect. 11.26).

More frequently, the stapedial artery can persist as an anastomosis between the external and the internal carotid arteries. It reaches the tympanic cavity through its floor and runs between the stapedial crura to the tympanic portion of the facial canal. It anastomoses with the posterior branch of the middle meningeal artery (Sect. 8.12), but it can also originate from the maxillary artery. According to Zuckerkandl [3], a rudiment of the stapedial artery may also be present in the adult and can be seen within the stapes during stapedectomy.

In addition to the stapedial artery, other large arteries can pass through the tympanic cavity (see Sect. 8.11). During middle ear surgery, large vessels found in the tympanum should not be coagulated because of possible disturbances in cranial nerve function, especially involving the facial nerve (Sects. 8.12, 8.19) but also affecting the III–VI cranial nerves.

Surgical Importance

Microsurgery of the middle ear, vascular decompression surgery.

References

1. Fischer AGT (1915)
2. Molsen HP, Winkelmann H (1980)
3. Zuckerkandl E (1882)

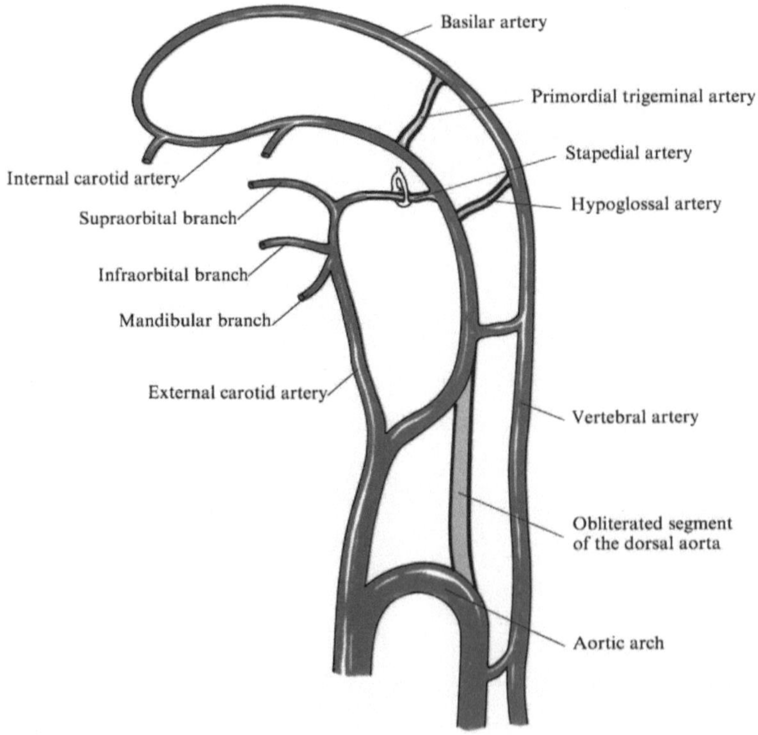

11.22 The Course of the Persistent Hypoglossal Artery

The persistent hypoglossal artery is a rare anomaly that passes with the hypoglossal nerve through its corresponding canal to form the basilar artery with a thin vertebral artery from the contralateral side. In the X-ray (tomogram) image, the normal hypoglossal canal is symmetrical and smooth. The presence of a persistent hypoglossal artery can result in enlargement of the hypoglossal canal. In certain cases, its smooth margins can be mistaken for a benign tumor (such as a neurinoma) [1–9].

Surgical Importance

Surgery of the hypoglossal nerve, basilar artery, and posterior cranial fossa.

References

1. Begg AC (1961)
2. Blain JG and Logothetis J (1966)
3. Keller HL and Weiss D (1973)
4. Khodadad G (1977)
5. Kodama N et al. (1976)
6. Krmpotić-Nemanić J (1978)
7. Oertel O (1922)
8. Shapiro R (1979)
9. Springer TD et al. (1974)

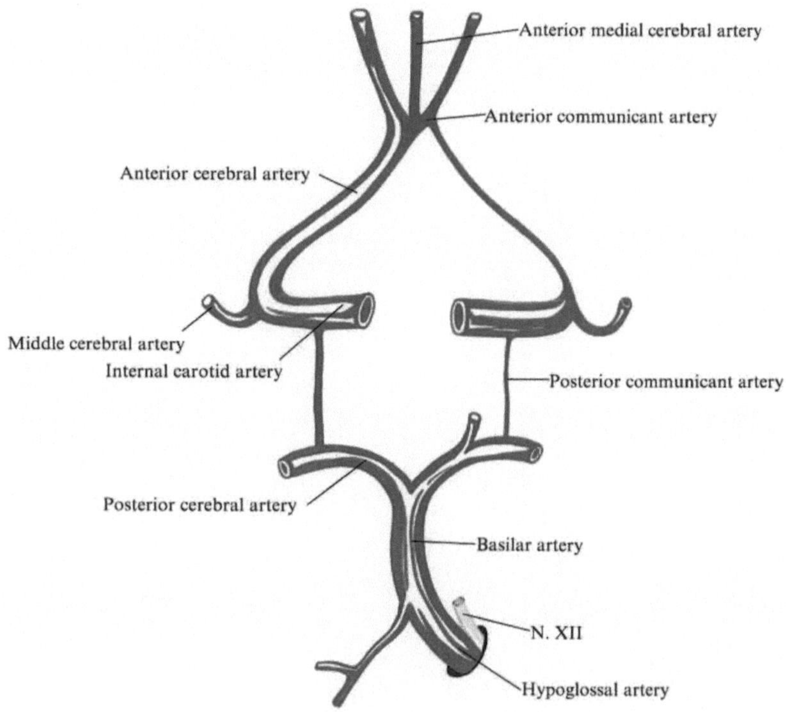

11.23 (Plates 1–5) The Normal Course and Typical Variations of the Middle Cerebral Artery

In the extracranial and intracranial vascular microsurgery made possible by improved radiological studies, knowledge of the possible variations of the cerebral vessels is indispensable. The normal vascular pattern of the middle cerebral artery is shown in Plate 1, while common variations are represented in Plates 2–5.

Surgical Importance

Radiological diagnoses, microvascular surgery involving intracranial and extracranial anastomoses.

Reference

1. Holbach KH et al. (1976)

Plates 1–5 adapted from Holbach et al. [1]

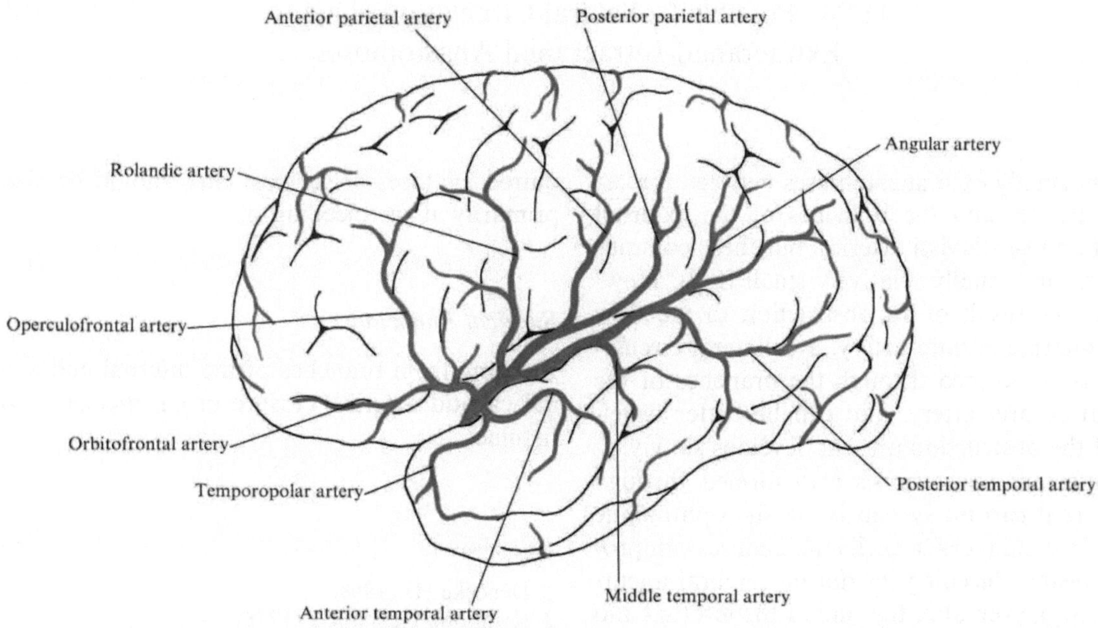

1

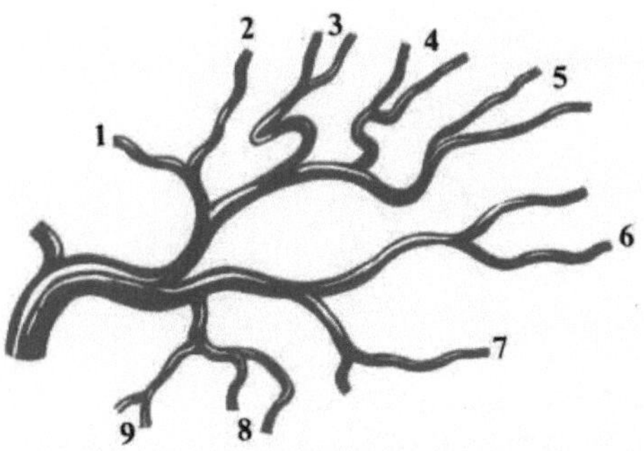

2

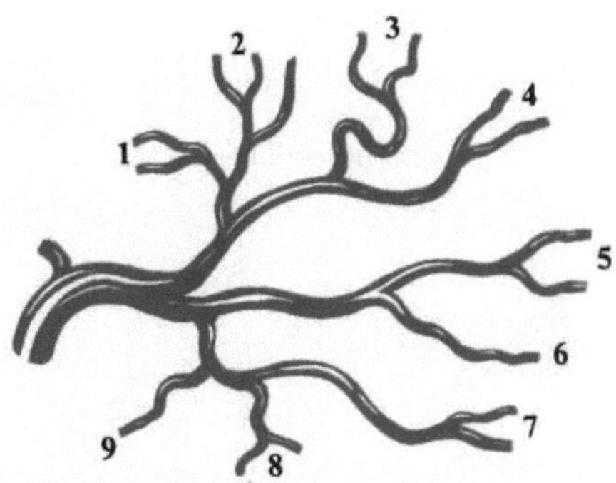

3

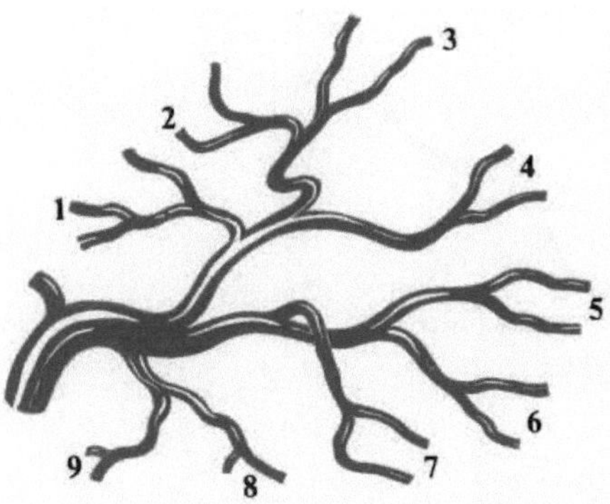

4

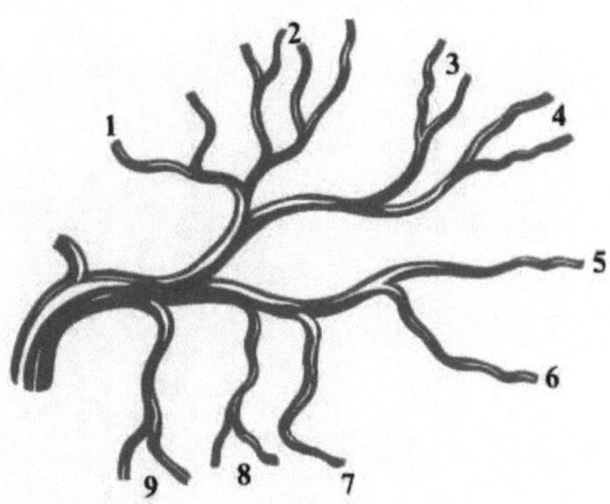

5

1 Orbitofrontal artery
2 Operculofrontal artery
3 Rolandic artery
4 Posterior and anterior parietal arteries
5 Angular artery
6 Posterior temporal artery
7 Middle temporal artery
8 Anterior temporal artery
9 Temporopolar artery

11.24 Possible Collateral Circulations Due to Extracranial-Intracranial Anastomoses

There normally exist anastomoses between the cerebral arteries and the branches of the external carotid and subclavian arteries, but these communications are usually relatively small [1–3]. However, as the result of an obstruction to the flow of the internal carotid artery, a collateral circulation may be formed through the branches of the external carotid artery, and can be better developed if the obstruction present develops slowly. An important anastomosis also formed through the internal carotid system is via the ophthalmic artery. In certain cases, such anastomoses can produce constant bleeding (as during cerebral aneurismectomy), even after ligature of the internal and external carotid arteries [1]. Consequently, any required ligatures or surgical clips should be placed primarily at the bleeding site.

Surgical Importance

Bleeding from branches of the internal and external carotid arteries, closure of caroticocavernous fistulae.

References

1. Denecke HJ (1968)
2. Krmpotić-Nemanić J (1978)
3. Quiring DP (1949)

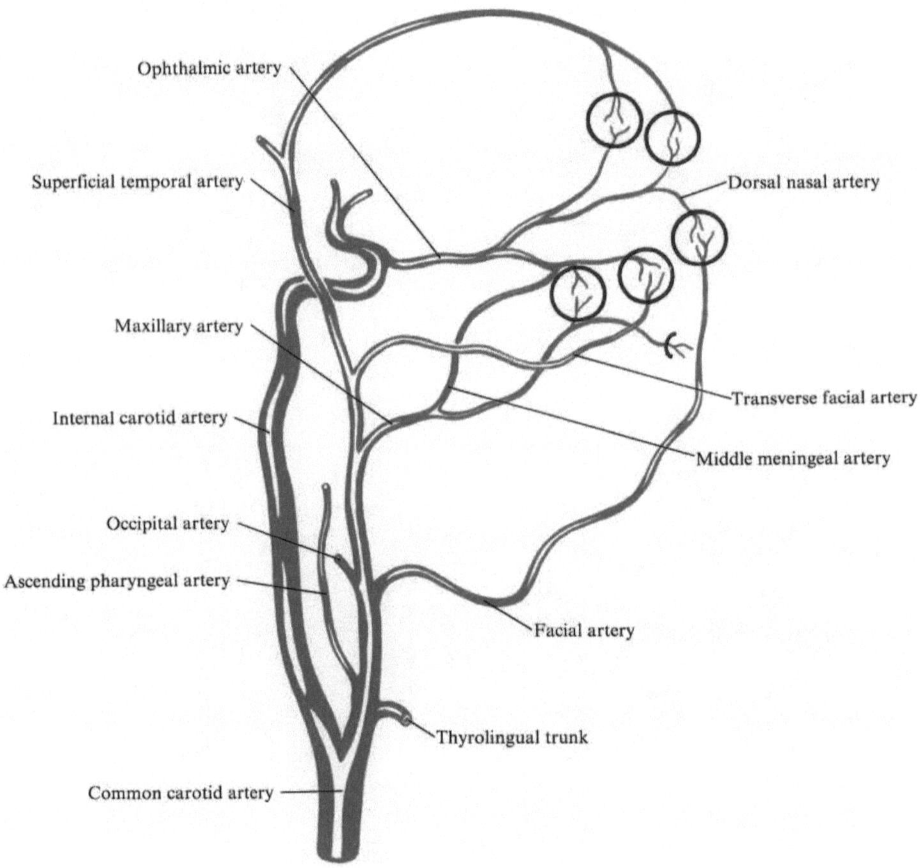

11.25 Possible Vascular Anastomoses Between the Central Nervous System and the External Carotid or Subclavian Arteries

Important communicating arteries in obstructions of the vessels of the central nervous system are the ophthalmic (Sect. 11.24), the anterior spinal, and the anterior tympanic arteries [1, 2]. The last two vessels are of lesser importance. Additionally, a large arterial anastomosis within the temporal bone can cause tinnitus.

Surgical Importance

Surgery for tinnitus, surgery of the orbit, surgery of the cervical spine.

References

1. Krmpotić-Nemanić J (1978)
2. De le Gars et al. (1980)

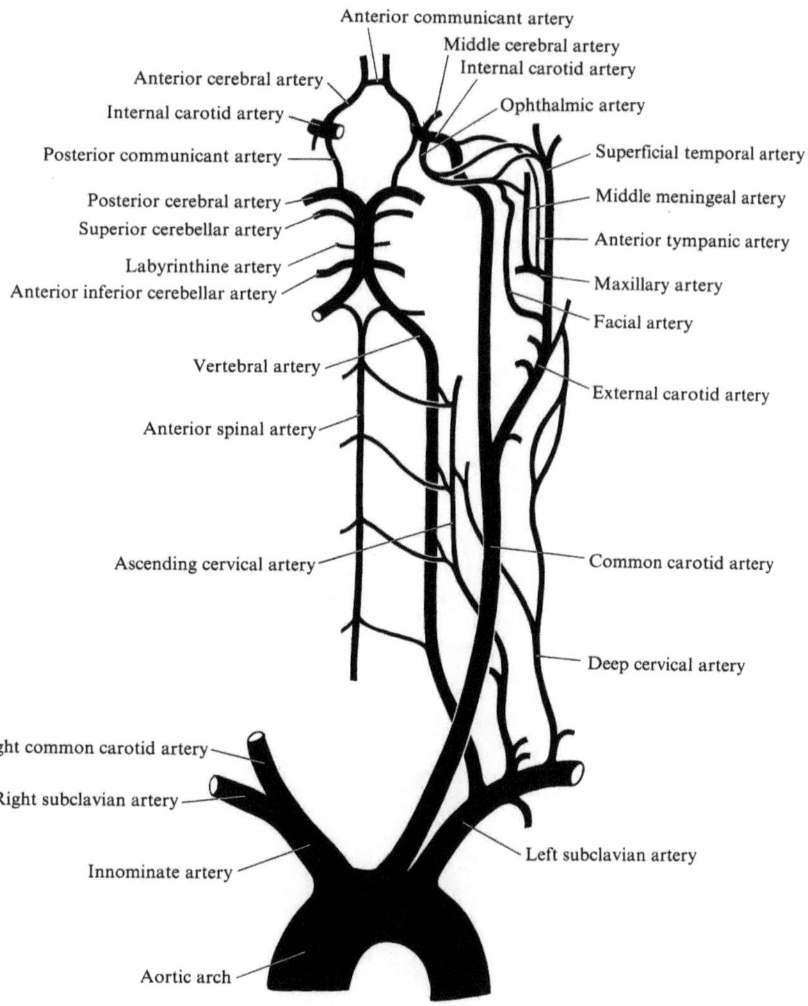

11.26 (Plates 1–8) The Formation of a Collateral Blood Supply in a Slowly Developing Obstruction of the Cerebral Arteries

The normal blood supply of the internal and external carotid arteries is shown in Plate 1. As illustrated in Plates 2–8, an obstruction of the internal carotid artery (*black*) can occur without significant clinical symptoms if the vascular anastomoses present are numerous [1]. In such circumstances (Plate 2), the following anastomoses (*violet*) can be useful: the anterior and posterior communicant arteries, the ophthalmic artery, the trigeminal arteries, and a *rete mirabile* between the dural branches of the middle meningeal artery and the anterior cerebral artery, as well as caroticotympanic and leptomeningeal anastomoses. In obstruction of the anterior cerebral artery (Plate 3), the following vessels can show compensatory dilatations: leptomeningeal anastomoses of the middle and posterior cerebral arteries, the anterior communicant artery, and the anterior accessory communicant artery. An obstruction of the middle cerebral artery can be only partially compensated by leptomeningeal anastomoses between the anterior and posterior cerebral arteries or a rete mirabile between the dural branches of the middle meningeal artery and pial branches of the obstructed middle cerebral artery (Plate 4). Obstruction of the posterior cerebral artery can be compensated by the posterior communicant artery, leptomeningeal anastomoses between the anterior and middle cerebral artery, choroidal anastomoses, a rete mirabile between the occipital artery and pial branches of the obstructed posterior cerebral artery (Plate 5). Obstruction of the basilar artery can only exceptionally be compensated for, even when slowly developing (Plate 6). Possible collaterals are the posterior communicant artery, a persistent trigeminal artery, and leptomeningeal anastomoses between the superior and posterior cerebellar arteries. Obstruction of the vertebral artery can be compensated by the contralateral vertebral artery, the posterior communicant artery, the trigeminal artery, the hypoglossal artery, the occipital artery, and the anterior spinal artery (Plate 7). A block of the vertebral artery close to its origin from the subclavian artery can be compensated by anastomoses between the anterior spinal arteries and the costocervical trunk (represented in an anterior view in Plate 8).

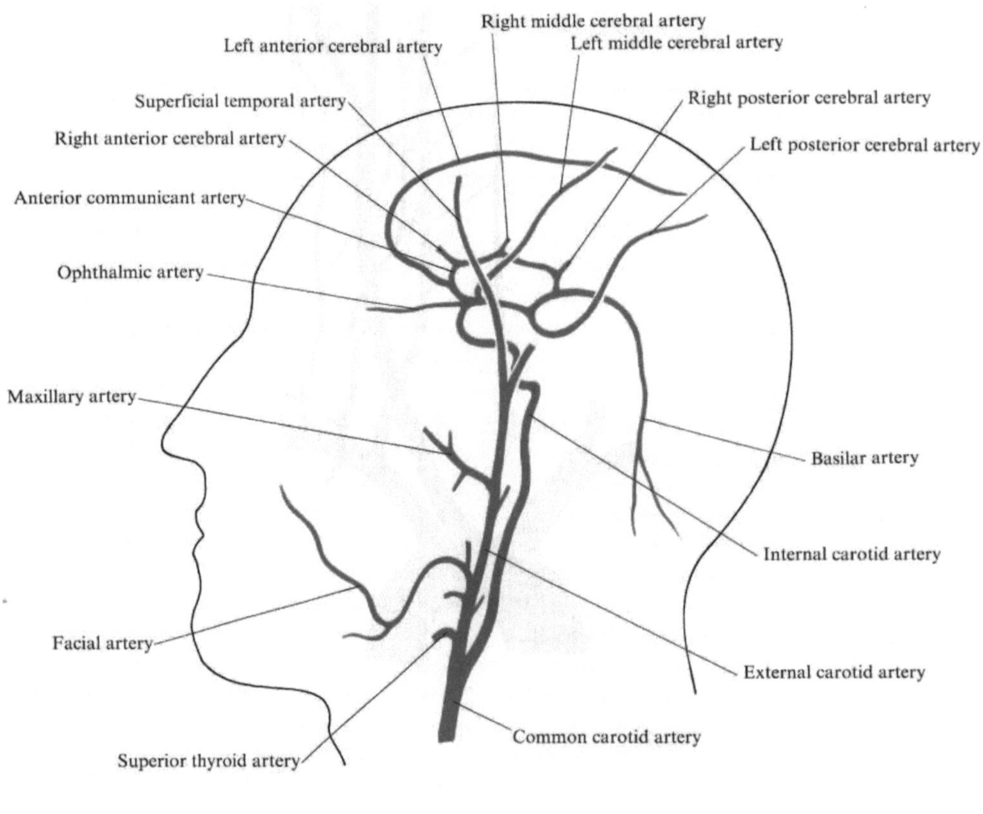

1

Surgical Importance

Diagnosis of cerebral vascular disturbances, superselective angiography, Doppler sonography, surgery to revascularize different cerebral areas, surgery of intra-extracranial vascular anastomoses.

Reference

1. Buccigrossi G et al. (1975)

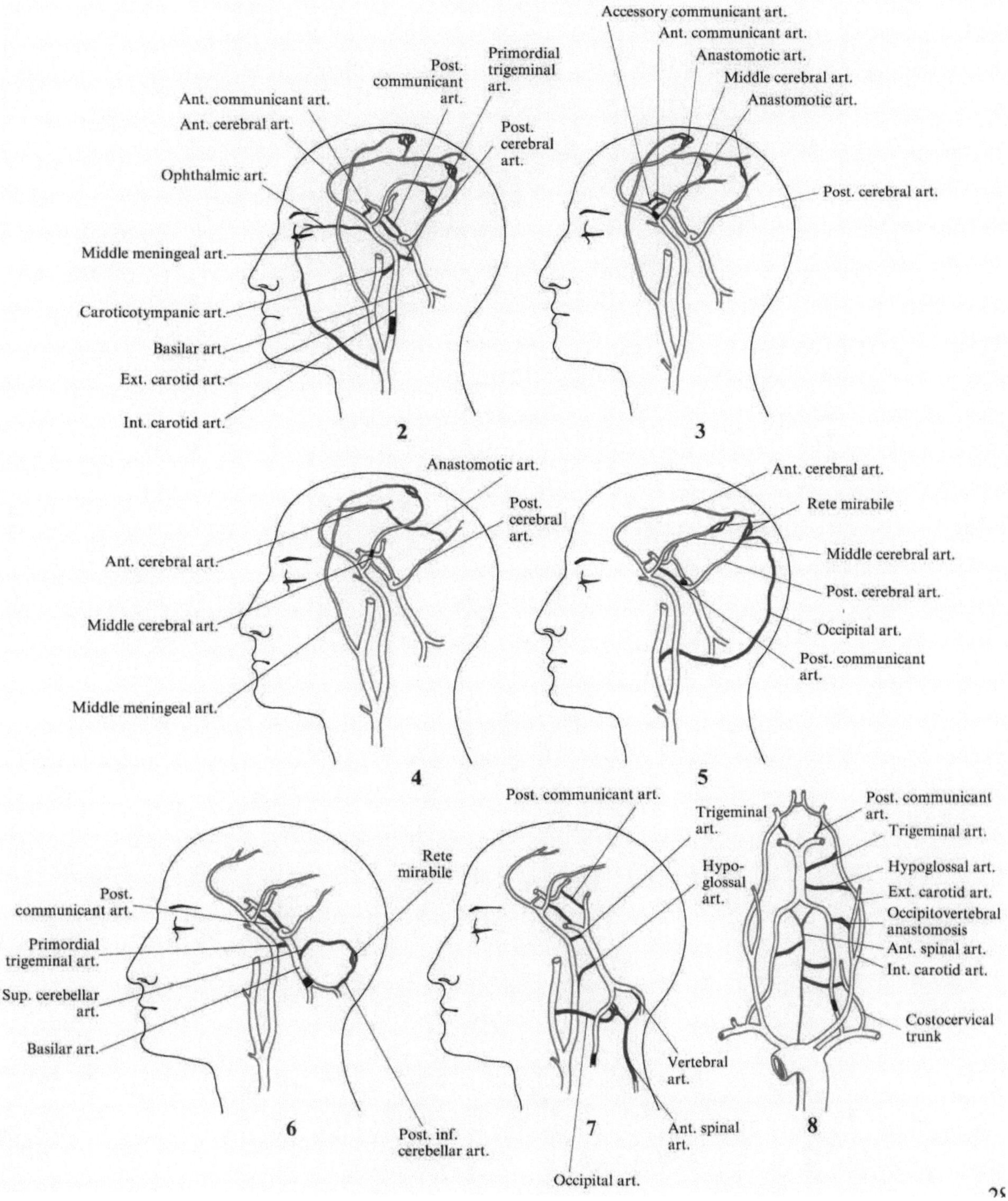

12 The Paraganglia

12.1 Locations of Non-Chromaffin Paraganglia in the Skull Base and in the Neck

The non-chromaffin paraganglia are located in the base of the skull (as examplified by the glomus tympanicum or jugular paraganglion), in the carotid bifurcation (the paraganglion caroticum or glomus caroticum), the larynx, the subclavian artery, the aorta (aortic paraganglion) and the pulmonary artery. These paraganglia are considered to be of branchiomeric origin.

The orbital and intravagal paraganglia constitute a special group of the paraganglia. These formations develop from cells of the neural crest. Their clinical importance derives from the possibility that tumors may develop in them (forming the paragangliomas or chemodectomas). Glomus tumors are characterized by infiltrative and destructive growths and a rich vascularization that can pose a significant problem when surgical removal is attempted [1, 2]. As a result of their growth characteristics, it is sometimes difficult to locate the site of origin of large tumors situated in the neck or in the base of the skull. Those tumors which occur most frequently are chemodectomas of the glomus caroticum type. True malignancy with metastases occurs in 10% of cases.

Surgical Importance

Surgery of chemodectomas.

References

1. Denecke HJ (1966)
2. Draf W and Samii M (1983)
3. Graf K and Fisch U (1979)

Plate from Graf and Fisch [3]

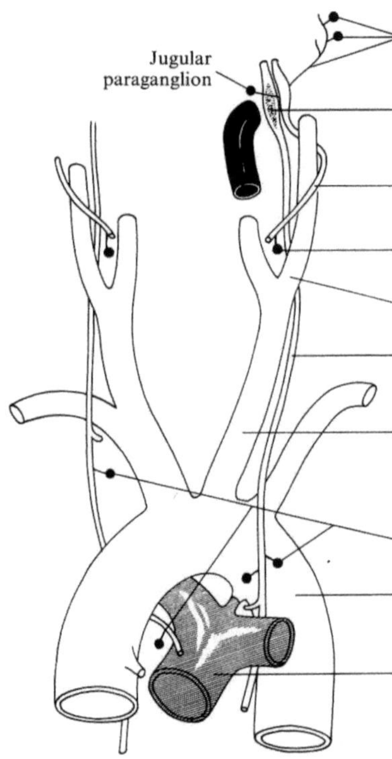

12.2 Possible Locations for Paraganglia in the Temporal Bone and in Lateral and Posterior Sections of the Skull Base

The *black dots* mark the possible locations of non-chromaffin paraganglia, from which invasive and destructive chemodectomas and paragangliomas can originate. A pulsating tinnitus is sometimes the first clinical symptom of such tumors. Paresis of the recurrent laryngeal nerve or other nerves can develop if the jugular foramen is invaded. These tumors can destroy bone and invade the middle and posterior cranial fossae, as well as the venous (transbasal) channels and the internal carotid artery [2, 4]. These tumors should be radically excised in order to avoid recurrences.

Paragangliomas are rare in the nasal cavity, the nasopharynx, and the larynx. In this latter site, laryngeal paragangliomas can be situated above the anterior rima glottidis, along the internal branch of the superior laryngeal nerve, between the thyroid and cricoid cartilages, along the recurrent laryngeal nerve, and in the transverse part (*pars transversa*) of the interarytenoideus muscle. The aryepiglottic folds are the preferred location for these tumors in the region of the larynx or the hypopharynx [1].

Surgical Importance

Chemodectomas of the temporal bone and the middle and lateral posterior skull base.

References

1. Batsakis J (1979)
2. Denecke HJ (1966)
3. Graf K and Fisch U (1979)
4. Teufel J (1964)

Plate from Graf and Fisch [3]

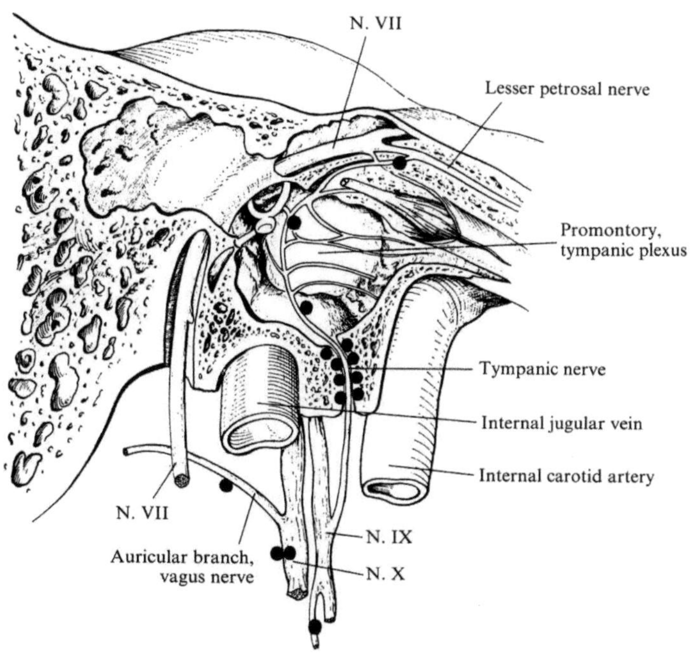

References

Aaron C, Chawaff AR (1967) Variations de la carotide externe et de ses branches. Etude anatomique et anatomoradiologique. CR Ass Anat 137:125–134

Abbé R (1898) A new plastic operation for the relief of deformity due to double harelip. Med Record 53:477

Abrahams P, Webb P (1978) Klinische Anatomie diagnostischer und therapeutischer Eingriffe. Springer, Berlin Heidelberg New York

Actis G, Marello M (1980) Anestesia loco-regionale in oftalmo-chirurgia plastica e ricostructiva, Minerva Anestesiol 46:1095–1098

Adachi B (1928) Das Arteriensystem der Japaner. Kyoto, Maruzen

Adachi B (1933) Das Venensystem der Japaner. Kyoto, Maruzen

Akazaku FK, Wyke B (1979) Innervation of the subglottic mucosa of the larynx and its significance. Folia Phoniatr (Basel) 31:271–283

Anderl H (1973) Reconstruction of the face through cross-face-nerve transplantation in facial paralysis. Chir Plastica (Berlin) 2:17–46

André M, Guerrier Y (1981) The anterior comissure of the larynx. Clin Otolaryngol 6:259–264

André P, Guerrier Y, Pinel J (1964) Oto-rhino-laryngologie et gros vaisseaux du cou. Paris, Arnette

Andrea H (1981) Vasculature of the anterior comissure. Ann Otol 90:18–20

Andrieu-Guitancourt J, Narcy P, Borde J, Melin Y, Manch Y, Dehesdin D, Desse P (1979) Duplications digestives de localisation pharyngée ou fistules de la quatrième poche endo-branchiale. A propos de 4 observations. Ann Otolaryngol Chir Cervicofac 96:863–879

Anson BJ, Donaldson JA (1973a) Surgical anatomy of the temporal bone and ear, II ed. Saunders, Philadelphia London Toronto

Anson BJ, Donaldson JA, Warpeha RL, Rensink MJ, Schilling BB (1973b) Surgical anatomy of the facial nerve. Arch Otolaryngol 97:201–213

De Antoni E, Murafori L, Buonocore F (1971) La preparazione del nervo laringeo inferiore nella chirurgia della tiroide; osservazioni anatomiche e casistica operatoria. Ann Ital Chir 46:479

Argenson C, Francke J, Sylla S (1979) The vertebral arteries (segments V_1 and V_2). Anat Clin 2:29–41

Ariyan S (1979) One-stage reconstruction for defects of the mouth using a sternomastoid myocutaneous flap. Plast Reconstr Surg 63:618

Ariyan S (1980) Pectoralis major, sternomastoid and other musculo-cutaneous flaps for head and neck reconstruction. Clin Plast Surg 7/1:89–109

Asamrah DK, Foy PM (1979) Some radiological abnormalities of the cervical carotid artery and their clinical significance. Clin Radiol 30:593–599

Askenazy M, Wisjenbeek EE, Herzberger E (1952) Absence de l'artère communicante postérieure. Du role favorisante possible de cette anomalie dans la pathogenie de l'obstruction de la carotide. Rev Neurol 86:274–275

Baldini N (1970) Su un caso della paralisi del faciale da otite media traumatica in soggeto portatore di deiscenza del canale di falloppio. Ann Laringol 69:197–203

Baldini M, Mosca L, Princi L (1980) The empty sella syndrome secondary to Rathke's cleft cyst. Acta Neurochir. (Wien) 53:69–78

Bardenheuer B (1931) Zit aus J Josef (1931) Nasenplastik und sonstige Gesichtsplastik, S. 638. Kabitzsch, Leipzig

Bartels P (1909) Lymphgefäßsystem. Fischer, Jena

Bassiounov A, Newlands WJ, Ali H (1982) Maxillary sinus hypoplasia and superior orbital fissure asymmetry. Laryngoscope 92:441–448

Batsakis J (1979) Tumors of the head and neck. Clinical and pathological considerations, II ed. Williams and Wilkins, Baltimore

Baumel JJ (1974) Trigeminofacial nerve communications, their function in facial muscle innervation and reinnervation. Arch Otolaryngol 99:34–44

Beauvillain C, Simon C, Wesoluch M (1982) Contribution à l'étude de l'anatomie du nerve facial du nouveau-né. 2e et 3e portions: application chirurgicale. Ann Otolaryngol Chir Cerficofac 99:223–230

Beck K (1933) Über die Behandlung von Stirnhöhlenentzündungen durch Drainage von außen. Laryngol Rhinol Otol (Leipz) 24:369–373

Beck K, Schneider P (1926) Missbildungen und Anomalien des Kehlkopfes und der Luftröhre und der großen Bronchien. In: Denker A, Kahler O (Hrsg) Handbuch der Hals-Nasen-Ohrenheilkunde, Bd II, Teil 2, S 423

Becker H, Newton TH (1978) Arteria carotis interna-Zuflüsse bei juvenilen Angiofibromen. Fortschr Roentgenstr 128/6:674–679

Bedford MA (1966) The cavernous sinus. Br J Ophthalmol 50:41

Begg AC (1961) Radiographic demonstration of the hypoglossal artery. A rare type of persistent anomalous carotid-basilar anastomosis. Clin Rad 12:187–189

Berendes J, Link R, Zöllner F (1977) Hals-Nasen-Ohrenheilkunde in Praxis und Klinik. Thieme, Stuttgart

Bernardi L, Dettori P (1975) Angiographic study of rare anomalous origin of the vertebral artery. Neuroradiol 9:43–47

Bernasconi M, Marasciuolo S (1972) L'ossificazione del legamento atlantooccipitale e la sindrome cervicale. Ann Laringol 71:43

Björk V (1969) Tracheal- und Bronchotrachealresektion. Langenbecks Arch klin Chir 322:886–899
Blain JG, Logothetis J (1966) The persistent hypoglossal artery. J Neurol Neurosurg Psychiatry 29:346–349
Boenninghaus G (1923) Die Operationen an den Nebenhöhlen der Nase. In: Katz L, Blumenfeld F (Hrsg) Handbuch der speziellen Chirurgie des Ohres und der oberen Luftwege, Bd III, 3. Aufl. Kabitzsch, Leipzig, S 91–286
Boisvert P, Tournier J, Larouche A (1980) Polypose nasale récidivante et neurectomie vidienne: rapport prélim. J Otolaryngol 9:235–240
Bollobas B (1980) Passive Pneumatisation. Nuovo Arch Ital Otol 8:11–24
Bose B (1982) Cervical auricle. Arch Surg 117:968–969
Bossy J (1980) Development of olfactory and related structures in staged human embryos. Anat Embryol (Berl) 161:225–236
Botmann JWM, Jongkees LBW (1955) Endotemporal branching of the facial nerve. Acta Otolaryngol 45:111
Bouchet A (1978) Les mégadolichoartère du système carotidien. A propos d'un cas opéré avec succès. Rev Otoneuroophthalmol 50:95–98
Boudin G, Laude M, Frèche Ch (1982) Anatomical bases for the surgical approach to the Vidian nerve (n. canalis pterygoidei). Anat Clin 3:239–242
Boutet JJ, Legent F, Beauvillain C (1982) Le ligament manubrial postérieur. Ann Otolaryngol Chir Cervicofac 99:263–264
Bremond GM, Garcin J, Magnan J, Pradour P, Pech-Gourg F (1976) Anatomie des acustico-facialen Komplexes. Acta Otorhinolaryngol Belg 30: 145–150
Bryce S, Choragiewicz B, Rakowski P, Vogue M (1979) Origine de l'artère vertébrale gauche de la carotide interne homolatérale: une variante anatomique inconnue. Rev Otoneuroophthalmol 51:421–423
Buccigrossi G, Mensorio C, Interligi M (1975) Sulla circolazione arteriosa collaterale del cervello in alcune sindromi neurologiche particolari. Chir Ital 23:41–51
Busuttil A, Davis BC, Maran AGD (1981) The soft tissue/catilage relationship in the laryngeal glottis. J Laryngol Otol 95:385–391

Cachin Y, Guerrier Y, Pinel J (1969) Les adénopathie cervicales néoplastiques. Soc française d'oto-rhino-laryng et de pathol cervico-faciale. Arnette, Paris
Caix M, Outrequin G (1979) Variability of the bony semicircular canals. Anat Clin 1:259–265
Calsao Hiosch (1925) Lehrbuch der Lokalanästhesie des Ohres und der oberen Luft- und Speisewege. Ewe, Stuttgart
Carlier M, Geertruyden J van (1980) Kyste Parathyroidien mediastinal. Ann Chir 34:99–102
Carter DR (1981) The anatomy of the Mediastinum. ENT J 60:153–157
Carter DR, Meyers AD (1979) Anatomy of the subglottic larynx. Otolaryng Head/Neck Surg 87:203–206
Chandler JR, Mitchell B (1981) Branchial cleft cysts, sinuses and fistulae Otolaryngol Clin North Am 14:175–186
Chevrel JP, Hurau J, Lassau JP (1965) Les voies de drainage veineux du corps thyroide, incidences chirurgicales. CR Ass Anat 124:457–463
Chidichino G, Pucci A, Rabitti G (1978) Il bypass temporaneo nella chirurgia carotidea. Chir Torac 31:374–377
Chouard CH, Charachon R, Morgon A, Cathala HP (1972) Anatomie pathologie et chirurgie du nerf facial. Masson, Paris
Citelli (1906) Solco glottideo. Internat Mschr Anat Physiol 23:421
Clara M (1942) Das Nervensystem des Menschen. Barth, Leipzig
Clemente CD (1975) Anatomy. A regional atlas of the human body. Lea u Febiger, Philadelphia
Combelles R, Bastide G, Juskiewenski S, Beçue J, Vaysse P, Roux P (1974) Vascularisation arterielle superficielle de la face. Bull Assoc Anat (Nancy) 58:833–841
Combelles R, Beçue J, Bastide G, Juskiewenski S (1974) Vascularisation arterielle des maxillaires. Bull Assoc Anat (Nancy) 58:85–94
Conley JJ (1970) Concepts in head and neck surgery. Thieme, Stuttgart. Gruner & Stratton, New York
Conley JJ (1971) Treatment of facial paralysis. Surg Clin North Am 51/2:403–416
Conley JJ (1975) Salivary glands and the facial nerve. Thieme, Stuttgart
Conley JJ (1976) Regional flaps of the head and neck. Thieme Stuttgart
Conso FJ, Reignier B, Salmon P, Honnart F, Gayral F, Calmat A (1974) Etude sur la position de la naissance et le trajet du tronc brachyocéphalique artériel par papport à l'axe médian et l'axe de la trachée. Ann Chir Thorac Cardiovasc 13/3:255–256
Converse JM (1942) New forehead flap for nasal reconstruction. Proc R Soc Med 35:811
Converse JM (1955) The cartilaginous structures of the nose. Ann ORL (St Louis)
Converse JM (1977) Deformities of the forehead, scalp and calvarium in reconstructive, plastic surgery. Saunders, Philadelphia
Conway H, Murray JE (1953) Indications for recostruction at the time of surgical excision of cancer of the oral cavity. Cancer 6:46
Corlieu P, Aaron C, Godefrov D (1981) Radioanatomie de la base du crâne, étage moyen et postérieur. Ann Otolaryngol Chir Cervicofac 98:173–179
Cormier JM, Ricco JB, Franceschi C (1979) 82 cas de reimplantation de la souclavière ou de la vertébrale dans la carotide primitive. Chirurgie 105:592–596
Correll RW, Jensen JL, Taylor JB, Rhyne RR (1979) Mineralisation of the Stylohyoid stylomandibular ligament complex. A radiographic study. Oral Surg 48:286–291
Cottle MH (1955) The structure and function of the nasal vestibule. Arch Otolaryngol 62:173
Counter RT (1980) A superior thyroid cornu anomaly: a report of a case. J Laryngol Otol 94:1087–1088
Czerny LJ (1973) Muscular dystopia in the facial nerve canal. Arch Otolarnygol 98:201–204

Davis RA, Anson BJ, Budinger JM, Kurth LE (1956) Surgical anatomy of the facial nerve and parotid gland based upon study of 350 cervicofacial halves. Surg Gynecol Obstet 102:384
Demard F, Charvet B Les sinus de la face. J Franç ORL 31:449–460

Denecke HJ (1961) Korrektur des Schluckaktes bei einseitiger Pharynx- und Larynxlähmung. HNO 9:351–353
Denecke HJ (1966) Zur Chirurgie ausgedehnter Glomustumoren im Bereich des Foramen jugulare. Arch Otorhinolaryngol (NY) 187:656–662
Denecke HJ (1968) Unfallchirurgie des Gesichtes und des Halses. Arch Otorhinolaryngol 191:217–404
Denecke HJ (1980) Die otorhinolaryngologischen Operationen im Mund- und Halsbereich. Springer, Berlin Heidelberg New York
Denecke HJ (1953) Die otorhinolaryngologischen Operationen. Die allgemein-chirurgischen Eingriffe am Hals. II Aufl. Springer, Berlin, Göttingen Heidelberg
Denecke HJ, Ey W (1984) Die Operationen an der Nase und im Nasopharynx. Allg. und spez. Operationslehre V/1. Springer, Berlin, Heidelberg, New York, Tokyo
Denecke HJ, Meyer R (1964) Plastische Operationen am Kopf und Hals. I Nasenplastik Springer, Berlin Göttingen Heidelberg
Denker A, Kahler O (1925) Handbuch der Hals-Nasen-Ohrenheilkunde Bd I Luftwege und Mundhöhle. Springer, Berlin. Bergmann, München
Dervaux JL, Bargy F, Labayle J (1976) A propos de la recherche du nerf récurrent droit dans la chirurgie thyroidienne à la base du cou. Ann Otolaryngol Chir Cervicofac 93:710–712
Desnos J, Andrieu-Guitrancourt J, Dehesdin D, Dubin J (1980) Vascular strictures of the respiratory tract in children. Internat J of Pediatric Otorhinolaryng 2:269–285. Biomedical Press, Elsevier/N Holland
Desnos J, Andrieu-Guitrancourt J, Dehesdin D (1981) Sténoses de la trachée et des bronches souches chez l'enfant. J Franc ORL 30:313–320, 323–333
Deutschmann H, Scharnagl E (1980) Skalpierungsverletzungen: Die Durchblutung der Kopfschwarte. Z Plast Chir 4:124–130
Diamant H, Wiberg A (1965) Does the chorda tympani in man contain secretory fibres for the parotid gland? Acta Otolaryngol 60:255–264
Domenech Mateu JM, Pueyo Mur FJ (1980) Development and arrangement of the tympanic plexus and the nerve of the pterygoid canal during human embryonic and fetal periods. Acta Morphol Neerl Scand 18:253–272
Domarus Hv, Weimer J, Friedrich HJ (1984) Die Hautarterien des Gesichtes. Variationen und mittlerer Verlauf dargestellt mit der Ultraschall-Doppler-Sonographie. Dtsch Z Mund-Kiefer-Gesichts-Chir 8:112–121
Donaghy RMP, Yasargil MG (1967) Microvascular Surgery. Thieme Stuttgart
Donaldson I (1980) Surgical anatomy of the tympanic nerve. J Laryngol Otol 94:163–168
Draf W (1978) Die Endoskopie der Nasenhöhlen. Springer, Berlin Heidelberg New York
Draf W (1978) Bedrohliche Zwischenfälle bei Trägern von Trachealkanülen. Notfallmedizin 4:14–23
Draf W (1982) Die chirurgische Behandlung entzündlicher Erkrankungen der Nasennebenhöhlen. Indikation, Operationsverfahren, Gefahren. Fehler und Komplikationen, Revisionschirurgie. Arch Otorhinolaryngol 235:133–305
Draf W (1983) Endoscopy of the paranasal sinuses. Springer, Berlin Heidelberg New York
Draf W, Samii M (1983) Fronto-basal injuries. Principles in diagnosis and treatment. In: Samii M, Brilaye J (eds) Traumatology of the skull base. Springer, Berlin Heidelberg New York, pp 61–68
Droulias C, Whitehurst JO (1976) The lymphatics of the tongue in relation to cancer. Am Surg 42:670–674
Droulias C, Tzinas S, Harlaftis N, Akin JT, Gray SW, Skandalakis JE (1976) The superior laryngeal nerve. Am Surg 42:635–638
Dupuch KM, Munier E, Lallemand D (1981) Kystes à localisation latérale cervicale du tractus aéro-digestif chez l'enfant. Rapport des investigations radiologiques. J Radiol 62:385–390

Eagle W (1958) Elongated styloid process. Arch Otolaryngol 67:172–175
Egami T, Sando I, Myers EN (1979) Temporal bone anomalies associated with congenital heart diseases. Ann Otol Rhinol Laryngol 88:72–78
Eickhoff W, Herberhold C (1968) Die Lymphbahnen der menschlichen Schilddrüse. Springer, Berlin Heidelberg New York
Eitschberger E, Rosslau A, Weidenbecher W (1981) Gefäßsystem und Präparationstechnik des pektoralen myokutanen Insellappens. HNO 29:79–82
Ercegovac N, Davidović R (1970) Foramen arcuale atlantis kao etiološki faktor vertebro bazilarne insuficijencije-dekompresija arterije vertebralis. Vojnosanit Pregl 10:435–441
Estlander JA (1872) Eine Methode aus der einen Lippe Substanzdefekte der anderen zu ersetzen. Langenbecks Arch Chir 14:622
Evans TH (1956) Carotid canal anomaly. Other instances of absent internal carotid artery. Med Times 84/10:1069–1072

Feldmann H (1975a) Supraganglionärer Verlauf des Nervus facialis durch Warzenfortsatz und Epitympanon; eine bisher nicht beobachtete Anomalie. Arch Ohr-Nas-Kehlkheilkd 210:346
Feldmann H (1975b) Anomalie des supraganglionären Facialisabschnittes Laryngol Rhinol Otol (Stuttg) 54:445–451
Feneis H (1982) Anatomisches Bildwörterbuch der internationalen Nomenklatur, 5. Aufl. Thieme, Stuttgart
Firbas W, Sinzinger H (1972) Über den Anfangsteil der Arteria cerebri anterior. Acta Anat (Basel) 83:81–86
Fisch U (1968) The surgical anatomy of the so-called internal auditory artery. 10. Nobel Symposium. Almquist and Wiksell, Stockholm, S 121
Fischer AGT (1914) A case of complete absence of both internal carotid arteries with a preliminary note on the development of the stapedial artery. J Anat Physiol 48:37
Fischer AGT (1915) Sigmoid tortuosity of the internal carotid artery and its relations to the tonsil and pharynx. Lancet 2:128
Fischer E (1938) Die Lageabweichungen der vorderen Hirnarterie im Gefäßbild Zentralbl Neurochir 3:300–313

Flament JB, Delattre JF, Pluot M (1982a) Vascularisation artérielle des glandes parathyroïdes. Incidence en chirurgie thyroïdienne. Anat Clin 3:16–20
Flament JB, Delattre JF, Pluot M (1982b) Arterial blood supply to the parathyroid glands: implications for thyroid surgery. Anat Clin 3:279–284
Fowler EP Jr (1961) Verlaufsanomalien des N. facialis im Schläfenbein. Z Laryngol Rhinol Otol 40:360
Francke JP, Macke J, Clarisse J, Libersa JC, Dobbelaero P (1982) The internal carotid arteries. Anat Clin 3:243–261
Freche C, Boudin G (1982) Les nerfs recurrents. J Franc ORL 31:51–64
Freeland AP (1974) Microphil angiography: a demonstration of the microvasculature of the larynx with reference to tumor spread. In: Albert PW, Bryce P (eds) Centenn. Conference on laryngeal cancer. Appleton-Cent-Grofts, New York, p 279
Fujii K, Chambers SM, Rhoton AL (1979) Neurovascular relationship of the sphenoid sinus. A microsurgical study. J Neurosurg 50:31–39

Gardner JW (1962) Concerning the mechanism of trigeminal neuralgia and hemifacial spasm. J Neurosurg 19:947–958
Gardner JW, Stowell A, Dutlinger R (1947) Resection of the greater petrosal nerve in treatement of unilateral headache. J Neurosurg 4:105
Gars de le, Lepage M, Deramond H (1980) Souffle intracrânien et malformations vasculaires. Rev Otoneuroophthalmol 52:135–141
Gerhardt HJ, Otto HD (1981) The intratemporal course of the facial nerve and its influence on the development of the ossicular chain. Acta Otolaryng (Stockh) 91:567–573
Gibo H, Lenkey C, Rhoton AL (1981) Microsurgical anatomy of the supraclinoid portion of the internal carotid artery. J Neurosurg 55:560–574
Glasscock ME, Dickins JRE, Jackson CG, Wiet JR (1980) Vascular anomalies of the middle ear. Laryngoscope 90:77–88
Godlewski G, Bossy J, Pradal D (1976) Contribution à l'étude des anastomoses des branches terminales du nerf facial et du nerf trijumeau. Bull Assoc Anat (Nancy) 60:677–693
Godlewski G, Lopez J, Pavart JC, Giraudon M, Dussaud J (1979) Etude anatomoradiologique de la vascularisation arterielle de l'articulation temporo-mandibulaire. Cah ORL Chir Cervicofac 14:397–402
Goethe JW v (1784) Über den Zwischenkiefer des Menschen und der Tiere. Zur Morphologie, Bd I Heft 2 (1820): Verh d. Kaiserlich Leopoldinisch-Carolinischen Akademie d. Naturforscher (1831) In Stapf P (Hrsg) Goethes Werke Bd 8. Naturwissenschaftl. Schriften, S. 879 Emil Vollmer Wiesbaden 1983
Golding-Wood PH (1962) Pathology and surgery of chronic vasomotor rhinitis. J Laryngol Otol 76:969–977
Godziewski S, Niserkowski C, Kindlik R (1979) Die morphologische Analyse des Canalis infraorbitalis und des Foramen infraorbitale beim Menschen. Anat Anz 145:517–527
Goodman RS, Cohen NL (1981) Aberrant internal carotid artery in the middle ear. Ann Otol Rhinol Laryngol 90:67–69
Gradenigo G (1907) Über die Paralyse des N abducens otitischen Ursprungs. Arch Hals-Nasen-Ohrenheilkd 74:249
Graf K, Fisch U (1979) Geschwülste des Ohres und des Felsenbeins. In: Berendes J. Link R, Zöllner F (eds) Hals-Nasen-Ohrenheilkunde in Praxis und Klinik, Bd V II. Thieme, Stuttgart
Grant JCB, Basmajian JV (1965) Grant's method of anatomy, VII. Williams and Wilkins, Baltimore
Grapulin G (1976) Le complicanze della chirurgia tiroidea, Acta Ital 32:347–389
Greenway RE, Hurst L, Fenton NA (1981) An unusual first branchial cleft cyst. J Otolaryngol 10:219–225
Grisoli F, Vincentelli F, Henry J (1982) Anatomical bases for the transsphenoidal approach to the pituitary gland. Anat Chir 3:207–220
Grünwald L (1925) Descriptive und topographische Anatomie der Nase und ihrer Nebenhöhlen. In: Denker A, Kahler O (Eds.) Hals-Nasen-Ohrenheilkunde. Springer, Berlin; Bergmann, München
Guerrero L, Leiva O (1979) Antrolitiasis maxilar. Rev ORL (Santiago) 39:69–70
Guerrier Y (1951) Les artères du nerf facial. Bull Sc ORL (Lyon) 2/4:83
Guerrier Y (1953) Terminaison du nerf laryngé externe. CR Assoc Anat 275
Guerrier Y (1981a) Essai d'explication anatomique de l'ophthalmoplégie et de l'amaurose après Videctomie. Cah ORL Chir Cervicofac 16:703
Guerrier Y (1981b) Les nerfs pétreux, le nerf vidien. Problèmes de nomenclature anatomique. Cah ORL Chir cervicofac 16:143–144
Guerrier Y (1981c) Essai d'explication anatomique de l'ophthalmoplégie et de l'amaurose après videotomie. Cah d'ORL J XVI, no 7:703
Guerrier Y, Andrea M (1977) La vascularisation des cartilages du larynx. Son importance clinique. Ann Otolaryngol (Paris) 94:273–289
Guerrier Y, Charissoux G (1981) L'artère pharyngienne ascendente. Cah ORL Chir cervico-fac 16:805–808
Guerrier Y, Franchebois P (1960) Phlébectasie jugulaire. Soc Chir Montpellier
Guerrier Y, Villaceque G (1949) Origine et compartament des artères cerèbelleuses moyennes CR Assoc Anat 65:377
Guleke N (1953) Die allgemein-chirurgischen Eingriffe am Halse. In: Guleke N, Zenker N (eds) Allgemeine und spezielle chirurgische Operationslehre, 2. Aufl, Bd 5.
Gupta SK, Sharma OP, Mohan V (1979) First and second branchial arch syndrome. ENT J 58:261–264
Guyader A le, Gayotte J, Baffont J, Renard M, Deteil C, Picard L, Brice M, Masson JP (1968) Variations morphologiques du polygone de Willis à propos de 170 observations anatomiques. Etud Anat Fac Med (Nancy) 2:45–55

Haar JG, Boulos EJ, Sadeghi NH (1981) Association of a thyroglossal duct cyst and a dermoid cyst in the neck. A case report. Ann 84 Otol (St Louis 90:181–182

Haecker V (1927) Goethes morphologische Arbeiten. G Fischer, Jena
Hafferl A, Thiel W (1969) Lehrbuch der topographischen Anatomie, 3. Aufl. Springer, Berlin Heidelberg New York
Halbrock KH (1960) Zweiteilung des N. facialis im Warzenfortsatz. Arch Ohr-Nas-Kehlkheilkd 194:465
Hall RM (1979) High tracheal bifurcation. Anaesthesia 34:503–505
Haller A v (1749) Icon Anat Fasc IV, Göttingae
Harner SG, McDonald TJ, Reese DF (1981) The anatomy of congenital choanal atresia. Otolaryngology (Rochester MN) 89:7–9
Hayek M (1915) Pathologie und Therapie der entzündlichen Erkrankungen der Nebenhöhle der Nase. 4. Aufl. Deuticke, Leipzig Wien
Healy JE, Seybold WD (1969) A synopsis of clinical anatomy. Saunders, Philadelphia London Toronto
Heiderich F (1934) Kopf, Hals, Bauch und Becken des Kindes. In: Peter K, Wetzel G, Heiderich F (eds) Handbuch der Anatomie des Kindes, Bd I. Bergmann, München
Helms J (1963) Über den Winkel zwischen Dens und Epistropheus. Diss München
Helms J (1978) Zur chirurgischen Anatomie des Felsenbeins in Kleinhirnbrückenwinkeltumoren. Plester D, Wende S, Nakayama N (Hrsg) Springer, Berlin Heidelberg New York
Helms J (1980) Variations of the course of the facial nerve in the middle ear and mastoid. Symposium on cranial nerves, Hannover 2/6
Helms J (1981) Variations of the course of the facial nerve in the middle ear and mastoid. In: Samii M, Janetta FJ (eds) The cranial nerves. Springer, Berlin Heidelberg New York
Helms J, Kimpel G (1978) Die Transposition der Zungengrundschilddrüse. Z Laryngol Rhinol Otol 57:252–256
Herlyn H (1943) zur Freilegung der A. vertebralis. Chirurg 15:713
Hermann F (1908) Gehirn und Schädel. Fischer, Jena
Himalstein MR (1979) Phylogeny of the cervical sinus with clinical correlation. ENT J 58:256–260
Himalstein MR (1980) Branchial cysts and fistulas. ENT J 59:23–29
Hiroto J, Hirano M, Tomita H (1968) Electromyographic investigation of human vocal cord paralysis. Ann Otol Rhinol Laryngol 77:296–304
Hiura A (1980) An anomalous ophthalmic artery arrising from the middle meningeal artery. Anat Anz 147:473–476
Hochstetter F (1937) Über einige Fälle einer bisher anscheinend noch nicht beobachteten Varietät der Arteria cerebralis posterior des Menschen. Z Anat Entwickl-Gesch 107/5:638–646
Hochstetter F (1938) Über eine Varietät der Vena cerebralis basialis des Menschen nebst Bemerkungen über die Entwicklung bestimmter Hirnvenen. Z Anat Entwickl-Gesch 108/2:311–336
Holbach KH, Wappenschmidt J, Bodosi M (1976) Angiographische Befunde bei extra-intrakraniellen Anastomosen. Radiologe 16:416–423
Holzapfel G (1899) Ungewöhnlicher Ursprung und Verlauf der Arteria subclavia dextra. Anat Hefte 12:369–523
House WF, Belal A (1980) Translabyrinthine surgery: Anatomy and pathology. Am J Otol 1:189–198
House WF (1964) Monograph I, transtemporal bone microsurgical removal of acustic neuromas. Arch Otolaryngol (Chic) 80:597–756

Iontov A (1962) Variationen im Verlauf der Arteria cerebri anterior beim Menschen. Tr Nauchnoissl Psikonevrol Inst Bekhterev (Leningrad) 28:45–68

Jacquemard C (1971) Le larynx, la trachée et les bronches de l'enfant. Rappel anatomique et physiopathologique. J Franc ORL 20:815–818
Jahrsdorfer RA, Cail WS, Cantrell RW (1981) Endolymphatic duct obstruction from a jugular bulb diverticulum. Ann Otol (St Louis) 90:619–623
James TM, Presley R, Steel F (1980) The foramen ovale and sphenoidal angle in man. Anat Embryol (Berl) 160:93–104
Janetta PJ (1977) Etiology and definitive microsurgical treatment of hemifacial spasm. J Neurosurg 47:321–328
Jo A, Trauzettel H (1974) Topographische Beziehungen der Venen in der Orbita. Verh Anat Ges 68:539
Johnson LG, Kingsley TC (1970) Herniation of the facial nerve in the middle ear. Arch Otolaryngol 91:598
Jost G (1975) Atlas of Aesthetic Plastic Surgery. Masson and Co. Paris
Jovanović S (1961) Supernumerary frontal sinuses on the roof of the orbit: their clinical significance. Acta Anat (Basel) 45:133
Jovanović S, Lotrić N, Radojević V (1954) Razvitak, morfologija i topografija sfenoidalnih sinusa kod djece. Acta Med Jugosl 8:232
Jumen-Lavillauroy C, Roux O, Reyt E (1981) Kystes du tractus thyreoglosse, réflexions à propos d'une série de 45 cas. J Franc ORL 30:257–261, 263–264
Juniper RP (1981) The superior pterygoid muscle. Br J Oral Surg 19:121–128
Juri J (1975) Use of parieto-occipital flaps in surgical treatment of baldness. Plast Reconstr Surg 55:456–460
Juri J, Juri C, Arufe HN (1978) Use of rotation skalp flaps for treatment of occipital baldness. Plast Reconstr Surg 61:23

Kaufman Arenberg I (1982) Abnormalities congenital anomalies and unusual anatomic variations of the endolymphatic sac and vestibular aqueduct: clinical, surgical and radiographic correlations. Group 3, 4 and 5 abnormalities. Am J Otol 3:221–240
Keller HL, Weiss D (1973) Ursprung einer das Vertebralisgebiet versorgenden Arterie aus dem Halsabschnitt der rechten Arteria carotis interna. Fortschr Roentgenstr 118:472–474
Kemmetmüller H (1911) Über eine seltene Varietät der Arteria vertebralis. Anat Hefte 44/2:307–360
Kempe LG (1968) Operative neurosurgery, vol 1 Cranial, cerebral and intracranial vascular diseases. Springer, Berlin Heidelberg New York
Kempe LG (1970) Operative neurosurgery vol II Posterior fossa, spinal cord and peripheral nerve diseases. Springer, Berlin Heidelberg New York

Keros P (1965) Über die praktische Bedeutung der Niveauunterschiede der Lamina cribrosa des Ethmoids. Z Laryngol Rhinol 41:808

Keros P, Nemanić G (1966) Research results in conduction anesthesia of the infraorbital nerve. Med J 88/2:4–9

Keros P, Nemanić G (1967) The terminal branching of the recurrent laryngeal nerve. Practica Otol Rhinol Laryngol 39:5–10

Khodadad G (1977) Persistent hypoglossal artery in fetus. Acta Anat 99:447–481

Kiesselbach W (1884) Über spontane Nasenblutung. Berlin Klin Wochenschr

Kiesselbach W (1885) Über Nasenbluten. Wien Med Zeitung

Killian G (1903) Die Nebenhöhlen der Nase in ihrer Lagebeziehungen zu den Nachbarorganen. Fischer, Jena

Killien FC, Wyler AR, Cromwell LD (1980) Duplication of the internal carotid artery. Neuroradiology 19:101–102

Kiyotaka Fujii, Chambers MS, Rhoton LA Jr (1979) Neurovascular relationsship of the sphenoid sinus. A microsurgical study. J Neurosurg 50:31–39

Klausberger EM, Samec P (1975) Foramen retroarticulare atlantis und das Vertebralisangiogram, Med Wochenschr 117:483

Kodama N, O'Hara H, Suzuki J (1976) Persistent hypoglossal artery associated with aneurisms. J Neurosurg 45:449–451

Kodama A, Sando I (1982a) Postnatal development of the vestibular aqueduct and endolymphatic sac, vol 91, no 4. Ann Otol Rhinol Laryngol [Suppl 96] 4:3–12

Kodama A, Sando I (1982b) Dimensional anatomy of the vestibular aqueduct and the endolymphatic sac (rugose portion) in human temporal bones. Statistical analysis of 79 bones, vol 91, no 4. Ann Otol Rhinol Laryngol [Suppl 96] 4:13–20

Koebke J (1976) Kompression der A carotis externa. Laryngol Rhinol Otol (Stuttg) 55:913–917

Konaschko PJ (1927) Die Arteria auditiva interna des Menschen und ihre Labyrinthäste. Z Anat Entwickl Gesch 83:241

Krause W (1905) Handbuch der Anatomie des Menschen. Hirzel, Leipzig

Krayenbühl H, Yasargil MG (1965) Die cerebrale Angiographie. Thieme, Stuttgart

Kreutzer EW, Iafek BW (1980) The vomeronasal organ of Jacobson in the human embryo and fetus. Otolaryng Head Neck Surg 88:119–123

Križan Z (1960) Beiträge zur descriptiven und topografischen Anatomie der Arteria maxillaris. Acta Anat (Basel) 41:319

Krmpotić J (1946) O izuzetnom lateralnom smještaju izvanje karotide prema stražnjem trbuhu mišića biventera u retromandibularnoj udubini. Liječ Vjesn 3/4:1–10

Krmpotić J (1960a) L'index neuromusculaire des nerfs relatifs à la phonation. Rev Laryngol Otol Rhinol (Bord) 81/3–4:232–343

Krmpotić J (1960b) Données anatomiques indispensables à connaître pour une chirurgie efficace du nerf statoacoustique. CR 57 Congrès franc à 'ORL 1959, Arnette, Paris

Krmpotić J (1963) L'index neuromusculaire chronométrique en otolaryngologie. Acta Otolaryngol (Stockh) 56/2–3:298–303

Krmpotić J (1965) Limfni sistem glave i vrata. Chirurgia Maxillofac Plast 5/1:4–19

Krmpotić J, Nikolić V (1964) Praktisch wichtige topographische Beziehungen des N. petrosus superficialis major in der mittleren Schädelgrube. Z Laryngol Rhinol Otol 43/12:748–753

Krmpotić J, Šercer A (1960) L'enfoncement profond du bulbe supérieur de la veine jugulaire dans la paroi inférieure de la pyramide comme cause de bourdonnements d'oreille et de surdité. Bull Acad Nat Med (Paris) 144:29–30, 701–704

Krmpotić J, Sunarić R (1963) Chirurgische Behandlung otogener und rhinogener Kopfschmerzen. Z Laryngol Rhinol Otol 4:260–266

Krmpotić J, Sunarić R (1963) Neue Gedanken zur operativen Ozaenatherapie. Arch Ohr-Nas-Kehlk-Heilkd, Z Hals-Nas-Ohr-Heilk 182/2:488

Krmpotić-Nemanić J (1977) Entwicklungsgeschichte und Anatomie der Nase und der Nasennebenhöhlen in Hals-Nasen-Ohrenheilkunde in Praxis und Klinik (1977) Berendes-Link-Zöllner, Bd 1. Obere und untere Luftwege. Thieme, Stuttgart

Krmpotić-Nemanić J (1978) Anatomie, Variationen und Mißbildungen der Gefäße im Kopf und Halsbereich. Arch Otorhinolaryngol 218:1–77. II Teil: Erläuterungen, Diskussionsbemerkungen und Schlußworte zu den Ref 219:285–330

Krmpotić-Nemanić J, Keros P (1973) Funktinale Bedeutung der Adaptation des Dens axis beim Menschen. Verh Anat Ges 67:393–397

Krmpotić-Nemanić J, Kostović I, Rudan P, Nemanić G (1972) Aging changes of the form and infrastructure of the external nose and its importance in rhinoplasty. Plast Reconstr Surg 1:84–91. Aesthetic Surgery 1972

Kümmerle F, Rothmund M, Diethelm L, Brünner H (1974) Primärer Hyperparathyroidismus mit mediastinaler Adenomlokalisation. Dtsch Med Wochenschr 9:983

Kwak S, Nagashima T, Kobayashi S (1980) Anomalous origine of the posterior meningeal artery. Neuroradiology 19:103–104

Lang J (1971) Zur Vascularisation der Dura mater cerebri. I. Z Anat Entwickl Gesch 135:20–34

Lang J (1979) Kopf. Teil B, Gehirn- und Augenschädel. In: Lanz-Wachsmuth (Hrsg) Praktische Anatomie, Bd I/1. Springer, Berlin Heidelberg New York

Lang J (1981) Klinische Anatomie des Kopfes. Springer Berlin Heidelberg New York

Lang J (1981) Neuroanatomie der Nn opticus, trigeminus, facialis, glossopharyngeus, vagus, accessorius und hypoglossus. Arch Otorhinolaryngol 231:1–69. Kongreßbericht Teil II 451–482

Lang J (1983) Macroscopic bony deficiency of the tegmen tympani in adult temporal bones. J Laryngol Otol 97:683–688

Lang J, Götzfried HF (1982) Über praktisch-ärztlich wichtige Masse an der Fossa cranialis media. Anat Anz 151:433–453

Lang J, Sakals LE (1981) Über die Höhe der Cavitas nasi, die Länge ihres Bodens und Masse sowie Anordnung der Conchae nasales und der Apertura sinus sphenoidalis. Anat Anz 149:297–318

Lang J, Schlehahn FA (1981) Ueber die postnatale Entwicklung der Fissurae orbitales. Gegenbaurs Morphol Jahrb 127:849–859

Lang J, Weigl M (1983) Nerve-vessel relations in the region of the jugular foramen. Anat Clin 5:41–56

Lang J, Schafhauser O, Hoffmann S (1983) Über die postnatale Entwicklung der transbasalen Schädelpforten: Canalis caroticus, Foramen jugulare, Canalis hypoglossalis, Canalis condylaris und Foramen magnum, Anat Anz 153:315–357

Lanz T v, Wachsmuth W (1955) Praktische Anatomie Bd I/2, Hals. Springer, Berlin Göttingen Heidelberg

Lamberty BGH, Živanović S (1973) The retroarticular vertebral artery ring of the atlas and its significance. Acta Anat (Basel) 85:113–119

Lasjaunias P (1981) Cranio-facial and upper cervical arteries, Vol. I Williams and Wilkins, Baltimore

Lasjaunias P (1983) Cranio-facial and upper cervical arteries, Vol. II Williams and Wilkins, Baltimore

Lasjaunias P, Manelfe C (1979) Arterial supply for the upper cervical nerves and the cervicocarotid anastomotic channels. Systematisation of radiological anatomy. Neuroradiol 10:125–131

Lasjaunias P, Moret J (1978) Normal and non-pathological variations in the angiographic aspects. Neuroradiol 15:213–219

Laudadio P, Casolino D, Lorenzo P (1981) Considerazioni su 34 casi di cisti branchiali latero-cervicali. Valsalva 57:44–52

Laude M, Gars de le (1980) Anatomie descriptive de la portion rétrogassérienne du nerf trijumeau chez l'homme. Anat Clin 2:223–228

Lautenschläger A (1912) Doppelbildung der Stimmbänder. Arch Laryngol Rhinol 26:706

Lautenschläger A (1934) Chirurgische Eingriffe an der Nase und ihren Anhangsgebilden. In: Kirschner M (Hrsg) Allgemeine und spezielle Operationslehre, 3. Bd, II. Teil. Springer, Berlin

Lazorthes G (1981) Les douleurs rapportées. Quelques exemples dans le domaine oto-neuro-ophthalmologique. Rev Otoneuroophthalmol 53:145–150

Lazorthes G, Gouaze A, Santini JJ, Salamon G (1979) The arterial circle of the brain circulus arteriosus cerebri. Anat Clin 1:241–257

Legent F, Beauvillain C, Kersaint-Gilly A (1981) Fosse pterygomaxillaire (fossa infratemporalis). Division topographique. Moyens d'exploration. Abord transmandibulaire de la region des pterygoidiens. Ann Oto-laryngol Chir Cervicofac 98:435–442

Leonardi HK, Nagger CZ, Ellis FH (1980) Dysphagia due to aortic arch anomaly Diagnostic and therapeutic considerations. Arch Surg 115:1229–1232

Lerault P, Freche C, Bouche J (1981) Note préliminaire sur l'anatomie et la physiologie du nerf pharyngien de Bock. Ann Otolaryngol 98:145–147

Lesoine W (1966) Experimentelle und klinische Untersuchungen über die Funktion des Zungenbeines beim Schluckakt. HNO 14:272–273

Lesoine W (1970) Die Morphologie des Zungenbeines. Z Laryngol Rhinol Otol 49:461–465

Lexer E (1908) General surgery. A presentation of the scientific principles upon which the practice of modern surgery is based. Appleton and Company, New York

Lexer E, Eden R (1911) Über die chirurgische Behandlung der peripheren Facialis-Lähmung. Beitr Klin Chir 73:116

Liston SL (1981) Fourth branchial fistula. Otolaryng. Head Neck Surg 89:520–522

Liston SL, Siegel LG (1979) Branchial cystes, sinuses and fistulas. ENT J 58:504–509

Little JR, Yamamoto YL, Feindel W, Meyer E, Hodge ChP (1979) Superficial temporal artery to middle cerebral artery anastomoses. Intraoperative evaluation by fluorescein angiography and Xenon 133 clearence J Neurosurg 50:560–69

Locatelli E, Catenacci N, Gentile G, Romagnino A (1977) Classificazione su base embriogenetica degli aneli vascolari periesofagotracheali. Chir Torac 30:457–462

Lood DM Mc, Karandy EJ (1981) Aberrant cervical thymus. A rare cause of acute respiratory distress. Arch. Otolaryngol 107:179–180

Loré (1973) Atlas of head and neck surgery, II. edn. Saunders, Philadelphia London Toronto

Madeira MC, Percinoto C, Silva MGM (1978) Signification clinique de l'innervation des incisives inférieurres: Etude du nerf mylohyoidien par dissection, Oral Surg 46:608–614

Maffei G, Zini C, Jemmi A, Botazzi D (1967) Ricerche sopra la vascolarisazione del nervo faciale (con particolare riguardo al decorso intratemporale). Arch Ital Otol [Suppl] 78:51

Malin JP (1979) Zur Aetiologie der Phrenicusparese. Bericht über 58 Fälle und Literaturüberblick, Nervenarzt 50:448–456

Malmgren LT, Gacek RR (1981) Acetylcholinesterase staining of fiber components in feline and human recurrent laryngeal nerve. Topography of laryngeal motor fiber region. Acta Otolaryngol 91:337–352

Mandiola E, Delgado A, Chatain I (1979) Contribucion al estudio de la fossa pterigopalatina/I Contenido vascular y tecnica de abordaje por la via transantral. Rev Otorhinolaryngol 39:39–46

Mangold U, Lierse W, Pfeifer G (1980) Die Arterien der Stirn als Grundlage des Nasenersatzes mit Stirnlappen. Acta Anat 107:18–26

Mann W (1980) Die Lymphdrainage der Kieferhöhle. Laryng Rhinol Otol (Stuttg) 59:782–786

Marquet J (1981) The incudo-malleal joint. J Laryngol Otol 95:543–565

Martin H, Helsper JP (1957) Spontaneous return of function following surgical section of VII cranial nerve in the surgery of parotid tumors. Ann Surg 146:715 (zit Miehlke, 1960)

Martin R, Saler K (1959) Lehrbuch der Anthropologie, Bd II. Fischer, Stuttgart

Martini A (1978) Considerazioni per una analisi filogenetica e modellistica dei canali semicircolari, Ann Laring (Pisa) 76:455–466

Masing H (1964a) Klinisch-anatomische Bemerkungen zur Cartilago septi nasi. Z Laryngol Rhinol Otol 43:604–612

Masing H (1964b) Ist die Killiansche Fensterresektion des Septums heute noch zeitgemäß? Z Laryngol Rhinol Otol 43:651–660
Mason J, Dugdale R, Price DJ (1979) Simplified theatre monitoring techniques to minimize the risks of carotid ligation, Acta Neurochir (Wien) [Suppl] 28:585–587
Maxwell GP, Manson PN, Hoopes JE (1979) Experiences with thirteen latissimus dorsi myocutaneous free flaps. Plast Reconstr Surg 4:1–18
Mazzoni A (1969) Internal auditory canal. Arterial relations at the porus acusticus, Ann Otol Rhinol Laryngol 78:797–815
Mazzoni A (1979) Vein of the vestibular aqueduct, Ann Otol Rhinol Laryngol E 3:759–767
Mazzoni A, Hansen CC (1970) Surgical anatomy of the arteries of the internal auditory canal. Arch Otolaryngol 91:128
McGregor (1963) The temporal flap in intra-oral cancer. Its use in repairing the postexcisional defect. Br J Plast Surg 16:329
Mehnert E (1901) Über topographische Altersveränderungen des Atmungsapparates, Jena
Mercier Ph, Cronier P, Mayer B, Billet J, Fischer G (1982) Microanatomical study of the arterial blood supply of the facial nerve in the pontocerebellar angle. Anat Clin 3:263–270
Merland JJ, Riche MC, Monteil JP (1980) Classification actuelle des malformations vasculaires, Ann Chir Plast 25:105–111
Messerklinger W (1978) Endoscopy of the nose. Urban & Schwarzenberg, Baltimore Munich
Mestdagh H, Drizenko A, Ghestem P (1981) Anatomical bases of suprascapular nerve syndrome. Anat Clin 3:67–71
Miehlke A (1960) Die Chirurgie des N facialis. Urban & Schwarzenberg, München Berlin
Miehlke A (1973) Surgery of the facial nerve, 2nd edn. Urban & Schwarzenberg, München Berlin Wien
Mihalkovics Vv (1899) Die Anatomie der Nase und ihrer Nebenhöhlen. In Hdb der Laryng und Rhinol, Wien
Miller-Fischer C (1965) The circle of Willis. Anatomical. Variations. Vasc Dis 2:99–105
Minatogawa T, Kumoi T, Hosomi H, Kokan T (1980) The blood supply of the facial nerve in the human temporal bone. Auris-Nasus Larynx (Tokyo) 7:7–18
Mitterwallner F (1955) Variationsstatische Untersuchungen an den basalen Hirngefäßen. Acta Anat (Basel) 24:51–88
Miura T, Grillo HC (1966) The contribution of the inferior thyroid artery to the blood of the human trachea. Surg Gynecol Obstet 123:99–102
Moffat DB (1962) The embryology of the arteries of the brain. Ann R Coll Surg Engl 30:368–382
Moretti A (1980) Rilevazione strumentale del piano glottico sulla cartilagine tiroidea nella laringectomia sopraglottica, Nuovo Arch Ital Otol 4:783–786
Moretti S, Babini L (1966) Studio anatomo-radiografico dei plessi venosi cervicali, Otorino Ital 35:134–158
Morita S (1902) Über eine seltene Varietät des Verlaufes und der Verästelung der Arteria carotis externa. Tokio Iggakwai-Zasshi 16:692
Most A (1917) Chirurgie der Lymphgefäße und der Lymphdrüsen. In: Küttner H (Hrsg) Neue Deutsche Chirurgie, Bd 24. Enke, Stuttgart
Müller F (1978) Die Arteria falcea anterior des Erwachsenen. Acta Anat (Basel) 102/1:1–12
Myerson CM (1932) The natural orifice of the maxillary sinus. Arch Otolaryngol 15:80

Nager GT, Proctor B (1982) II anatomical variations and anomalies involving the facial canal. Ann Otol 91:45–61
Nakajima H, Martyama Y, Koda E (1981) The definition of the vascular skin territories with Prostaglandin E_1. The anterior chest, abdomen and tigh-inguinal region. Br J Plast Surg 34:258–263
Narcy P, Mansch Y, Gandiche O, Brasnu D (1979) Tronc artériel brachiocéphalique compressif et reflux gastro-oesophagien dans les suites opératoires d'atresie de l'oesophage. Ann Otolaryngol 96:841–847
Natvig P, Sether LA, Gingrass RP, Gardner WD (1971) Anatomical details of the osseous cartilaginous framework of the nose. Plast Reconstr Surg 48:528
Naumann HH (1976) Kopf und Halschirurgie, Bd 3. Thieme, Stuttgart
Nemanić G (1968a) Kritički osvrt na dijagnostiku i operativnu terapiju paranazalnih sinusa u svijetlu novih morfoloških spoznaja, Diss Biološki Glasnik 21:53
Nemanić G (1968b) The reflection of some anatomical structures of the maxillary sinus on the clinical and x-ray diagnostic. Ref XII. Kongres UAJ Novi Sad
Nikolić V (1961) O kolateralnim granama gornjeg dijela zajedničke karotide, Rad Med Fak Zagrebu 3:269–278
Nikolić V (1963) O odnosima etmoidnih ćelija i čeonog sinusa i o mjehurastim izbočenjima tih ćelija u frontalni sinus. Rad Med Fak Zagrebu 11:141
Nikolić V (1967) Variations du trou sphenopalatin, Acta Anat 68:189–198
Nikolić V, Jo A (1967) Öffnungen und Dehiszenzen an der Vertikalplatte des Gaumenbeins. Anat Anz 121:272–282
Nikolić V, Rudež V (1961) Variation der Lage des oberen Knies des sigmoiden Sinus. Liječn Vjesn 88:381–388
Nitzsche H (1980) Kombination einer seltenen Aortenbogenmißbildung (cervical aortic arch) mit einer A lusoria. Fortschr Roentgenstr 132/1:93–94
Noyek AM, Friedberg J (1981) Thyroglossal duct and ectopic Thyroid disorders, Otolaryngol Clin North Am 14:187–201

Oertel O (1922) Über die Persistenz embryonaler Verbindungen zwischen der Arteria carotis interna und der Arteria vertebralis cerebralis, Anat Anz 55:281–291
Ogawa A, Sando I (1982) Spatial occupancy and facial nerve in the facial canal. Ann Otol 91:14–19
Ogura YJ, Clemis D (1971) A study of the gross anatomy of the human vestibular aqueduct, Ann Otol 80:813
Okano Y, Sando I, Mayers EN (1980) Branch of the singular nerve (posterior ampullary nerve) in the otic capsule. Ann Otol 89:13–19
Olsen KD, Maragos NE, Weiland LH (1980) I branchial cleft anomalies. Laryngoscope 90:423–436

Onishi T (1981) Bony defects and dehiscences of the roof of the ethmoid cells. Rhinology 19:195–202
Onodi A (1922) Die topographische Anatomie der Nasenhöhle und ihrer Nebenhöhlen. In: Katz L, Blumenfeld F (Hrsg) Handbuch der speziellen Chirurgie des Ohres und der oberen Luftwege, Bd 1, 1. Hälfte. Kapitzsch, Leipzig, S 61
Osborn AG (1980) The vidian artery: Normal and pathological anatomy. Radiology 136:373–378
Owsley JQ (1981) Functional upper eyelid anatomy as applied to aesthetic blepharoplasty, Aesh Blast Surg 5:115–121

Palmer FJ (1977) Origin of the right vertebral artery from the right common carotid artery: angiographic demonstration of three cases. Br J Radiol 50:185–187
Papadatos D (1978) Deux observations anatomiques d'un nerf récurrent – non récurrent. Anat Anz 144:97–106
Papangelou L (1972) Study of the human internal auditory canal. Laryngoscope 82:125
Papangelou L (1974) Volumetric study of the human internal auditory canal. J Laryngol. 88, 349
Parry RH (1904) A case of tinnitus and vertigo treated by division of the auditory nerve. J Laryng (London) 19:402–406
Pech A, Cannoni M (1969) La chirurgie correctrice de la pointe du nez. Cah d'ORL 4 La Simarre, Tours, 317
Pech A, Cannoni M, Latil G, Gitenet P (1976) La chirurgie thyroidienne et les anomalies de la region recurrentielle. A propos de 100 dissections chirurgicales, Les Cah d'ORL 11/2:125–131
Penzholz H, Kühner A (1981) Critical remarks on different surgical methods in trigeminal neuralgia. In: Samii M, Jannetta PJ (Eds) The cranial nerves. Springer, Berlin Heidelberg New York, p. 341
Pérez Olivares F, Schuknecht HF (1979) Width of the internal auditory canal. A histological study, Ann Otol 88:316–323
Perneczky A, Tschabitscher M (1976) Kleinhirnarteriensegmente: Verh Anat Ges 70:385
Pernkopf E, Ferner H (1963) Atlas der topographischen und angewandten Anatomie des Menschen, Bd I. Kopf- und Hals. Urban & Schwarzenberg, München Berlin
Perović D (1954) Beiträge zur Kenntnis der Entwicklung und des Baues des menschlichen Oberkiefers sowie des Sinus maxillaris. Bull Sci Cons Acad RPF (Yougosla) 11:5
Perović D (1961) Über den Agger nasi. Wien Klin Wochenschr 54/7:1–12
Perrin Cl, Truong-Mink E, Long FX (1979) Réflexions sur la technique de Nomura pour la résection du nerf vidien, J Franc d'ORL 28/6:345–351
Pesch HJ, Glas Wv, Stephan M, Thull R (1981) Zum Ossifikationsprinzip des Schildknorpels. Arch Otorhinolaryngol (NY) 231:829–831
Peter K (1925) Entwicklungsgeschichte der Nase und ihrer Nebenhöhlen. In: Denker A, Kahler O (Hrsg) Handbuch der Hals-Nasen-Ohrenheilkunde, I. Bd. Luftwege und Mundhöhle. Springer, Berlin. Bergmann, München
Peter K, Wetzel G, Heiderich F (1934) Handbuch der Anatomie des Kindes, Bd 1. Bergmann, München
Pfiffer CR, Zorzetto NL (1980) Course and relations of the abducens nerve, Anat Anz 147:42–46
Pisk E (1914a) Über eine seltene Varietät im Verlaufe der Arteria carotis externa. Z Angew Anat Konstitut 1:99
Pisk E (1914b) Über eine seltene Varietät im Verlaufe der Arteria carotis externa beim Menschen und beim Hunde, Anat Anz 45:373
Platzer W (1962) Verlauf und Varietäten der Arteria carotis interna als Grundlage der Karotisangiographie. Wien Klin Wochenschr 74:245–248
Plester D, Wende S, Nakayama N (1978) Kleinhirnbrükkenwinkeltumoren. Springer, Berlin Heidelberg New York
Poncet P, Miller P (1979) Le trajet abérrant de la carotide interne intrapétrause dans la caisse du tympan. Analyse d'un cas personel. Revue de la littérature. Ann Otolaryngol (Paris) 96:793–804
Portmann M, Sterkers JM, Charachon R, Chouard CH (1975) The internal auditory neatus. Anatomy, pathology and surgery. Livingstone, Edinburgh London New-York
Portmann, M, Guillen G, Chabrol A (1982) „How I do it": head and neck. A targeted problem and its solution. Elektrocoagulation of the Vidian nerve via the nasal passage. Laryngoscope 92:453–455
Proctor B, Nager ST (1982) The facial canal: normal anatomy, variations and anomalies. I. Normal anatomy of the facial canal. Anm Otol 1:44
Proctor B, Nielsen E, Proctor C (1981) Petrosquamisal suture and lamina. Otolaryngol Head Neck Surg 89:482–495
Prot D, Narcy F, Manach N (1979) Faux aspects d'arc aortique au cours de la laryngomalacie. Ann Otolaryngol (Paris) 96:913–918
Puchades-Orts A, Nombela-Gomez N, Ortuño-Pacheco G (1976) Variations in Form of Circle of Willis. Some Anatomical and Embryological Considerations. Anat Rec 185/1:119–123

Quiring DP (1949) Circulation collaterale. Lea and Febiger, London
Quisling RS (1980) Intrapetrous carotid artery branches: pathological application. Radiology 131:109–113

Rabattu JP (1981) L'abscêss de la bourse de Tornwald. L'organe vomeronasal de Jacobson. Pathologie méconnue, pathologie oubliée, Cah ORL Chir Cervicofac 16:571–573
Rabischong P, Guerrier J, Vignaud J, Traserra J, Tolosa F (1981) Anatomical bases for the surgical acces to the pterygopalatine fossa. 2:209
Rácz L, Maros T (1981) Anatomische Variationen des Nervus lingualis beim Menschen. Anat Anz 149:64–71
Rácz L, Maros T, Geres-Sturm L (1981) Die anatomischen Variationen des N. alveolaris inferior und ihre Bedeutung in der Praxis. Anat Anz 149:329–333

Radojević S, Negovanović B (1963) La gouttière et les anneaux osseux de l'atlas. Acta Anat (Basel) 55:183

Ray J, Bouche J (1980) Le rettrapage des echecs de la desensibilisation par la chirurgie vidienne dans les allergies nasosinusiennes. Ann Otolaryng 97:375–383

Ricbourg B, Mitz V, Lassen JP (1975) Artère temporale superificielle. Etude anatomique et deductions prattiques. Ann Chir Plast: 20:197–213

Roch JB, Bouchayer M, Cornut G (1981) Le sulcus glottidis, Rev Laryngol Otol Rhinol (Bord) 102:336–346

Romeo G, Falco A, Gaini RM, Vecchi E (1980) Il cordoma rinofaringeo (Profilo anatomo clinico di 5 osservazioni. Nuovo Arch Ital Otol 8/I:25–37

Rose KG, Ortmann R, Seegers D (1979) Vidian neurectomy: neuroanatomical considerations and a report on a new surgical approach. Arch Otorhinolaryngol 224:157–168

Rothmund M (1980) Hyperparathyreoidismus. Thieme, Stuttgart New York

Rowe LD, Lowry LD, Keane WM, Manoucher Fallenjad (1978) Aortic arch anomalies in adult disorders of deglutition. Ann Otol 87:498–508

Rudež V (1960) Topografski odnošaj etmoidnog labirinta prema maksilarnom sinsusu. Diss Med Fakultet Zagreb

Runne J (1957) Strumaresektion und Stimmbandlähmung. Zentralbl Chir 82/44:1841

Saad EF (1980) The underdeveloped palate in ear, nose and throat practice. Laryngoscope 90:1371–1377

Sadeghi EM, Bell WA (1980) Developmental cyst of floor of the mouth: soft tissue variety of median mandibular cyst. J Oral Surg 38:841–843

Saito H (1976) On the tracheoesophageal arteries of the human cervical and upper thoracic regions. Acta Anat Nippon 51:409 (cit Staindl)

Salassa JR, Pearson BW, Payne WS (1977) Gross and microscopical blood supply of the trachea. Ann Thorac Surg 24:100

Salvi G (1901) Anom Lar hum. Arch Psichiatr 22:369

Samii M (1975) Modern aspects of peripheral and cranial nerve surgery. In: Advances and technical standards in Neurosurgery, vol 2. Springer, Berlin Heidelberg New York, S 35–85

Samii M, Janetta PJ (1981) The cranial Nerves. Springer, Berlin Heidelberg New York

Saraniti G, Cusimano F, Ferrara P (1976) Aspects oto-rhino-laryngologiques de l'anomalie de Kimmerle. Rev Laryngol Otol Rhinol (Bord) 97:411–416

La Scala G, Feltrin G, Liessi G, Della Costa F (1978) Agnesi della carotide int. Acta Chir Ital 34:343–347

Schaefer SD, Byrg HS, Holmes RE (1980) Forehead and scalp reconstruction after wide-field resection of skin Carcinoma Arch Otolaryngol 106:680–684

Schewitsch I, Stalsberg H, Schroder KE, Mair IWS (1980) Cysts and sinuse of the lateral head and neck. J Otolaryngol 9:1–6

Schimmelbusch C (1931) (zit aus Joseph J) Nasenplastik und sonstige Gesichtsplastik. Kabitzsch, Leipzig, S 643

Schlemmer M, Navratil J, Stiskal A, Wimmer M (1979) Trachealkompression durch Knickung des Truncus brachiocephalicus. Paediatr Paedol 14:455–462

Schlungbaum (1921) Über die Bildung mehrerer Nebenräume der Nasenhöhle im Oberkiefer und ihre Deutung, mit Berücksichtigung der Beziehungen der Zähne zu ihnen. Vierteljahrhundertsschrift für Zahnheilkunde 43.

Schumacher S (1925) Histologie der Luftwege und der Mundhöhle: In: Denker A, Kahler O (Hrsg) Handbuch der Hals-Nasen-Ohrenheilkunde, Bd 1/I. Springer, Berlin Bergmann München, S 295

Schuman DM (1981) Nasolabial cysts: mechanism of development. ENT J 60:389–394

Sciascia R, Coppini L, Ronandelli R (1937) La vascolarisacione del nervo faciale con particolare riguardo all'arteria petrosa. Otorinolaryngol Ital 35:231–242

Schwemmle K (1980) Die allgemeinchirurgischen Operationen am Halse. Springer, Berlin Heidelberg New York

Scotti G, Grandi C de, Colombo A (1978) Ectasie of the intracranial arteries diagnosed by computed tomography. Neuroradiology 15:183–184

Sedgwick C, Cady B (1980) Surgery of the thyroid and parathyroid glands. II. edn. Saunders, Philadelphia

Seiffert A (1928) Unterbindung der A maxillaris (sh Denecke 1953)

Sellars I (1981) An anatomical study of the bursa subhyoidea. J Laryngol Otol 95:487–491

Šercer A (1951) Otolaringologija. Medicinska Knjiga (Beograd-Zagreb)

Šercer A (1960) Communication préliminaire sur la chirurgie du nerf statoacoustique. Comptes rendus 57 Congrès français d ORL, 1959. Arnette, Paris

Šercer A, Krmpotić J (1958) Über die Ursache der progressiven Altersschwerhörigkeit (Presbyacusis). Acta Otolaryngol Suppl 143:1–36

Šercer A, Mündnich K (1962) Plastische Operationen an der Nase und an der Ohrmuschel. Thieme, Stuttgart

Seriat-Gautier B, Collet S (1981) The anatomical basis for percutaneous puncture of the internal jugular vein in the newborn infant Anat Clin (Berl) 2:87–90

Shapiro R (1979) Enlargement of the hypoglossal canal in the presence of a persistent hypoglossal artery. Radiology 133:395–396

Shugar MA, Healy GB (1980) The fourth cleft anomaly. Head Neck Surg 3:72–75

Silver AJ, Mawad ME et al. (1984) Computed tomography of the carotid space and related cervical spaces, Part I, Anatomy. Radiology 150:723–728

Šimunić V (1974) Dobne promjene arterija unutrašnjeg uha. Magisterarbeit Universität Zagreb

Sistrunk WE (1920) The surgical treatment of cysts of the thyreoglossal Tract. Ann Surg 71:121–122

Sistrunk WE (1928) Technique of removal of cysts and sinuses of the thyroglossal Duct. Surg Cynecol Obstet 46:109–112

Skandalakis JE, Droulias C (1976) The recurrent laryngeal nerve. Am Surg 42:629–634

Skerlev M, Mačković M (1981) Die Variationen der Dimensionen des Saccus endolymphaticus. Medicinar

Smith D, Larsen JL (1979) On the symmetry and asymmetry of the bifurcation of the common carotid artery. A study of bilateral carotid angiograms in 100 adults. Neuroradiology 17:245–247

Springer TD, Fischbone G, Shapiro R (1974) Persistent

hypoglossal Artery Associated with Superior Cerebellar Artery Aneurysm. J Neurosurg 40:397–399

Staindl O (1984) Zur Pathogenese und Therapie cervikaler Trachealstenosen; Habilitationsschrift aus der Hals-Nasen-Ohrenabteilung der Landeskrankenanstalten Salzburg, Universität Wien

Stamm WK (1981) Anatomy of the pterygopalatine foramen and the fontanella in the lateral nasal wall. Rhinology 19:87–91

Stennert E (1979) Combined approach in extratemporal facial nerve reconstruction. Clin Plast Surg 6:481–486

Štern-Padovan R (1977) Promjene unutrašnjeg slušnog hodnika u tijeku postnatalnog života. Magisterarbeit, Universitat Zagreb

Stewart GR, Mountain JC, Colcock BP (1972) Nonrecurrent laryngeal nerve. Br J Surg 59:379–381

Stock AL, Collins HP, Davidson TM (1980) Anatomy of the superficial temporal artery. Head Neck Surg 2:466–469

Storrs LA (1976) Vascular anomalies of the Middle Ear. Laryngoscope 86:247–250

Su WY, Marion MS, Hinojosa R (1982) Anatomical measurement of the cochlear aqueduct, round window membrane, round window niche and facial recess. Laryngoscope 92:483–486

Subotić R (1979) The high position of the jugular bulb. Acta otolaryngol (Stockh) 87:340–344

Sunarić R (1963) O osteomima paranazalnih sinusa. Diss Medicinski Fakultet (Zagreb)

Sutton D (1950) Anomalous Carotid-basilar Anastomosis. Br J Radiol 23:617–619

Tabb HG, Becker FF (1971) Bilateral benign mixed tumor of the deep lobe of the parotid gland. Arch Otolaryngol 94:158–161

Tabb HG, Tannehill JF (1973) The tympanomastoid fissure: A reliable approach to the facial nerve in parotid surgery. South Med J 66/11:1273–1276

Tabb HG, Scalco AN, Fraser SF (1970) Exposure of the facial nerve in parotid surgery. Laryngoscope 80/4:559–567

Taguchi K (1898) Über das Foramen clinoideo-opthalmicum mit Berücksichtigung der A. ophthalmica beim Menschen. Jap Tokyo Igakkwai Zashi 12:117

Taillens JP (1960) Etudes anatomoclinique des chaines ganglionnaires lymphatiques du cou et de leurs ganglion sattelites bucco-linguopharyngés, vol 22. Practica ORL 1:44–58

Tan WS, Spigos DG (1979) Right vertebral artery originating from the right common carotid. Radiologe 19:155–156

Terracol J, Ardouin P, Truffet P (1965) Anatomie des fosses nasales et des cavités annexes Libr Maloine, Paris, S 1

Terrier G (1978) L'endoscopie rhinosinusale moderne. Morell Officime Grafica, Osnago

Teufel J (1964) Einbau der A carotis in den canalis caroticus unter Berücksichtigung des Venenabflusses. Gegenbauers Morphol Jahrb 106:188–274

Thomas JR (1979) Thyroglossal-duct cyst, Ent J 58:510–514

Thomas JR (1980) Tympanic neurectomy and chorda tympani section, Aust NZJ Surg 50:352–355

Thomassin JM (1982) Le nerf laryngé supérieur. J Franç ORL 31:139–150

Thomsen J, Reiter S, Borum P (1981) Tomography of the internal acoustic meatus. A critical evaluation of the radiological appearance in normals and in patients with acoustic neuromas. J Laryngol Otol 95:1191–1204

Tiraboschi R, Grupi G, Locatelli G, Siew Yen Ho, Parenzan L (1980) Cervical aortic arch with aortic obstruction: report of two cases. Thorax 35:26–30

Tobin GR, Schusterman M, Peterson GH, Nichols G, Bland KI (1980) The intramuscular neurovascular anatomy of the latissimus dorsi muscle: the basis for splitting the flap. Plast Reconstr Surg 637–641

Toldt C, Hochstetter F (1979) Anatomischer Atlas, 27. Aufl. Urban u. Schwarzenberg, München Berlin Wien (J Krmpotić-Nemanić)

Tornwaldt GL (1885) Über die Bedeutung der Bursa pharyngea für die Erkennung und Behandlung gewisser Nasenrachenkrankheiten. Bergmann, Wiesbaden

Tschabitscher M, Perneczky A (1974) Über die Beziehungen von Kleinhirnarterien zum Meatus acusticus internus. Acta Anat (Basel) 88:231–234

Tucker AG (1981) A case of postauricular sinus. J Laryngol Otol 95:619–621

Tucker HM (1978) Human laryngeal reinnervation: Long-term experiences with nerve-muscle pedicle technique. Laryngoscope 88:598–604

Turvey TA, Fonseca RJ, Hill Ch (1980) The anatomy of the internal maxillary artery in the pterygopalatine fossa: its relationship to maxillary surgery. J Oral Surg 38:92–95

Underwood AS (1910) An inquiry into the anatomy and pathology of the maxillary sinus. J Anat Physiol 44

Vergnon L, Gotlieb O (1979) Ophtalmoplégie totale après section du nerf vidien. A propos de deux cas. Ann Otolaryngol (Paris) 96:407–410

Vidyasagen C (1979) Persistent embryonic veins in arteriovenous malformations of the posterior fossa. Acta Neurochir (Wien) 48:67–82

Vogel G, Bauer G (1978) Über eine Varietät des Circulus arteriosus. Gegenbaurs Morphol Jahrb 124: 360–363

Vogt K, Schrade F (1979) Anatomische Varianten des Ausführungsgangsystems der Stirnhöhle. Laryngol Rhinol Otol 58:783–794

Vuillard P, Bouchet A, Gouillat C, Armand D (1978) Le nerf laryngé inférieur non recurrent. A propos de 15 cas opératoires. Bull Assoc Anat 62:497–505

Wang CA (1976) The anatomic basis of parathyroid surgery. Ann Surg 183–271

Wang CA (1977) Parathyroid reexploration: A clinical and pathological study of 112 cases. Ann Surg 186–140–145

Warfel JH (1976) The head, neck and trunck: muscles and motor points, 4th edn. Lea Febiger, Philadelphia

Watanabe K, Watanabe J (1980) The ultrastructural characteristics of the capillary walls in human nasal mucosa. Rhinology 18:183–195

West JM, Bandy BC, Jafek BW (1974) Aberrant jugular bulb in the middle ear cavity. Arch Otolaryngol 100:370–372

Wigand ME, Haid T, Berg M, Rettinger G (1982) The enlarged transtemporal approach to the cerebellopontine angle: Technique and indications. Acta otorhinol. ital. 2:571–582

Wigand ME, Haid T, Berg M, Rettinger G (1983) Hirnnerven bei cochleovestibulären Störungen über einen erweiterten transtemporalen Zugang. Mikrochirurgische Neurologie des VIII. HNO 9:295–302

Wilbrand HF (1974) Multidirectional tomography of minor detail in the temporal bone. Acta Univ Upsaliensis, Abstracts of Uppsala Dissertationis 187, Uppsala

Wilson JSP, Galvas MSL, Brough MD (1980) The application of hairbearing flaps in head and neck surgery. Head and Neck Surgery, May-June, S 386–406

Winerman I, Nathan H, Arensburg B (1980) Posterior ligament of the incus: variations in its components. ENT J 59:227–231

Yasargil MG (1969) Microsurgery applied to Neurosurgery. Thieme, Stuttgart

Zanolli L, Viada J (1979) Hemorrhagia tarda post traqueostomia. Rev ORL (Santiago) 39:77–81

Zehm S (1966) Die Topographie des retromaxillären Raumes in chirurgischer Sicht. Habil Schrift Würzburg

Zehm S (1970) Der retromaxilläre Raum. Thieme, Stuttgart

Zelnik J, Gingrass RP (1979) Anatomy of the alar cartilage. Plast Reconstr Surg 64:650–653

Ziarah HA, Atkinson ME (1981 a) The surgical anatomy of the cervical distribution of the facial nerve. Br J Oral Surg 19:171–179

Ziarah HA, Atkinson ME (1981 b) The surgical anatomy of the mandibular distribution of the facial nerve. Br J Oral Surg 19:159–170

Zuckerkandl E (1882) Normale und pathologische Anatomie der Nasenhöhle und ihrer pneumatischen Anhänge, Bd I. Braunmüller, Wien Leipzig

Ždanov DA (1952) Šeinaja čast grudnogo protoka, obščaja anatomij i fiziologija limfatičeskoj sistemi. Lenjingr Otdel 273–281

Index of Operations and Surgically Important Anatomic Relationships

Most of the operative procedures are listed under the heading "Surgery" or the affected organ being treated. The numbers indicate pages.

Subject Index

Page numbers in *italies* indicate illustrations

MIX
Papier aus verantwortungsvollen Quellen
Paper from responsible sources
FSC® C105338

If you have any concerns about our products,
you can contact us on
ProductSafety@springernature.com

In case Publisher is established outside the EU,
the EU authorized representative is:
Springer Nature Customer Service Center GmbH
Europaplatz 3, 69115 Heidelberg, Germany

Printed by Libri Plureos GmbH
in Hamburg, Germany